WATER POLLUTION

WATER POLLUTION

Editor

Kilian Schreier

scitus
academics

Water Pollution

Edited by **Kilian Schreier**

ISBN: 9781681179735
Library of Congress Control Number: 2017949110

SCITUS Academics LLC,
616, Corporate Way, Suite 2, 4766,
Valley Cottage, NY 10989

www.scitusacademics.com

Notice

Printed in United States of America.

Preface

Over two thirds of Earth's surface is covered by water; less than a third is taken up by land. As Earth's population continues to grow, people are putting ever-increasing pressure on the planet's water resources. In a sense, our oceans, rivers, and other inland waters are being "squeezed" by human activities—not so they take up less room, but so their quality is reduced. Since water is the source of life and essential for all life on the planet, the use of this resource is a highly important issue. Poorer water quality means water pollution. Water pollution is any contamination of water with chemicals or other foreign substances that are detrimental to human, plant, or animal health. Water pollutants include fertilizers and pesticides from agricultural runoff; sewage and food processing waste; lead, mercury, and other heavy metals; chemical wastes from industrial discharges; and chemical contamination from hazardous waste sites. Worldwide, nearly 2 billion people drink contaminated water that could be harmful to their health. Water pollution is a major global problem that requires ongoing evaluation and revision of water resource policy at all levels.

This volume provides detailed information about different topics and gives a general overview of the current issues in water pollution and its management. The chapters, contributed by eminent scientists and experts, cover a wide range of current issues, reflecting on current problems and demonstrating the complexity of water pollution. It also reports on water pollution management technology that is low-tech, cheap and above all ecologically friendly. This book will be useful to environmental scientists, water professionals, researchers, academics and students.

Table of Contents

CHAPTER 1

Point Source Pollution and Variability of Nitrate Concentrations in Water from Shallow Aquifers

Jasna Nemcˇic´-Jurec[1] Anamarija Jazbec[2]

[1]*Institute of Public Health of Koprivnica-Krizevci County, Trg Tomislava Bardeka 10/10, 48 000 Koprivnica, Croatia*

[2]*Department for Forest Inventory and Management, Faculty of Forestry, University of Zagreb, Svetosˇimunska 25, 10 000 Zagreb, Croatia*

ABSTRACT

Agriculture is one of the several major sources of nitrate pollution, and therefore the EU Nitrate Directive, designed to decrease pollution, has been implemented. Point sources like septic systems and broken sewage systems also contribute to water pollution. Pollution of groundwater by nitrate from 19 shallow wells was studied in a typical agricultural region, middle Podravina, in northwest Croatia. The concentration of nitrate ranged from <0.1 to 367 mg/l in water from wells, and 29.8 % of 253 total samples were above maximum acceptable value of 50 mg/l (MAV). Among regions R1–R6, there was no statistically significant difference in nitrate concentrations ($F = 1.98$; $p = 0.15$) during the years 2002–2007. Average concentrations of nitrate in all 19 wells for all the analyzed years were between recommended limit value of 25 mg/l (RLV) and MAV except in 2002 (concentration was under RLV). The results of the repeated measures ANOVA showed statistically significant differences between the wells at the point source distance (proximity) of <10 m, compared to the wells at the point source distance of >20 m ($F = 10.6$; $p < 0.001$). Average annual concentrations of nitrate during the years studied are not statistically different, but interaction between proximity and years is statistically significant ($F = 2.07$; $p = 0.04$). Results of k-means clustering confirmed division into four clusters according to the pollution. Principal component analysis showed that there is only one significant factor, proximity, which explains 91.6 % of the total variability of nitrate. Differences in water quality were found as a result of different environmental factors. These results will contribute to the implementation of the Nitrate Directive in Croatia and the EU.

Keywords Nitrate Pollution Point source Water Shallow well Croatia

INTRODUCTION

The most important and abundant source of drinking water in the European Union (EU) is groundwater constituting over 80 % of the supply (Council Directive 2000). Pollution of groundwater is caused by different pollutants, among which nitrate plays the most important role. Nitrate pollution from agriculture is an issue of major concern for the UN (Addiscott et al. 1991), and a series of environmental policies have been implemented in the EU to decrease nitrogen (N) emissions from agriculture. The Nitrate Directive is one of the main policies to reduce nitrate leaching from agriculture primarily through good agricultural practise and the application of organic and mineral fertilizers according to regulations (Velthof et al. 2014).

As nitrogen is the key factor in a high and stable cultivated plant production, it is no surprise that modern agriculture affects nitrogen content in water, as it is a high nitrogen user (Addiscott et al. 1991; Bašić and Herceg 2010). Nitrate concentration in natural waters has been increasing during the last few decades, coinciding with the increased use of mineral fertilizers and a higher yield of cultivated plants in agriculture. This problem has become even more significant by introducing the stricter standards concerning the nitrate content of drinking water in the EU. The maximum acceptable value (MAV) is 50 mg/l NO_3^- (Regulation 2013), but the recommended limit value (RLV) is only 25 mg/l NO_3^- (Council Directive 1991).

Agriculture is the biggest source of pollution because of the so-called diffuse source pollution that does not originate from a single discrete source. Pollution is often a cumulative effect of small amounts of contaminants gathered from a large area (US EPA 1987). Numerous scientists have shown nitrate leaching from agricultural areas. Elhatip et al. (2003) showed that agricultural activities contribute to an increased nitrate concentration in groundwater either directly by nitrate leaching from agrochemicals or by agrochemicals that affect the processes in the soil and increase nitrate leaching from the soil. Nitrate leaching depends on a variety of environmental factors. Mild climate and favorable winter temperatures allow the extension of the growing season of plants, which reduces nitrate leaching (Hooker et al. 2008). Benson et al. (2006) also evaluated the importance of the excessive use of agrochemicals. They point out that the use of fertilizers in doses of 500 kg/ha increases the nitrate concentration in the water up to a distance of 2 km from the well.

Point source pollution refers to contaminants that enter a waterway from a single, identifiable source such as livestock farms, slurry lagoons or manure depots with an inappropriate building or location (US EPA 1987). Dense populations and discharges from point sources like septic systems or broken sewer systems contribute significantly to water pollution by nitrate in urban and suburban areas (Nemčić-Jurec et al. 2007, 2013; Drake and Bauder 2005; Melian et al. 1999; US EPA 1987).

Effects of point sources of pollution are usually localized and limited (Krapac et al. 2002; Nemčić-Jurec et al. 2013). Ciravolo et al. (1979) showed that point source (lagoon with liquid pig manure) affected the concentration of nitrate in the water at the distance of 36 meters from the well. However, they point out that the contamination occurred in the wells located on sandy soil, while at locations where there is clay soil, there was no significant pollution. In similar studies, Richards et al. (1996) showed that the concentration of nitrate in the water from the well in sandy soil, at the distance of 6 m from the point source pollution (animal feedlot or barnyard), was twice as high (3.6 mg/l) compared to the concentration of nitrate (1.8 mg/l) from the wells located 60 m from the pollution. Nemčić-Jurec et al. (2013) showed by comparison of nitrate concentration in two different geological areas that the nitrate concentration occasionally or regularly exceeds MAV in wells close to point sources of pollution (within 10 m of the well). In both areas of research, however, the frequency of unsafe samples was significantly higher in Prigorje area compared to Podravina area. Nitrate concentration in water depends on the vertical and horizontal transports of nitrate leaching (Houben et al. 2001). They point out that the highest nitrate concentrations are determined in the shallow aquifers that are richer in oxygen at a depth of 12 m. Katz et al. (2004) showed that the concentration of nitrate in the shallow aquifer is spatially and vertically highly variable. Although the average nitrate content between the shallow and deep zone was not significantly different (0.13 and 0.17 mg NO_3-N/l), the maximum content was much higher in the shallow zone (7.3 mg NO_3-N/l) compared to the deeper zone (1.6 mg NO_3-N/l). The variability of nitrate concentration was higher in shallower layers.

Podravina is an area of relatively intensive agriculture where, according to modern agricultural demands, higher doses of organic and mineral fertilizers are applied, causing leaching of the soil and nitrate input in aquatic ecosystems. Considering the large share of arable land in the River Drava basin, the highest possible nitrate pollution of water in this region can be expected. Due to the increasing trend of application of mineral and organic fertilizers in the world (Woo-Jung et al. 2007; Pacheco et al. 2001; Bouwman et al. 2005; Cetindag 2005; Widory et al. 2003) and, on the other hand, due to stricter criteria concerning reduction in nitrate pollutions in water (EU Nitrate Council Directive 1991), the aim of this research was to investigate the annual nitrate concentration variability in the area of an intensive agriculture and dense population of the middle Podravina region. The areas especially prone to pollution are easily permeable soils in drinking water protection zones in shallow unconfined aquifers (Burkart and Stoner 2007; Kraft and Stites 2003). Therefore, the aim was to determine to what extent point sources of pollution from agriculture and septic systems contribute to total pollution, as well as to determine the nitrate concentration variability in individual wells depending on the environmental factors.

Study areas

Location

The locations of the wells within the regions (R1–R6) are illustrated in Fig. 1. The regions are based and defined on political (municipality) and geographical boundaries. The study area (20 × 30 km) located in the water catchment of middle Podravina in northwest Croatia extends from Legrad in the east to Sokolovac in the west. From the north, the area is bordered by the state border with the Republic of Hungary and in the south with the northern slopes of Bilogora and Papuk. The old alluvial terraces area is composed of drift of the River Drava and its tributaries, as well as terraced sediment.

Climate and hydrogeology

From the hydrogeological point of view, the quaternary sand–gravel aquifer is the most important. The top part of the aquifer is built of sand and clay, with a significant proportion of quicksand and mostly of swampy loess in the southern and eastern part of water catchment (Mayer et al. 1996). A mild-continental climate dominates, and the average annual precipitation is about 820 mm. The average annual temperature is about 11 °C which indicates mild-warm climate (data obtained from the Meteorological and Hydrological service). These climate conditions favor agricultural production and increased population density.

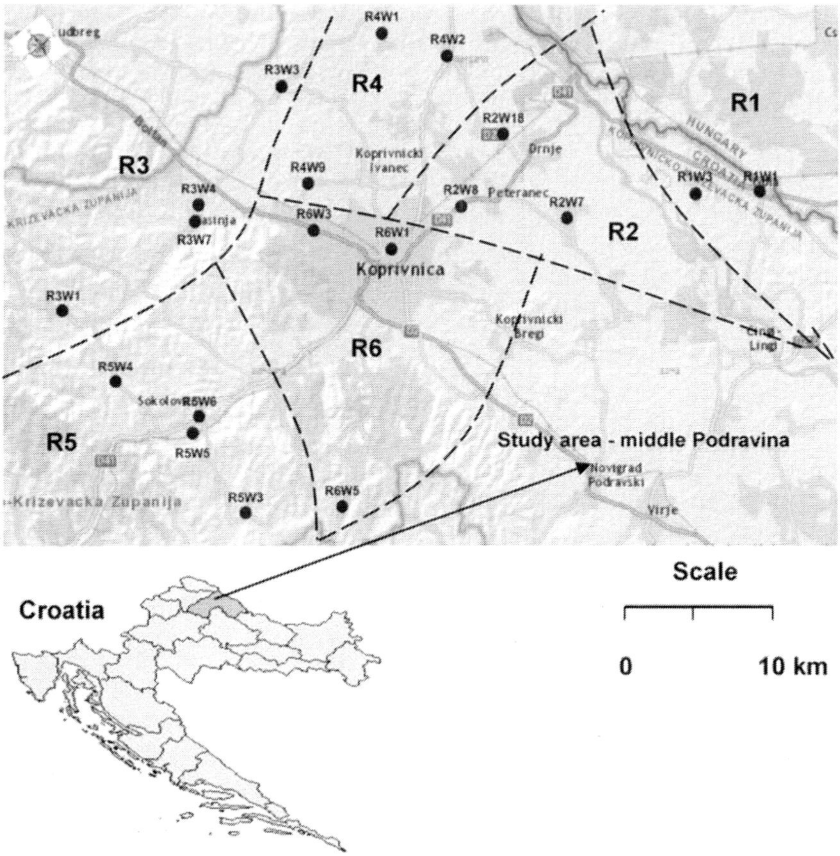

Figure. 1 Locations of the wells within the regions in middle Podravina (*R* region, *W* well)

Land use and point sources of pollution

The research area is inhabited by about 60,000 people with traditionally developed plant and animal production. The total research area covers 177,464 ha, out of which agricultural land covers 59 %. The remaining area consists of forests (32 %), arid land, settlements and water surfaces (9 %). Of the agricultural land, 99 % is arable (gardens, orchards, vineyards and meadows) and 1 % is pastures.

According to the data of the Faculty of Agriculture, average doses of organic and mineral fertilizers applied in middle Podravina (152 kg N/ha agricultural area) are higher in comparison with the average doses of fertilizers applied in Croatia (54 kg N/ha agricultural area). In the research area, point sources of pollution of agricultural origin such as farms and barns, and landfill of organic fertilizers depots are often very close to the wells (5–20 m). Most of the population in urban areas, up to 80 %, is connected to a sewerage system, while the population of suburban areas use septic systems which are also often very close to the wells.

On the sandstone outcrop area, the soil is freely draining so the nitrate input from point sources of pollution can be expected to accumulate in the aquatic ecosystems including the River Drava basin.

MATERIALS AND METHODS

Sampling

Nitrate concentration variability was tested in drinking water from 19 wells originating from the shallow aquifer. Characteristics of the wells are shown in Table 1. The locations of the wells were chosen in such a way to represent a range of end uses. The wells that supply water to public buildings, such as schools, kindergartens, shops and restaurants, in densely populated and agricultural areas of middle Podravina were selected. The surroundings of the wells were examined, the sources of pollution were identified, and the distance of the pollution source from the wells was measured by a digital meter (m). The wells were located at a greater or lesser distance from the point sources of pollution of agricultural origin (farms and organic fertilizers depots) and septic systems. Information about the location of septic systems was obtained from the owner of the well, and the distance was measured by a digital meter. During a 6-year study (2002–2007), a total of 253 samples of water were taken during all seasons (2–3 times per year) at the same time (through a week) from each well.

For quality control purposes, duplicate samples were taken and analyzed at a rate of 1 in 20. These were treated as completely separate samples. Precision judged (% RSD) from the duplicates is on average in the range of 2–8 % for nitrate concentration.

The sampling was done according to HRN ISO 5667-5:2000. The water sampling from wells was done after the releasing of water for 2–3 min on the outlet of a water supply system, in a clean polyethylene vessel of 1 l volume. The samples were stored in a portable fridge, in the dark at 1–5 °C for transportation. After delivery to the laboratory, the samples were stored at the same temperature until the analysis (within 24 h).

Table 1 Characteristics of the wells by regions (R1–R6)

Region	Well	Well depth (m)	Pollution source	Proximity (m)	Soil type
R1	W1	6	Ag	>20	Gravel–sand
R1	W3	5	Ag, SS	>20	Gravel–sand
R2	W7	5	Ag	<10	Silt–gravel–sand
R2	W8	6	Ag	>20	Silt–gravel–sand
R2	W18	5	Ag	<10	Silt–gravel–sand
R3	W1	10	Ag	>20	Clay–silt–gravel
R3	W3	5	Ag	>20	Silt–gravel–sand
R3	W4	10	Ag	>20	Clay–silt–gravel
R3	W7	7	Ag, SS	10–20	Clay–silt–gravel
R4	W1	5	Ag	>20	Silt–gravel–sand
R4	W2	5	Ag	>20	Silt–gravel–sand
R4	W9	7	SS	>20	Gravel–sand
R5	W3	10	SS	10–20	Clay–gravel
R5	W4	10	Ag	>20	Clay–gravel
R5	W5	10	SS	10–20	Clay–gravel
R5	W6	15	Ag	<10	Clay–gravel
R6	W1	20	SS	>20	Gravel–sand
R6	W3	10	Ag, SS	<10	Gravel–sand
R6	W5	5	Ag, SS	<10	Clay–gravel

R region, *W* well, *Ag* agriculture, *SS* septic systems

Water analysis

Nitrate concentration in drinking water was determined using the ion chromatography method (HRN ISO 10304-1:1998). Prior to the injection into the analyzer, ion chromatograph ICS 3000, Dionex (USA), samples were filtered through a membrane filter (of pore 0.45 μm) to remove any particulate matter. After establishing the calibration function, samples were injected into the ion chromatograph and the peaks were measured in accordance with the manufacturer's instructions.

The chromatograph system consists of the following components: auto sampler device, sample injection system (incorporating sample loop of 50 μl), gradient pump, liquid chromatography device, conductivity detector and eluent generator. The AS 15 anion separator column and appropriate precolumn were used for the separation. ASRS ULTRA II-4 mm performed chemical suppression. 38 mM NaOH was used as an eluent, and the flow was 0.5 ml/min. The whole system was supervised by Dionex Chromeleon software.

Data analysis

Descriptive statistics was done for all the data ($N = 253$ samples; $N = 114$ data), and it was compared to the MAV (Regulation 2013), i.e., recommended level value (RLV) of Nitrate Directive (Council Directive 1991). Average nitrate concentrations in each of 19 wells through the 6 years were calculated. Data from each year and location are averaged in order to determine the general water quality in a particular well due to nitrate. Box plots of annual nitrate concentration (mg/l) indicating MAV and RLV were log-transformed (ln) to provide a more symmetric (normalized) distribution.

The difference in the nitrate concentration during the years 2002–2007 among regions (R1–R6) was tested with an analysis of variance (ANOVA).

In order to compare the concentrations of nitrate by wells and depending on the distance from the point sources of pollution in the analyzed period, we used the repeated measure analysis of variance. Since we wanted to detect precisely which group differences were statistically significant, we used the Tukey multiple comparison post hoc test (Davis 2002).

In order to confirm the grouping of wells according to the average annual values and distance from the point source pollution, we used k-means clustering.

In order to analyze and present wells in respect of their annual average concentration of nitrate, we did principal component analysis (PCA).

The correlation between the concentrations of nitrate and depth of wells was done as well as the correlation between the concentrations of nitrate and precipitation.

Descriptive statistics and the repeated measure ANOVA were done using the SAS 9.2. PCA was done using the STATISTICA 8.0 (2011), and graphical presentations were done using both statistical packages.

RESULTS

Nitrate concentration in groundwater

Concentration of nitrate ($N = 114$) in the research area of middle Podravina ranged from <0.1 to 367 mg/l in groundwater from shallow wells, and 29.8 % of the samples were above the MAV.

If we analyze the regions (R1–R6), shown in Fig. 1, we see that there is no statistically significant difference in the nitrate concentration during the years 2002–2007 among regions ($F = 1.98$, $p = 0.15$), and the interaction between regions and years is not statistically significant too ($F = 0.99$, $p = 0.49$).

The box plot of an annual nitrate concentration (mg/l) indicating MAV and RLV is shown in Fig. 2. Box plots, log-transformed to provide a more symmetric distribution, show that the average concentration of nitrate in all 19 wells for all the analyzed years was between RLV and MAV except in 2002 when the concentration was under the RLV. The distribution of the concentration of nitrate per year (mean, median, min and max) shows that the 2002 distribution was negatively asymmetric which means that most of the wells (over 50 %) had values below the RLV and almost 70 % below the MAV; 2003 and 2004 distribution was symmetric, i.e., about half of the wells had average concentration values up to the RLV. In 2005 and 2006, half of the wells had a concentration higher than the average and higher than the RLV, in 2005 almost reached the maximum (MAV), and after that value slightly decreased until 2007 (half of the wells had the average value that is slightly higher than the RLV). Nitrate pollution is present in wells of all regions, and during a 6-year study, there are year-to-year nitrate concentration differences. Figure 2 also shows that the highest variability of nitrate concentration was established in 2002. It is assumed that extremely low concentrations (detection limit—DL) of one well (R3W1) increase the total variability (negatively asymmetric) in 2002. Considering that the nitrate concentrations in the well are below the RLV and that the well is not exposed to contamination, it is difficult to explain the reason for the variability.

Point sources of nitrate pollution

The results of the repeated measure ANOVA show that there is a statistically significant difference in the concentration of nitrate due to the point source distance (Table 2). Results of Tukey post hoc tests show that there is statistically significant difference between the wells within 10 m from the point sources of pollution compared to the wells within 20 m from the source of pollution ($F = 10.6$; $p < 0.001$), while the wells within 10–20 m from the point sources are not statistically different from wells within 10 m and within 20 m.

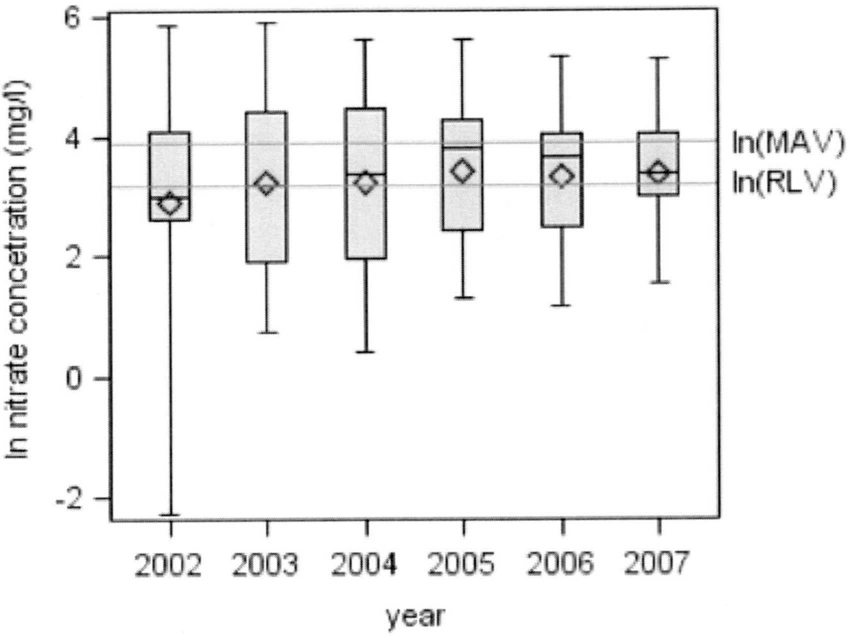

Figure. 2 Box plot of annual nitrate concentration (mg/l) indicating RLV (recommended level) and MAL (maximum allowed level) (*filled circle* mean, *hyphen* median, Q1–Q3 box, min–max whiskers)

Table 2 Results of repeated measure analysis of variance for ln of nitrate concentration

Source of variation	**DF**	**Mean square**	***F* value**	**Pr > *F***
Between subject				
Point source distance	2	46.50	10.58	0.001
Error	16	4.39		
Within subject				
Year	5	0.06	0.23	0.956
Point source * year	10	0.52	2.07	0.036
Error	80	0.25		

Average values of nitrate during the years are not statistically different, but the interaction between point source distance (proximity) and years is statistically significant ($F = 2.07$; $p = 0.04$), which indicates that the concentration of nitrate during the monitoring does not behave the same with respect to the distance (Fig. 3). Average values of nitrate concentration for wells whose distance is >20 m increase over the years, while those whose distance is <10 m decrease. In the group of the wells at the distance of point sources >20 m, nitrate concentration is generally below the MAV, and only in one well (R4W9), a concentration above the MAV in 2005 (74.2 mg/l) and 2006 (57.5 mg/l) was determined. After that, the concentration in that well decreases again.

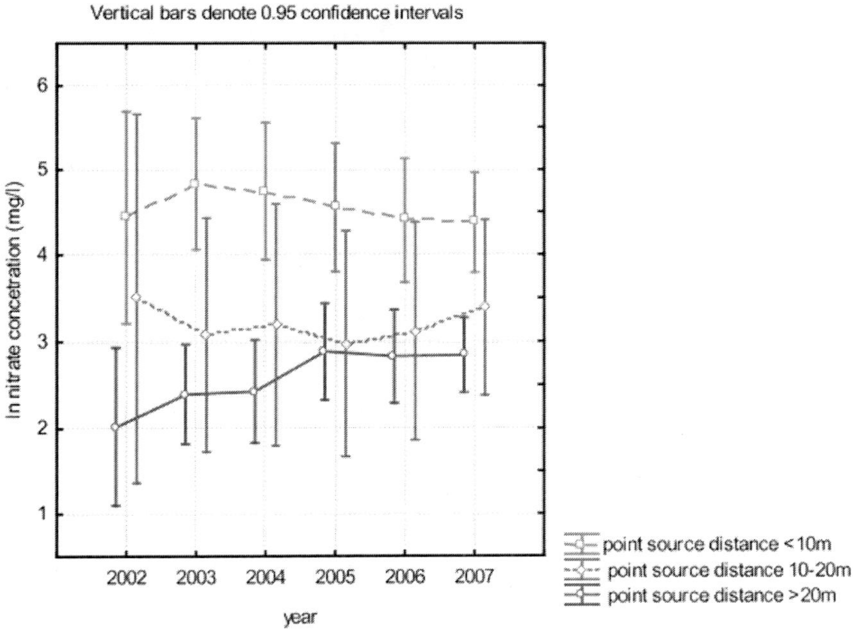

Figure. 3 Average of ln nitrate concentration (mg/l) and 95 % confidence intervals according to point source distance (m)

Figure 3 also shows that the average concentration of nitrate at the distance of point sources <10 m is the highest in 2003. In a well (R2W7) from this group, concentration ranged from 43.6 to 186 mg/l. The maximum nitrate concentration (186 mg/l) was determined in 2003. In 2004, the concentration decreased (157 mg/l), and a decreasing trend continued until 2007. The variability of the above-mentioned well suggests an average increase in nitrate concentration in 2003. The figure shows that there are differences in the average concentration from year to year and that the variability of individual wells influences the average nitrate concentration through a period of research from 2002 to 2007. The reason behind the variability of the above-mentioned well is unknown.

It is evident that the proximity of point sources of pollution affects the average nitrate concentration. Figure 4 shows the grouping of the wells, considering the type of pollution (of agricultural origin—Ag, septic systems—SS) and a comparison with RLV and MAV. Results of k-means clustering confirmed a division into four clusters according to the pollution. Cluster 1 (R6W3AgSS) is the highest one with an average value ranging from 201 to 367 mg/l during the years and is far higher than the MAV. Cluster 2 (R2W7Ag, R2W8, R2W18Ag, R4W9SS, R5W5SS, R5W6Ag, 6W5AgSS) clustered wells whose average values are over the MAV, and it is evident that this includes all wells that are exposed to Ag, SS or both. Cluster 3 (R1W1, R3W3, R3W4, R3W7, R5W3SS, R6W1, R4W1) has the nitrate concentration values a little less than the RLV, while cluster 4 (R1W3, R3W1, R4W2, R5W4) is the lowest.

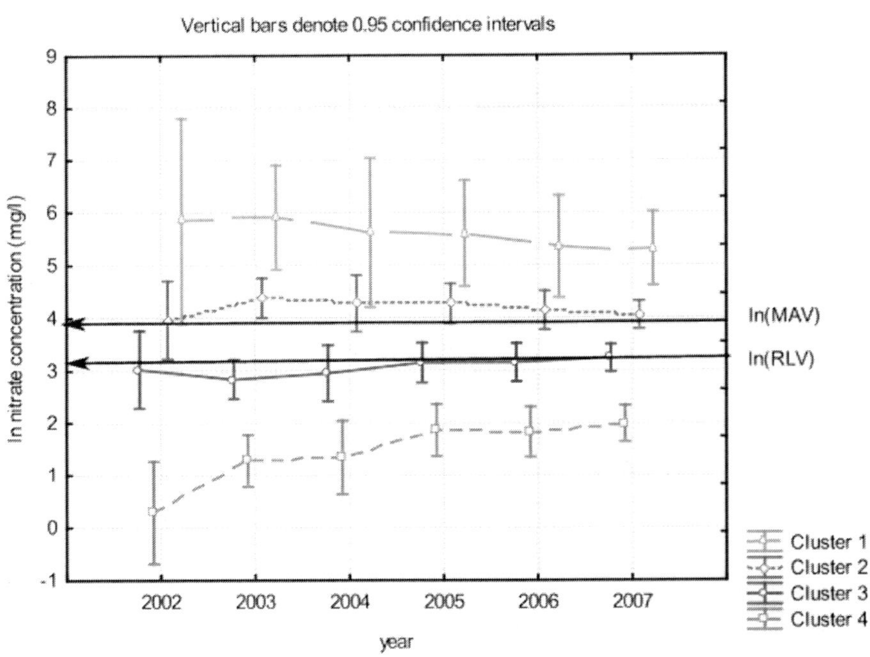

Figure. 4 Average of ln nitrate concentration (mg/l) of 4 clusters according to k-means clustering. RLV (recommend level) and MAL (maximum allowed level)

According to the results, it is clear that the wells that are close to one type of point source of pollution (agriculture or septic systems) show an increased nitrate concentration, while the wells close to both types of pollution are most likely to have nitrate problems.

In the wells far from pollution (Cluster 3 and Cluster 4), established nitrate concentrations are lower (mostly below the RLV). However, according to Fig. 4, it can be seen that the difference in nitrate concentrations was also determined in these wells.

Given that the other environmental factors have not been studied in detail, we determined how many point sources of pollution contribute to the total variability of nitrate in the wells. Principal component analysis projections of the wells with respect to the first two factors according to the average annual nitrate concentration are shown in Fig. 5. Our results show that the first factor explains 91.6 % of the total variability, which means that the distance point source (proximity) is a very important predictor, and another factor accounts only for 5.67 %. We can see that the clusters are lined up one after the other and that the concentration of nitrate following point source distance varies from highest to the left to lowest to the right. The figure also shows that cluster includes two wells away from point sources of >20 m. In these wells, concentrations are close to the MAV (R2W8) or occasionally above the MAV (R4W9SS). Similar characteristics of these wells are such that nitrate concentrations are high even though the distance of point source is large or it is not determined at all.

Effects of multiple factors

The characteristics of the wells are very significant for contamination in shallow wells. For this reason, the correlation between the depth of the wells and nitrate concentrations was examined. There is no statistically significant correlation between the concentrations of nitrate and depth of wells (mean = 8.32 m, SD = 4.08 m) for any observed year (r2002 = −0.02, r2003 = 0.14, r2004 = 0.11, r2005 = 0.11, r2006 = 0.07, r2007 = 0.07).

In addition to depth, several other factors, for example precipitation, may be associated with more frequent contamination problems. The effect of some other factors (precipitation) can be related to the soil characteristics. According to the results, there was no statistically significant correlation ($r = -0.12$) between the average amount of rainfall (mm) per year and the average nitrate concentration.

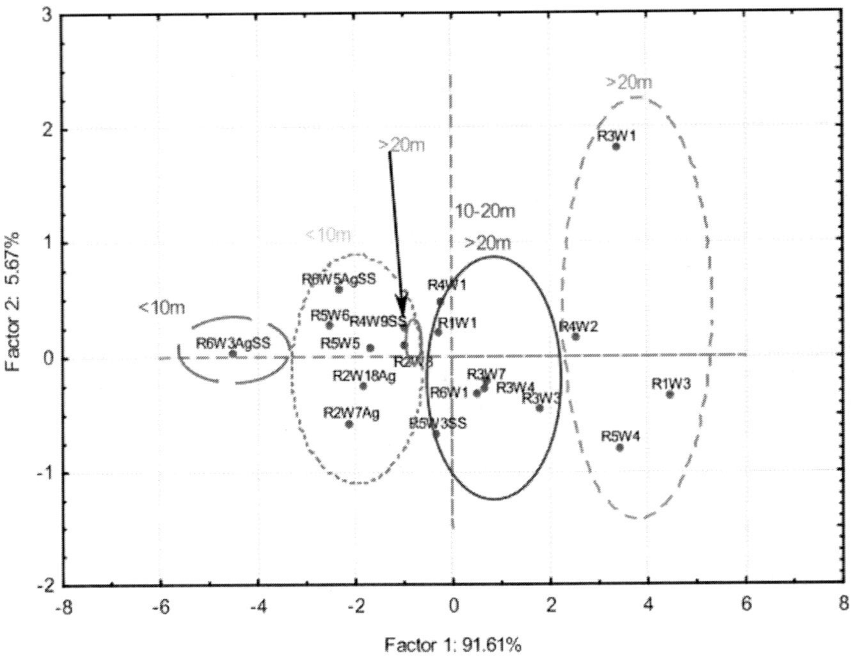

Figure. 5 Projections of the wells with respect to the first two factors (PCA) according to the ln of average annual nitrate concentrations indicating cluster membership and point source distance (Ag—agricultural source, SS—septic systems source, AgSS—both sources)

DISCUSSION

In the region of middle Podravina, a relatively high nitrate concentration (to 367 mg/l) was found in water wells from the shallow aquifer consistent with results of previous research. Kattan (2001) found a concentration of nitrate in shallow wells ranging from 11 to 80 mg/l and some other researchers (Obeidat et al. 2007) ranging up to 330 mg/l in the wells, which was recognized as contamination. Other researchers reported an even lower nitrate concentration (8.5 mg/l) in drinking water (Hallberg 1989) that they already considered as pollution, while Kazemi (2004) and Jalali (2005) point out that the concentration of 14 mg/l is satisfactory because it is far below the MAV. Aelion and Conte (2004) found a low concentration (0.94 mg NO_3-N/l) in their research area. However, after an increase in the intensity of use of agricultural land, an increase in nitrate concentration up to 3 mg NO_3-N/l followed. Although scientists have different opinions about the concentrations considered as pollution, it is evident that 29.8 % of samples (above the MAV) in middle Podravina are contaminated by nitrate.

Contaminations of some wells do not necessarily reflect a generalized contamination of local or regional aquifers. Consistent local patterns of contamination probably indicate contamination of the aquifer locally (Richards et al. 1996; Bouwman et al. 2005). It is evident that the nitrate problem exists in all regions, but the nitrate concentration does not depend on the region but on the location of the wells within the region. This difference could be explained as a result of different specific habitat factors (Singh et al. 2015).

Although ammonia and nitrite can cause water quality problems, nitrate is most often associated with the contamination of drinking water (Follett and Hatfield 2001). In the earlier Podravina research, ammonia was not detected in the analyzed samples and it very rarely occurs in waters from shallow wells, whereas nitrate is very often a cause of water contamination. The water from the aquifer in the research area is rich in oxygen, and if ammonia occurs, it probably rapidly oxidizes to nitrate by redox processes (Houben et al. 2001; Follett and Hatfield 2001; Gates et al. 2008).

The most common sources of nitrate contamination include fertilizers, livestock manure, feedlots, septic tanks and land application of municipal sludge (Addiscott et al. 1991; Elhatip et al. 2003; Obeidat et al. 2007). Plant and livestock production in the middle Podravina region is well developed, but the infrastructure (water supply and sewage systems) in settlements that are supplied with water from shallow aquifers, i.e., wells, has not been built yet. Also, point sources of pollution like septic systems can often be found near the wells so these results were expected. Woo-Jung et al. (2007) compared the concentrations of nitrate during 3 years of monitoring in the agricultural area with various sources of pollution, such as sown agricultural land, farms of livestock or dense population, in order to determine types of source contaminations. They found a different variability and in some areas an increase in nitrate concentration above the MAV. In accordance with our results, they determined the lowest concentration of nitrate in the natural area where there are no sources of pollution close to the well, following the crop area. The largest concentrations were found in densely populated areas as well as in the areas with the farms close to the wells. The monitoring of the nitrate concentration in middle Podravina during a 6-year period has also shown that water from the wells far from pollution (determined by surveying of the well surroundings) was not contaminated by nitrate. The contamination by nitrate was determined in densely populated areas in the vicinity of point sources of agricultural origin (leaching from organic waste of private farms) or septic systems. Such results are shown by

Krapac et al. (2002) as well, but alongside the type and vicinity of the source of contamination, the type of soil is also important for the leaching of nitrate (Ciravolo et al. 1979). By monitoring groundwater over 8 years, Ciravolo et al. showed contamination from the source of contamination in the area with sandy soil, while they did not determine significant water contamination in the area of clay soil. They emphasize the importance of the distance of point sources of pollution of the wells, but the effect of the impact depends on the permeability and the type of soil. Given that in Podravina there is sandy and gravelly soil (less clay soil), the obtained results are consistent with the hypotheses.

In this study, it was found that when the distance between the point sources and the wells was within 10 m, nitrate concentrations were above the MAV. In contrast, when the distance was greater (more than 10 m), effect on water quality was much rarer. Richards et al. (1996) also showed that the concentration of nitrate is significantly higher in well close to point source (within 6 m) of agricultural origin or septic systems from the well, compared to the concentration far from point source (within 60 m). It is obvious that the distance of the point source contributes to the final concentration of nitrate and nitrate variability. Gardner and Vogel (2005) also pointed out that distance and numbers of septic tanks and point sources of agricultural origin contribute to an increased nitrogen concentration in water even at distances up to 300 m and that the proximity of these types of pollution is a prerequisite for poor water quality due to nitrate.

In the study area, four clusters are shown according to the level of pollution. Water from wells close to point source pollution was worst quality. In wells that are far from point source, water quality is better, but nitrate concentrations are different. Factors that contribute to the difference in the nitrate concentration are not identified in this paper, but are assumed to be the result of the geological structure and porosity of the soil. The physical and chemical weathering of minerals, leaching and runoff are the main factors responsible for groundwater quality deterioration with some geogenic contribution from soil (Singh et al. 2015). The structure and depth of the wells and the proximity of diffuse pollution of agricultural origin or the use of agrochemicals on intensive agricultural areas affect the concentration of nitrate in groundwater in some locations of wells (Elhatip et al. 2003; Burkart and Stoner 2007; Kraft and Stites 2003). This result emphasizes the fact that vulnerability factors bear only a probabilistic relationship to the concentration of contaminants in a well and that even under conditions that combine several factors associated with a low concentration, a certain proportion of wells are contaminated.

Nemčić-Jurec et al. (2007), Melian et al. (1999), Nemčić-Jurec et al. (2013) identified multiple effects of different sources of pollution including agricultural point sources, since their samples had been mostly analyzed in connection with other point sources of pollution. However, some of the previous researches showed the pollution effect of different point sources like septic systems (Drake and Bauder 2005) and agricultural point sources (Widory et al. 2003). In this study, PCA showed that the distance is the main reason for the total variability, which means that it is a very important predictor in middle Podravina. Concentration of nitrate following point source distance ranges from highest (within 10 m, wells are exposed to Ag and SS) to lowest (wells are not exposed to contamination) which is in accordance with above-mentioned studies.

Concentrations of nitrate have shown different trends during the study. A higher nitrate concentration from year to year may be related to lower organic C content in older soils, and consequently less removal of nitrate by reduction or because of older wells, deteriorating structure and higher permeability (Richards et al. 1996). According to the results (data obtained from the Meteorological and Hydrological service) from 2004 to 2007, the annual precipitation is higher (753–991 mm) on average compared to the initial period of the study (740 mm–748 mm), which may contribute to the difference in the concentration of nitrate in the wells, in whose vicinity there are no point sources. However, there is a possibility of influencing underground water through diffuse sources of pollution, which can in general contribute to increasing concentrations in most of the wells. Nitrate concentration changes over time, and the interpretation of research results can be affected by the duration of the research period. Brian et al. (2004) emphasized that geocemical processes are too slow. They showed increase in nitrate concentration in groundwater during 3 years of monitoring. They pointed out that longer period of monitoring includes processes and the influence of potential hazards on groundwater quality. Variability or trends of water quality can be determined by 3 or more years of monitoring. The monitoring of trends during a period longer than 6 years was done by other researchers (Thorburn et al. 2003) reporting 4 categories of trends according to nitrate concentration: varying, steady, decreasing and increasing categories. They showed the effect on water quality during a longer period of time, as well as different effects of point sources near wells and influence of other agroecological factors.

Changes in the soil can affect the transport of nitrate. Alluvial areas which are usually formed by the deposition of soil or sediments have high porosity and permeability. They are considered good for groundwater recharge, and hence, they are also very vulnerable to groundwater pollution in comparison with confined aquifers (Singh et al. 2015). The concentration and variability of nitrate in water depend on the vertical and horizontal transports (Houben et al. 2001). According to Houben et al., vertical transport depends on the distribution of the reactive material in the aquifer, and it is estimated that the velocity of reactive vertical transport of pollutants is on average a few cm/year, whereas nonreactive is much faster, 1 m/year. They found the highest concentrations of nitrate in shallow layers of the aquifer that are richer in oxygen, and transport of nitrate is the fastest. According to Katz et al. (2004), the average concentrations of nitrate were similar in the shallow and deep zones, but the maximum content was much higher in the shallow zone compared to the deep zone. They point out that the concentration of nitrate is highly variable and that the variability of nitrate concentration is greater in shallow layers. As previously mentioned, the variability of nitrate concentrations in Podravina was also recorded in some wells close to pollution.

According to the studies of the Geological Survey, the hydraulic conductivity values range on average between 1×10^{-3} and 3×10^{-3} m/s. The analysis of anisotropy has found that the maximum anisotropy is determined in the shallowest part of the soil (2–7 m depth) made of gravel and sand. In the research area, the anisotropy coefficient usually varies from 10 to 35, while the vertical hydraulic conductivity is 10–35 times smaller than the horizontal. The results showed that the maximum horizontal conductivity is in a 5-m-deep layer (the most common depth of the monitored wells), which further explains the transfer of nitrate from the point source of pollution to the well. The nitrate of agricultural origin was leached for years from point sources into deeper layers, while septic systems are also often several meters deep and are often to be found at the level of the wells. For this reason, contamination of the wells through higher horizontal conductivity of nitrate was expected. Considering that the wells are shallow (mostly 5–10 m depth) and that in the research area, the soil is easily permeable, horizontal hydraulic conductivity values are high and a rapid transfer of nitrate from pollution sources to the well is expected, contributing to variability in individual wells. These shallow wells are more often irregularly, unsafely and spontaneously constructed. We assume that such kind of wells construction contributes to variability of water quality in the wells. Nemcic-Jurec et al. (2013) had examined nitrate concentration in part of Podravina not far away from the study area examined in this paper. Both areas showed similar geological structure and similar average nitrate concentrations (about 25 mg/l). However, proportions of samples above the MAV are different. According to Nemčić-Jurec et al. (2013), only 6 % was above the MAV, while at the nearby area examined in this paper, proportion of samples above MAV was significantly higher, about 30 %. We assume that similar geology and hydrogeology contribute to similar average nitrate concentrations in both areas, while higher proportion of samples above MAV shows periodically higher nitrate concentrations and higher variability. Differences are probably caused by micro-location and by safe environmental conditions, by proximity of source of pollution and by well construction itself. Bad and unsafe well construction and bad environmental conditions like proximity of source of pollution contribute to periodically higher nitrate concentrations, and they are an important prerequisite for poor water quality. Actually, bad and unsafe wells are indicators of proximity of pollution.

The perennial accumulation of nitrate in the soil near the source of pollution contributes to nitrate leaching into groundwater and drinking water in wells (Katz et al. 2004).

Richards et al. (1996) and McLay et al. (2001) have shown that the characteristics of the wells are very important for the contamination of groundwater from shallow wells. In the study area, the wells are usually shallow, and the difference in depths is not significant (90 % of the monitored wells are at 5–10 m depth), it is expected that there is no relationship between nitrate concentration and depth of the wells. Such a relationship should be investigated in different geological areas and in wells with significantly greater differences in depth (range 5–30 m or more). Previous research (Richards et al. 1996; McLay et al. 2001) determined the negative correlation between nitrate concentration and significantly different depth of wells (between wells <15 m deep and wells >30 m deep).

As previously noted, the effect of precipitation on wells depends on the characteristics of the soil and wells in sandy soils more frequently exceed the MAV. The alluvial aquifer is particularly sensitive to pollution (Burkart and Stoner 2007; Singh et al. 2015). Podravina predominantly has alluvial aquifers, but correlation between rainfall and nitrate concentration has not been established. Whatever the exact explanation for the unexpected relationships with well water quality shown by these two factors (precipitation and type of soil), it is clear that if they have any tendency to contribute to an increase in nitrate concentration, this tendency is small enough to be masked by confounding factors acting in the other directions. Similar results were obtained by other researchers (Burkart and Stoner 2007; Kraft and Stites 2003). The largest proportion of wells show varying trends over a longer period of time as a result of all environmental factors, and in these wells, the correlation between nitrate and precipitation was not found. They emphasize that groundwater is particularly sensitive to pollution in areas with a humid climate. More precipitation contributes to the variability of nitrate. However, according to some researchers, one of the more significant environmental factors which influence the concentration of nitrates in wells is precipitation.

Pacheco et al. (2001) found the lowest nitrate concentration during the rain season due to dilution of water in wells caused by rain which led to low nitrate concentrations, and they also found a relationship between nitrate and precipitation.

CONCLUSIONS

Nitrate-contaminated groundwater has been observed in shallow wells in a typical agricultural region, middle Podravina. Nitrate concentration in some wells exceeded the maximum accepted value of 50 mg/l required by the EU Drinking Water Standards; 29.8 % of the samples were above the MAV.

There is no statistically significant difference in the nitrate concentration during the years 2002–2007 among regions (R1–R6) nor was the interaction between regions and years statistically significant, which indicates that during the monitoring, all regions behaved approximately equally. The nitrate concentration problem was observed in all regions with the key predictor of concentration being the well location within the region.

In the study area, there is a significant difference in the concentration of nitrate due to the point source distance (proximity). Water quality is better in the wells at the proximity of <20 m compared to the wells at the proximity of >10 m. Average values of nitrate during the years are not statistically different. In the wells close to one type of point source pollution (agricultural origin or septic systems), there is an increased nitrate concentration, while in the wells close to both types of contamination, the water is more frequently unsafe for health. Point source distance (proximity) explains 94 % of the total variability of nitrate, which means that the proximity is a very important predictor. Bad and unsafe well construction also contributes to periodically higher nitrate concentrations. It is an indicator of pollution proximity, and it is an important prerequisite for poor water quality.

Because high nitrate concentrations in water can cause different health problems, it is necessary to protect drinking water in the study area. High nitrate concentrations are associated with improperly located wells. It is necessary to locate new wells uphill and least 20 m from feedlots, septic systems, barnyards and chemical storage facilities. Monitoring of nitrate in the tested wells should be continued, and specific rehabilitation measures of wells to end users should be proposed. Wherever possible, it is necessary to build a public water supply system to ensure safe drinking water.

It is also important to manage nonpoint sources of water pollution (fields, lawns) to limit the loss of excess water and plant nutrients and match fertilizer applications to precise crop uptake needs in order to minimize groundwater contamination. Careful fertilizer management can reduce nitrate leaching to groundwater.

These measures will contribute to the implementation of the Nitrate Directive in Croatia and the EU, thus providing pollution control and agricultural product safety.

REFERENCES

1. Addiscott TM, Whitmore AP, Powlson DS (1991) Farming, fertilizers and the nitrate problem. CAB International, Wallingford, Oxon, UK
2. Aelion CM, Conte BC (2004) Susceptibility of residential wells to VOC and nitrate contamination. Environ Sci Technol 38:1648–1653
3. Bašić F, Herceg N (2010) Principles of agronomy. Synopsis, Zagreb, Croatia
4. Benson VS, VanLeeuwen JA, Sanchez J, Dohoo IR, Somers GH (2006) Spatial analysis of land use impact on ground water nitrate concentrations. J Environ Qual 35:421–432
5. Bouwman AF, Van Drecht G, Van der Hoek KW (2005) Global and regional surface nitrogen balances in intensive agricultural production systems for the period 1970–2030. Pedosphere 15:137–155
6. Brian GK, Angela RC, Thomas RP (2004) Use of chemical and isotopic tracers to assess nitrate contamination and ground-water age, Woodville Karst Plain, USA. J Hydrol 289:36–61
7. Burkart MR, Stoner JD (2007) Nitrate in aquifers beneath agricultural systems. Water Sci Technol 56:59–69
8. Cetindag B (2005) Investigation of discharge and groundwater contamination characteristics around the Karasu Spring, Mus Area, Turkey. Environ Geol 47:268–282
9. Ciravolo TG, Martens DC, Hallock DL, Collins ER, Kornegay ET, Thomas HR (1979) Pollutant movement to shallow ground water tables from anaerobic swine waste lagoons. J Environ Qual 8:126–130
10. Clean Water Act (CWA), Section 502 (1987). General definitions. US EPA
11. Davis CS (2002) Statistical methods for the analysis of repeated measurements. Springer, New York
12. Drake VM, Bauder JW (2005) Ground water nitrate-nitrogen trends in relation to urban development, Helena, Montana, 1971–2003. Ground Water Monit Remediat 25:118–130
13. Drinking water safety Regulation (2008) NN 47/2008
14. Elhatip H, Afsim M, Kuscu I, Dirik K, Kurmac Y, Kaviurmaci M (2003) Influences of human activities and agriculture on groundwater quality of Kayseri-Incesu-Dokuzpinar springs, central anatolian part of Turkey. Environ Geol 44:490–494
15. European Union(EU) (1991) Council Directive 91/676/EEC of 12 December 1991 concerning the protection of waters against pollution caused by nitrates from agricultural sources. Off J 375:0001–0008
16. European Union(EU) (2000) Council directive 2000/60/EEC of 23 October 2000 establishing a framework for the Community action in the field of water policy. Off J 327:0001–0073

17. Follett RF, Hatfield JL (2001) Nitrogen in the environment: sources, problems and management, 1st edn. Elsevier science, Amsterdam, The Netherlands
18. Gardner KK, Vogel RM (2005) Predicting ground water nitrate concentration from land use. Ground Water 43:343–352
19. Gates JB, Bohlke JK, Edmunds WM (2008) Ecohydrological factors affecting nitrate concentrations in a phreatic desert aquifer in northwestern China. Environ Sci Technol 15:3531–3537
20. Hallberg GR (1989) Nitrates in groundwater in the United States of America. In: Follet RF (ed) Nitrogen management and groundwater protection: developments in agriculture and managed-forest ecology. Elsevier, Dordrecht
21. Hooker KV, Coxon CE, Hackett R, Kirwan LE, O'Keeffe E, Richards KG (2008) Evaluation of cover crop and reduced cultivation for reducing nitrate leaching in Ireland. J Environ Qual 37:138–145
22. Houben GJ, Martiny A, Basler N, Langguth HR, Pluger WL (2001) Assessing the reactive transport of inorganic pollutants in groundwater of the Bourtanger Moor area (NW Germany). Environ Geol 41:480–488
23. HRN ISO 10304-1:1998. Water quality—Determination of dissolved anions by liquid chromatography of ions-Part 1: Determination of bromide, chloride, fluoride, nitrate, nitrite, phosphate and sulfate
24. HRN ISO 5667-5:2000. Water quality-Sampling-Part 5: Guidance on sampling of drinking water and water used for food and beverage processing
25. Jalali M (2005) Nitrates leaching from agricultural land in Hamadan, western Iran. Agric Ecosyst Environ 110:210–218
26. Kattan Z (2001) Use of hydrochemistry and environmental isotopes for evaluation of groundwater in the Paleogene limestone aquifer of the Ras Al-Ain area (Syrian Jezireh). Environ Geol 41:128–144
27. Katz BG, Chelette AR, Pratt TR (2004) Use of chemical and isotopic tracers to assess nitrate contamination and ground water age, Woodville Karst plain, USA. J Hydrol 289:36–61
28. Kazemi GA (2004) Temporal changes in the physical properties and chemical composition of the municipal water supply of Shahrood, northeastern Iran. Hydrogeol J 12:723–734
29. Kraft GJ, Stites W (2003) Nitrate impacts on groundwater from irrigated-vegetable systems in a humid north-central US sand plain. Agric Ecosyst Environ 100:63–74
30. Krapac IG, Dey WS, Roy WR, Smyth CA, Storment E, Sargent SL, Steele JD (2002) Impact of swine manure pits on groundwater quality. Environ Poll 120:475–492
31. Mayer D, Urumović K, Hlavnjak B, Gold H (1996) Hydrogeological study of the development of water source Ivanšćak. Faculty of mining, geology and petroleum Engineering, Zagreb, Croatia
32. McLay CDA, Dragten R, Sparling G, Selvarajah N (2001) Predicting groundwater nitrate concentrations in a region of mixed agricultural land use: a comparison of three approaches. Environ Poll 115:191–204
33. Melian R, Myrlian N, Gouriev A, Moraru C, Radstake F (1999) Groundwater quality and rural drinking-water supplies in the Republic of Moldova. Hydrogeol J 7:188–196
34. Nemčić-Jurec J, Mesić M, Bašić F, Kisić I, Zgorelec Ž (2007) Nitrate concentration in drinking water from wells at three different locations in northwest Croatia. Cereal Res Commun 35:845–848
35. Nemčić-Jurec J, Konjačić M, Jazbec J (2013) Monitoring of nitrates in drinking water from agricultural and residential areas of Podravina and Prigorje (Croatia). Environ Monit Assess 185:9509–9520
36. Obeidat MM, Massadeh AM, Al-Ajlouni AM, Athamneh FS (2007) Analysis and evaluation of nitrate levels in groundwater at Al-Hashimiya area, Jordan. Environ Monit Assess 135:1–3

37. Pacheco J, Marin L, Cabrera A, Steinich B, Escolero O (2001) Nitrate temporal and spatial patterns in 12 water-supply wells, Yucatan, Mexico. Environ Geol 40:708–715
38. Regulation about the parameters of assessment and methods of analysis of water for human consumption. National newspaper 125 (2013), Zagreb, Croatia
39. Richards RP, Baker DB, Creamer NL, Kramer JW, Ewing DE, Merryfield BJ, Wallrabenstein LK (1996) Well water quality, well vulnerability and agricultural contamination in the Midwestern United States. J Environ Qual 25:389–402
40. SAS 9.2 (http://support.sas.com/documentation/92/index.html). Accessed 25 April 2008
41. Singh SK, Srivastava PK, Singh D, Han D, Gautam SK, Pandey AC (2015) Modeling groundwater quality over a humid subtropical region using numerical indices, earth observation datasets, and X-ray diffraction technique: a case study of Allahabad district, India. Environ Geochem Health 37:157–180
42. Statistica 8.0 (Electronic version): StatSoft, Inc (2011). Electronic statistics textbook. Tulsa, OK: StatSoft. WEB:http://www.statsoft.com/textbook/. Accessed 1 Oct 2011
43. Thorburn PJ, Biggs JS, Weier KL, Keating BA (2003) Nitrate in groundwaters of intensive agricultural areas in coastal Northeastern Australia. Agric Ecosyst Environ 94:49–58
44. Velthof GL, Lesschen JP, Webb J, Pietrzak S, Miatkowski Z, Pinto M, Kros J, Oenema O (2014) The impact of the nitrates directive on nitrogen emissions from agriculture in the EU-27 during 2000-2008. Sci Total Environ 468:1225–1233
45. Widory D, Kloppman W, Chery L, Bonnin J, Rochdi H, Guinamant JL (2003) Nitrate in groundwater: an isotopic multi-tracer approach. J Contam Hydrol 72:165–188
46. Woo-jung C, Gwang-Hyun H, Sang-mo L, Goon-Taek L, Kwang-sik Y, Soo-Myung C, Hee-Myong R (2007) Impact of land-use types on nitrate concentration and ó15N in unconfined groundwater in rural areas of Korea. Agric Ecosyst Environ 120:259–268

CHAPTER 2

Factors Affecting Water Pollution: A Review

Chaudhry FN* and Malik MF

Department of Zoology, University of Gujrat, Hafiz Hayat Campus, Gujrat, Pakistan

ABSTRACT

This study was conducted at University of Gujrat during 2016 to 2017 as a term paper for Master of Philosophy. The data regarding effect of ozone depletion on human was reviewed and compiled as a review paper from various published articles of international reputed journals annual/environmental reports of recognized organization and ebooks. Factors which are effecting water pollution are addressed. Water pollution is very important problem of 21st century. Due to water pollution pure water is becoming less scare day by day. The biggest cause of water pollution is industrialization and increase in population. By drinking polluted water people becoming more and more ill.

Keywords: Water pollution; Pesticides and herbicides; Surface water quality; Chemical pollution; Agricultural pollution; Herbicides; Pesticides; Chemical pollution; Sediment pollution; Urban water runoff; Salt water intrusion

INTRODUCTION

Water pollution has become a global problem now a day's ongoing evaluation of water resource policy is needed to counter this problem.

Deaths and diseases are caused worldwide due to water pollution and approximately 14000 people die every day due to water pollution [1-3].

Both developed as well as developing countries are facing water pollution problems [4]. Water quality is influenced by many factors like precipitation, climate, soil type, vegetation, geology, flow conditions, ground water and human activities. The greatest threat to water quality is posed by point sources of industries and municipalities. Activities like mining, Urban development and Agriculture also effect water quality. Non-point source pollution also includes nutrients, sediments and toxic contaminants [5].

POLLUTANTS

It is a substance which when introduced into environment causes undesirable effects or spoils resources. Long or short term damage may be caused due to pollutant. Biodegradable pollutants only cause short term damage. Some pollutants like DDT again produce pollutants upon degradation like DDD and DDE.

Pollutants may be of different types and having different properties like Stock pollutants which include non-biodegradable plastics, synthetic chemical and heavy metals have no or very little absorptive capacity. These pollutants accumulate in environment with the passage of time. Their damage increases as their quantity increases. For future generations stock pollutants are burdens. Similarly Fund pollutants have some absorptive property in environment. They only cause problem when their quantity increases beyond environment absorbance capacity. E.g., Carbon dioxide only causes problem when its amount increases. These pollutants can only be diluted to reduce their toxicity or recycled into non harmful substances [6].

POINT SOURCE POLLUTION

When source of water pollution is known or pollutants that are entering into water are from identifiable source like ditch, pipe industry, storm drain and sewage treatment plants etc. pollution is known as point source pollution [7]. It can be distinguished from other pollution sources [8].

NON-POINT SOURCE POLLUTION

When source of water pollution is not known or pollution does not come from single discrete source pollution is known as non-point source pollution [9]. It is very difficult to control and may come from different sources like pesticides, fertilizers industrial wastes etc. [10]. Non-point source pollution is the main and leading cause of water pollution in USA [11,12].

GROUND WATER POLLUTION

When pollutants which are present on ground enter the water bodies under earth they cause ground water pollution. When fecal water containing pathogens reaches under earth it makes it unfit for drinking. Pathogen polluted ground water may contain viruses, protozoa and bacteria and rarely in some cases helminth eggs. Consumption of this water causes diseases like diarrhoea and cholera [13,14]. Similarly nitrates also causes ground water pollution causing disease in children called blue baby syndrome in rural population of Bulgaria and Romania. It is observed that when nitrates concentration exceeds above 10 mg/L (10 ppm) in ground water chances of blue baby syndrome increases [15,16]. Excessive use of nitrate fertilizers can also cause water pollution because very small amount of nitrates is utilized by plants most of it accumulates in soil which later on reaches to ground water by leaching and contaminate it [17-21]. Ground water polluted with high levels of fluoride causes dental and skeletal problems [22].

URBAN STORM WATER RUNOFF

It is due to highly populated cities. It comes from homes and office places [1]. In suburban and urban areas pavement and buildings covers much of land surface so whenever there is snow melt or rain the water does not soak into ground. This storm water carries much type of pollutants like dirt, oil, lawn fertilizers and chemicals directly to rivers and streams where they cause water pollution [1]. In the case of natural landscape these pollutants are trapped into pores soil and water is filtered but in cities as water is not able to soak into ground so it wash away all of these pollutant's into water bodies thus polluting them. Moreover this storm water has high speed of flowing which erodes more sediment from embankments of water bodies thus causing water pollution.

AGRICULTURAL POLLUTANTS

As in rural areas population is less so it mostly contains fertilizers, pesticides and eroded soil and these pollutants reach to water bodies through runoff after rain and flood [1]. Agricultural runoff cases fresh water body's eutrophication. Half of lakes in US are eutrophic. Phosphate is the main contributor to eutrophication its high concentration promotes Cyanobacteria and Algae growth which ultimately reduces dissolved oxygen in water [23]. Harmful toxins which accumulate in food chain are produced by cyanobacterial blooms [24]. Nitrogen rich fertilizer compounds causes dissolved oxygen deficiency in rivers, lakes and coastal zones which have devastating effects on oceanic fauna. In America and Northwest Europe nitrogen fertilizer use is controlled from 2006 [25,26]. Nitrogen fertilizers have high water solubility and increased runoff and leaching rate which results in ground water pollution [27-29].

Similarly pesticides are used to control pests these pesticides leaches to ground water thus polluting ground water. Water soluble pesticides leach more. Sandy soil also favours leaching [30,31].

Selenium (Se) is a heavy metal that occurs naturally in soil but due to irrigation practices it accumulates in the soil. This accumulated selenium reaches to water reservoirs and is very toxic for animals and humans [32].

ATMOSPHERIC POLLUTANTS

It is due to small particles which are present in air which it reaches to water bodies through rain. It includes carbon dioxide which produced by burning of fossil fuels its quantity is increasing which it combines with water molecules its forms sulphuric acid. Sulphur dioxide produced from volcanoes and industries also combines with water molecules to form sulphuric acid. Sulphur dioxide is also produced by combustion of coal and petroleum products. Similarly Nitrogen dioxide also combines with water to form nitric acid. Particulates also play very important role in effecting water pollution these particulates reach to water bodies through rain [1,9].

PATHOGENS

Pathogens are the microorganisms which causes disease. Most bacteria in nature are non-pathogenic or beneficial but few are pathogenic and these pathogenic bacteria also pollute drinking water. Coliform bacteria are a bacterial indicator species used for the identification of water pollution. Disease causing bacterial species includes Cryptosporidium parvum, Burkholderia pseudomallei, Giardia lamblia, Norovirus, Salmonella and Parasitic worms like Schistosoma [33-35].

PESTICIDES AND HERBICIDES

Herbicides and pesticides are used to control weeds and pests. Both of them also contribute to water pollution [36]. Their leaching also pollutes ground water. Leaching is influenced by soil texture, pesticide properties, irrigation and rain fall. If soil is sandy and pesticide is water soluble more will be the leaching. Similarly pesticides and herbicides also reach natural water bodies through runoff. These pesticides residues when reach to natural water bodies they disturb flora and fauna there. Pesticides which don't degrade easily or take time to degrade are more harmful [37-39].

CHEMICAL POLLUTANT

It comes from waste of harmful chemicals factories it is a material which is left as a by-product during manufacturing process and it also plays a big role in polluting water bodies [40-42]. Hazardous chemical waste may be in solid, liquid or in gaseous form. The characteristics which make material hazardous are corrosively, Ignitability, toxicity and reactivity [41]. It started with the start of industrial revolution [42]. Industrial waste chemicals can only be treated by using special waste treatment plants they cannot be treated by sewage treatment plants [43].

SEDIMENT POLLUTION

Sedimentation due to runoff effects water quality. It decreases the capacity of streams, ditches, navigation channels and rivers. It decreases the penetration of light into water due to which due to under water flora is disturbed. So the fishes and other fauna feeding on that flora are also disturbed and whole food chain is disturbed. Pollutants like pesticides and phosphorus are transported and accumulated due to sedimentation [38]. Sediment particles also attach to fish gills so fishes feel difficulty to respire in this way they causes fish death. Similarly sediments carry dangerous chemicals like pesticides and petroleum products to water bodies thus polluting them [1].

SALTWATER INTRUSION

Salt water intrusion is another very important factor which pollutes ground water. It occurs when saline water from sea enters into ground water near coastal areas. It occurs naturally but some human activities like pumping of fresh groundwater also increases salt water intrusion. Navigation channels, drainage channels and agriculture channels also play important role in salt water intrusion [44].

CONCLUSION

Water is polluted by many factors among which industrial wastes are the most important. Beside industrial wastes other factors include herbicides, pesticides and atmospheric pollutants. Pathogen in polluted water causes serious diseases in humans. The whole ecosystem of water bodies is disturbing due to water pollution.

RECOMMENDATIONS

To treat industrial wastes there should be special industrial waste treatment plants with every industry. Similarly there should also be urban runoff pounds to remove pollutants from runoff and to prevent floods. Toxic pesticides and Herbicides should be replaced with nontoxic ones or Pesticides should be replaced with biological control.

REFERENCES

1. Letchinger M (2000) Pollution and Water Quality, Neighbourhood water qualityassessment. Project oceanography.
2. Pink DH (2006) Investing in tomorrow's liquid gold. World Journal of Analytical Chemistry 2: 42-46.
3. Larry w (2006) World water day: A billion people worldwide lack safe drinking water.
4. National Water Quality Inventory Report to Congress (2009) Washington, D.C: United States Environmental Protection Agency (EPA). EPA 841-F-08-03.
5. Florescu, Ionete RE , Sandru C,Iordache A, Culea M (2010)The influence of pollution monitoring parameters in characterizing the surface water quality from Romania southern area.Rom Journ Phys 56: 7-8.
6. Tietenberg T (2006) Economics of Pollution Control, Chapter 15 in Environmental and Natural Resource Economics, (7th edn.), Pearson, Boston.
7. Claudia C (2016)Clean Water ActSection .United States.
8. Hogan CM(2010) Water pollution. Encyclopedia of Earth, Topic ed. Mark McGinley, ed., in chief C.Cleveland, National Council on Science and the Environment, Washington DC.
9. Brian M (2008) "Water Pollution by Agriculture" (PDF). Phil. Trans. Royal Society B363: 659-666.
10. Texas Commission on Environmental Quality (2013) Municipal Solid Waste in Texas: A Year in Review.
11. U.S. Environmental Protection Agency (EPA). Washington, D.C. Cycle. "(October 2007). Document No. EPA-841-R-07-001.
12. (2003) National Management Measures to Control Non-point Source Pollution from Agriculture."Document No. EPA 841-B-03-004.
13. WolfL, Nick A, CroninA (2015)How to keep your groundwater drinkable: Safer siting of sanitation systems – Working Group 11 Publication. Sustainable Sanitation Alliance, pp: 1-7.

14. Jennyfer W, Ustün P, Annette,Cumminget al.(2014)"Systematic review: Assessing the impact of drinking water and sanitation on diarrhoeal disease in low- and middle-income settings: systematic review and meta-regression". Tropical Medicine & International Health19: 928-942.
15. Buitenkamp M , StintzingAR(2008)Europe's sanitation problem – 20 million Europeans need access to safe and affordable sanitation. Women in Europe for a Common Future (WECF), The Netherlands.
16. KnobelochL,Salna B ,Hogan A, Postle J,Anderson H(2000) Blue Babies and Nitrate-Contaminated Well Water.EnvironHealth Perspect 108: 675-8.
17. Khan MN, Mohammad F (2006) Eutrophication: Challenges and Solutions. In: Ansari AA, Gill SS (eds.), Eutrophication: Causes, Consequences and Control, Springer Science Business Media Dordrecht 2014.
18. Rosen CJ ,Horgan BP(2009) Preventing Pollution Problems from Lawn and Garden Fertilizers. Extension.umn.edu.
19. Singh B, Singh Y, Sekhon GS(2012) Fertilizer-N use efficiency and nitrate pollution of groundwater in developing countries.Journal of Contaminant Hydrology20: 167-184.
20. Schonbeck M(2004) NOFA Interstate Council: The Natural Farmer. Ecologically Sound Nitrogen Management.
21. Jackson LE, Burger M, Cavagnaro TR (2008) Roots, Nitrogen Transformations,and Ecosystem Services.Annual Review of Plant Biology59: 341-363.
22. Lennon MA, Whelton H, O'Mullane D, Ekstrand J (2004) Fluoride in drinking-water. World Health Organization.
23. Werner W (2002) Fertilizers, 6. Environmental Aspects. Ullmann's Encyclopedia of Industrial Chemistry, Wiley-VCH, Weinheim.
24. Schmidt JR, Shaskus M, Estenik JF, Oesch C, Khidekel R, et al. (2013) Variations in the microcystin content of different fish species collected from a eutrophic lake. Toxins (Basel) 5: 992-1009.
25. Van Grinsven HJM, Ten Berge HFM, Dalgaard T, Fraters B, Durand P, et al. (2012) Management, regulation and environmental impacts of nitrogen fertilization in northwestern Europe under the Nitrates Directive; a benchmark study. Biogeosciences 9: 5143-5160.
26. State-EPA Nutrient Innovations Task Group (2009)An Urgent Call to Action – Report of the State-EPA Nutrient Innovations Task Group.
27. Rosen CJ, Horgan BP (2009) Preventing Pollution Problems from Lawn and Garden Fertilizers.
28. NOFA Interstate Council: 9 (2004). The Natural Farmer. Ecologically Sound Nitrogen Management. Mark Schonbeck.
29. Singh B, Singh Y, Sekhon GS (2006) Fertilizer-N use efficiency and nitrate pollution of groundwater in developing countries. Journal of Contaminant Hydrology. 20: 167-184.
30. Environmental Fate of Pesticides(2015) Pesticide Wise. Victoria, BC: British Columbia Ministry of Agriculture.
31. McBride DK (1989) Managing pesticides to prevent groundwater contamination. North Dakota State University Extension Service, Publication E-979.
32. Ganje TJ (1966)Selenium In: ChapmanHD(ed.) Diagnostic Criteria for Plants and Soils: 394-404.
33. EPA (2003) U.S. Environmental Protection Agency Nonpoint Source Control Branch (4503T) 1200 Pennsylvania Avenue, NW Washington, DC 20460.
34. USGSRestonVA(2014) A Primer on Water Quality.U.S. Department of the Interior, U.S. Geological Survey Feb 18 2014.
35. Thomas RS (2000) Microbes and Urban Watersheds: Concentrations, Sources, & Pathways. Center for Watershed Protection. Ellicott City, MD.
36. Environmental Databases: Ecotoxicity Database. (2006) Pesticides: Science and Policy. Washington, D.C.: U.S. Environmental Protection Agency (EPA).

37. Environmental Fate of (2015)Pesticide Wise. Victoria, BC: British Columbia Ministry of Agriculture.
38. Dudal R (1981) An evaluation of conservation needs. In: Morgan RPC (ed.),Soil Conservation, Problems and Prospects, Chichester, U.K.: Wiley.
39. Pope CA, Bhatnagar A, McCracken JP, Abplanalp WT, Conklin DJ, et al.(2016)Exposure to fine particulate air pollution is associated with endothelial injury and systemic inflammation. Circulation Research 119: 1204-1214.
40. Bill H (2010) Techniques for Efficient Hazardous Chemicals Handling and Disposal. Pollution Equipment News, pp: 13.
41. Laboratory chemical waste management guidelines (2016) Environmental Health and Radiation Safety University of Pennsylvania.
42. Maczulak A (2010) Pollution: Treating Environmental Toxins. New York: Infobase Publishing,pp: 120.
43. https://www3.epa.gov/npdes/pubs/pretreatment_program_intro_2011.pdf
44. Teddy J (2007) Battling Seawater Intrusion in the Central and West Coast Basins (PDF). WRD Technical Bulletin. Journal of Geoscience and Environment Protection 4.

CHAPTER 3

Water Related Health Problems in Central Asia—A Review

Zakir Bekturganov [1], Kamshat Tussupova [2], Ronny Berndtsson [2,*], Nagima Sharapatova [3], Kapar Aryngazin [4] and Maral Zhanasova [5]

[1] *Department of Public Health, Karaganda State Medical University, Gogol Street 40, Karaganda 100008, Kazakhstan*

[2] *Department of Water Resources Engineering & Center for Middle Eastern Studies, Lund University, Box 118, Lund SE-221 00, Sweden*

[3] *MROO "Educational-health Center", Petrova str. 30/1, 13, Astana 010000, Kazakhstan*

[4] *Department of Professional Training and Protection of the Environment, Pavlodar State University, Lomov Street 64, Pavlodar 140008, Kazakhstan*

[5] *Department of Surgical Disease, Karaganda State Medical University, Gogol Street 40, Karaganda 100008, Kazakhstan*

ABSTRACT

The present paper provides an extensive literature review on water related health issues in Central Asia. Even though the per capita amount of available freshwater is substantial in all Central Asian states the uneven distribution in time and space creates problems for water availability. Due to this, the Central Asian economies are developing under increasing water deficiency. The degradation of water supply systems and sewage treatment plants is often severe leading to potentially high water loss rates and inadequate accessibility to safe water supply. In this context, rural areas are the most affected. Low tariffs in combination with absent metering and low collection rates for water fees mean that operation and maintenance costs for basic services of water supply and sanitation are not covered. Unsafe water supply contains both microbiological and non-microbiological contaminants. Helminthiasis and intestinal protozoa infections are of considerable public health importance in Central Asia. Agricultural and industrial pollution is especially affecting downstream areas of Amu Darya and Syr Darya rivers. In large areas copper, zinc, and chromium concentrations in water exceed maximum permissible concentration. Thus, there is an urgent need to strengthen the environmental monitoring system. Small-scale water supply and sanitation systems need to be developed in line with more efficient public spending on these.

Keywords: Central Asia; water supply; public health; safe drinking water; sanitation; pollution

1. INTRODUCTION

Central Asia refers to the five former Soviet Union states Kazakhstan, Turkmenistan, Uzbekistan, Tajikistan, and Kyrgyzstan [1]. Central Asia stretches from the Caspian Sea in the west to China in the east and from Afghanistan in the south to Russia in the north (Figure 1).

Consequently, all five Central Asian states have experienced the crumbling of the Soviet Union in 1991, the following collapse of the economic system, and the subsequent socioeconomic upheaval [2]. The transition from state planned to market economy has meant changing patterns for basic services such as water supply and sanitation [3]. At present Central Asia is home to a population of about 66 million (Kazakhstan 17.0 million, Kyrgyzstan 5.7 million, Tajikistan 8.0 million, Turkmenistan 5.2 million, and Uzbekistan 30.0 million) [4]. By 2040, the total population is expected to have increased to about 86 million. This together with countryside migration to urban areas will put an enormous stress on water and infrastructure. Water supply availability in Central Asia is complicated due to that the region's two major rivers, the Syr Darya and Amu Darya, are transboundary. Thus, continual conflicts over the region's water resources have characterized the Central Asian states since the collapse of the Soviet Union [4].

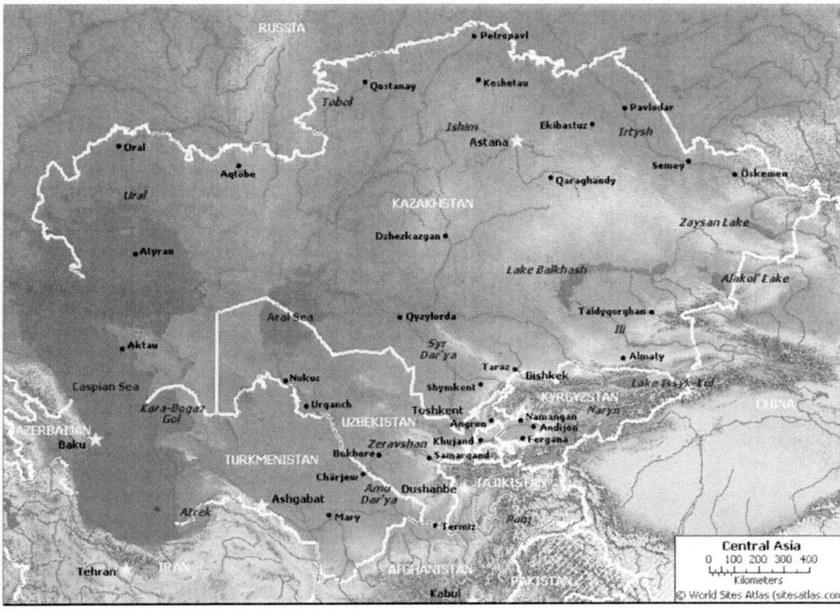

Figure 1. Outline of Central Asia and the five former Soviet Union states Kazakhstan, Turkmenistan, Uzbekistan, Tajikistan, and Kyrgyzstan (courtesy of World Sites Atlas).

The Central Asian economies are developing under increasing water deficiency [5,6,7,8,9]. According to I.V. Severskiy [2], 70% of developmental problems in the region are caused by freshwater shortage. The main reasons for this are increasing political tensions and worsening ecological and socioeconomic conditions. Freshwater shortage, however, also affects the public health and providing continuous access to safe drinking water is fundamental to protecting the population from microbiological disease [10]. Before 1991, public health service mainly followed the Soviet model with priority on regulatory hygienic and sanitary control measures for combatting infectious diseases [11]. After 1991, the notion of "New Public Health" including a broader range of evidence-based scientific, technological, and management based systems for improving the health of individuals became more widely accepted [12,13,14]. The new approach, that also stressed chronic non-communicable diseases, health policy, health promotion, health systems management, and public health practice, eventually lead forward to the reformulation of educational programmes and legislation. At present the main challenge for Central Asian states public health sectors is in clarifying, coordinating, and streamlining the roles and responsibilities of different agencies responsible for public health and health promotion activities [15].

In view of the above, there is a remarkable absence of basic information on water related health issues of Central Asia in the English scientific literature. The lack of information may reflect a general absence of investigation efforts during the period following the collapse of the Soviet Union. It is believed that this collapse also was followed by a breakdown in public health infrastructure. On the other hand, much of this information may be delimited to the Russian scientific literature. Relevant data and information may also be available in unpublished documents and reports. The authors are, however, not aware of any previous scientific publication trying to cover the state of art situation on water related health problems in the Central Asia. Along this line, we try to summarize all available scientific literature and publically available reports on this subject, synthesise, and draw general conclusion from the state of water and health in Central Asia. The introductory chapter is followed by a description of distribution of freshwater resources of Central Asia. The chapter after this describes threats to water quality. In the following chapter, access to safe water is outlined. The final chapter treats occurring and possible future health impacts. We close with a conclusion and a discussion on possible ways forward.

2. DISTRIBUTION OF FRESHWATER RESOURCES

Central Asia is located in the centre of the Eurasian continent occupying an area of about 4 million km^2 [16]. The main part of the Central Asia is arid and thus, there is a general deficit of freshwater. The climate is strongly continental arid to semiarid with hot cloud-free summers and humid temperate winters in the south and cold winters in the north [17]. Average annual precipitation for Central Asia is about 273 mm. It varies from about 161 mm in Turkmenistan to 691 mm per year in Tajikistan. The by far largest country Kazakhstan receives on average about 250 mm per year. The plains and deserts receive less than 70 mm per year and the high mountains of central Tajikistan up to 2400 mm per year [18,19,20,21,22]. Potential evaporation similarly varies from above 2250 mm per year in the most arid region to less than 500 mm per year in the mountainous areas.

Water resources in Central Asia are constituted by surface water and groundwater. The major part of potentially usable water comes from the large rivers of the region. All these rivers, however, are transboundary. The largest, the Amu Darya and the Syr Darya, flow through more than three countries. The Amu Darya and Syr Darya account for 90% of

Central Asia's river water. Especially, Turkmenistan and Uzbekistan are vulnerable due to that a major part of the water resources is generated outside of their respective territory. The Central Asian water problem is thus, complex and also involving other transboundary stakeholders such as Russia, China, Afghanistan, Iran, and Pakistan. Other large rivers are the Irtysh and Ishym located in the east parts of Central Asia, the Chu and Talas located in the south, in the west the Ural can be found, and finally the Ishim and Tobol are found in the north [2]. The by far largest withdrawal of water for irrigation is done from the large rivers of the region. However, the area also contains countless smaller river and creeks that as well contribute to irrigated agriculture. The information on this, however, is often absent.

Totally available water resources the respective country are given in Table 1 [23,24,25]. The groundwater resources generally constitute about 10%–15% in comparison to the surface water resources [23]. Turkmenistan, though, has reminiscent groundwater only. As seen from the table, there are substantial potentially available water resources as a whole in Central Asia. The smallest per capita amount of water is represented by Uzbekistan with 1870 m^3/capita and year. In general, an amount of less than 3000 m^3/capita and year may be regarded as economic water scarcity and less than 1000 m^3/capita and year as physical water scarcity [26]. According to this criterion, only Uzbekistan falls below the threshold for economic water scarcity. Most part of the surface water in Table 1 can be regarded as annually renewable water. However, it must be remembered that there is a great spatial and temporal variability of surface water. Groundwater as well, especially deeper groundwater, will not be replenished as quickly as the surface water. As seen from the table, especially Uzbekistan and Turkmenistan are dependent on water sources located outside of the country borders. This makes them more vulnerable in case of political conflict and hydropolitical issues.

Table 1. Totally renewable water resources in Central Asian countries (surface and groundwater, after [23,24,25]).

Country	Total Renewable Water Resources (km^3/Year; within Brackets is Given Share of Outside Sources)	Total Renewable Water Resources (m^3/Capita and Year)	Average Precipitation (mm/Year)	Main Agricultural Production
Kazakhstan	117 (34)	6490	250	Wheat, livestock
Kyrgyzstan	58 (0)	8480	530	Livestock
Tajikistan	99 (16)	13,500	690	Cotton, wheat
Turkmenistan	25 (23)	4090	160	Fruit, vegetables
Uzbekistan	59 (34)	1870	265	Cotton, wheat, fruit

During the last decades of the 20th century, Central Asia suffered an enormous ecological crisis. The Aral Sea, the fourth largest lake in the world, started to dry up [27]. The reason for this was the doubling of irrigated agricultural area from 4.3 to 8.2 million ha. Due to the changed ecology, several hundred thousand square kilometers with a population of several million have been damaged [28].

As seen from the above, the Central Asian states are not water scarce in terms of total water supply per capita. The uneven distribution in space and time in combination with excessive and often uncontrolled withdrawal for irrigation, however, create water scarcity especially in areas to the south. These problems can only be solved at the international level and involving integrated water resources management. This, however, requires a political will and the involvement of all stakeholders, including an environmentally aware public.

3. THREATS TO WATER QUALITY

3.1. Human Waste

Just before the Soviet Union collapse, about 70% of all cities and 20% of all villages had their own wastewater collection systems [23]. Wastewater treatment plants were usually designed with mechanical and biological treatment. When the Soviet State collapsed in 1991, the majority of wastewater treatment plants halted their operation. During the following period, lack of investment and maintenance resulted in serious degradation of the majority of sewage treatment plants. Still, most wastewater treatment plants do not operate efficiently. Their present technical standard is insufficient due to lack of equipment and spare parts, chemicals, and trained personnel [23]. Almost everywhere, biological treatment stages are not functioning efficiently. Mechanical treatment is only operating at a few city wastewater treatment plants.

According to Unicef and the MDG (Millennium Development Goals) the portion of the 2015 population who gained access to improved sanitation since 1990 for Caucasus and Central Asia was 24% [29]. All Central Asian countries except for Turkmenistan (no data) achieved 93%–100% improved sanitation. Reviews indicate that improved sanitation can reduce diarrhoeal diseases by 32%–37% [30,31,32]. However, improved sanitation needs to be combined with good operation and maintenance of water supply systems. Worn-out water pipe-lines are often the reason of deteriorated potable water quality and higher incidence of intestinal infections.

One gram of fresh faeces from an infected person can contain 10^8 viral pathogens, 10^8 bacterial pathogens, 10^4 protozoan cysts, and 10^4 helminth eggs [33,34]. The lack of sanitation, as well as occurrence of polluted water, cause different consequences in the health of the population [35]. Drinking water provided by old water pipes can be the cause of acute gastrointestinal illnesses such as typhoid, dysentery, and cases of viral Hepatitis A [35,36,37]. Diarrhoeal disease accounts for an estimated 4.1% of the total DALY (disability-adjusted life year) global burden of disease. It has been estimated that 88% of that burden are attributable to unsafe water supply, sanitation, and hygiene according to World Health Organization [28]. In the harsh climate of Central Asia, rural villagers often obtain water from shallow wells [38,39]. These wells are typically located near the house and the outdoor toilet. The prevalence of *Heliobacter pylori* infections is inversely related to living standard and sanitary practice [40,41]. *H. pylori* infection is linked with gastritis and associated diseases such as peptic ulcer, gastric adenocarcinoma, and primary gastric lymphoma [36]. In Kazakhstan, the morbidity and mortality associated with gastrointestinal diseases are high and gastric cancer is the second largest cause of cancer death [38,42]. Results by Nurgalieva *et al.* [38] suggest that transmission of *H. pylori* can be water borne, related to poor sanitary practices, or both.

3.2. Agriculture

Central Asia has been one of the world's fastest growing regions since the end of the 1990s [18]. Thus, the area has shown a strong development potential. The reasons for this are natural resources such as oil, gas, and gold together with cotton and agricultural production combined with acceptable infrastructure and human capital. The added value of agriculture to the GDP is about 10% for the Central Asian region (2010) [18]. It generally ranged from 5% in Kazakhstan to about 21% in Kyrgyzstan and Tajikistan. On average, about 30% of the economically active population are engaged in farming (ranging from 14% in Kazakhstan to 29% in Turkmenistan) [18]. Cultivated area per person economically active in agriculture varies from 1.1 to 1.7 ha/person in Tajikistan and Uzbekistan respectively, 2.7 ha/person in Kyrgyzstan and Turkmenistan to almost 20 ha/person in Kazakhstan [18]. This gives an average for the region of 3.3 ha/person.

Requirements to maximise irrigated agriculture during the Soviet era resulted in a general degradation of water resources. After independence, most countries in the region adopted national policies regarding water supply and sanitation. Still, about 90% of the total water supply are used for irrigation [23]. Intensified irrigation increases the water pollution by chemicals used in agriculture. Increased agricultural production leads to an increased usage of mineral fertilizers and chemicals protecting plants against pests, weed, and diseases. As a result, many chemical substances, pesticides, and herbicides are discharged to the environment. Some of the substances such as DDT are bio-resistant and have a reported half-life of about 15 years in soil.

The drying up of the Aral Sea and the decrease of the water surface area combined with increasing contamination levels in the remaining water bodies, constitute a threat for environment and humans [27]. Irrigation return contains salts, fertilizers, pesticides, herbicides, and cotton defoliants [43]. However, at present significant efforts are being made to restore wetlands, improve habitat conditions, and reduce pollution in the Aral Sea region [44]. Even so, I.V. Severskiy [2] notes that salinization and pollution problems are increasing due to irrational land and water use. About 50% of the irrigated area in the Aral Sea basin are experiencing increased salinization. This has resulted in up to 50% reduction in agricultural productivity [45,46]. Agricultural pollution is especially affecting downstream areas of Amu Darya and Syr Darya [2]. Copper, zinc, and chromium concentrations exceed maximum permissible concentration [47]. More than 70% of the area within the Amu Darya Basin in Uzbekistan have a water quality that is dangerous to health. More than 10% of the water are extremely dangerous [48,49].

The above clearly materializes two areas of strong concern, safe sanitation and safe management of chemical elements in agriculture. Large investments are needed to rehabilitate both water supply and sanitary systems in Central Asia. The public need to be made aware of the dangers with using open unprotected water sources and the direct benefits of paying for a safe water supply. For irrigated agriculture in the south Central Asia, wasteful irrigation techniques need to be curbed. If possible, re-use techniques of water should be developed. Alternative methods to the excessive use of herbicides and pesticides need to be established. At the same time, environmental control and management need to be put in place involving all concerned stakeholders.

3.3. Industry

Surface and groundwater quality in the Central Asian region, is commonly affected by agriculture, municipal wastewater as well as industry. In Kazakhstan, water sources are often classified as unsatisfactory [18]. Common types of water pollution is chemical, oil, manufacturing and metallurgical industry contamination. The Kazakhstan Hydrometeorology Service Bureau classified less than 30% as clean out of 44 investigated water sources in 2002. Also, besides the common polluters such as industrial, mineral extracting and refinery enterprises, urban buildings, farms, irrigated fields, waste containers, and storage facilities for liquid and solid waste and oil products contribute to the general pollution level [18]. River water in Kyrgyzstan, however, usually displays good water quality due to that it is fed by glacial melt. Nitrate, organic matter content, and nutrient contents are usually low. There are a few cases of water pollution from storage facilities and use of fertilizers and chemicals, industrial waste, non-compliance of the sanitary code, improper conditions for sewerage systems, cattle breeding, and industrial effluent. The water source for drinking water is usually groundwater. Nuclear tailing dumps, however, are a serious problem in Kyrgyzstan. Also in Tajikistan, water supply is in general satisfactory, except for a few lakes and groundwater sources. In Turkmenistan, however, river water and drainage networks are often polluted. The river water repeatedly contains high concentrations of salts and pesticides both from domestic sources and upstream international basins. This is especially the case for the Aral Sea Basin where the human pressure on surface water is high. In Uzbekistan, some rivers that receive discharge of sewage and municipal wastewater display high pollution levels. Pollution from petroleum industry may reach 0.4 to 8.2 MAC (maximum allowable concentration), phenols pollution may reach 6 MAC, nitrates 3.7 MAC, and heavy metals 11 MAC [18]. This also threatens the groundwater.

4. ACCESS TO SAFE WATER DRINKING WATER

4.1. Drinking Water Supply

In general, during the last 10–15 years of independence, the quality of the water supply services has dramatically deteriorated [23]. The main cause for this deterioration is the marked reduction in public spending due to the general economic recession. In 2010, about 74% of the total population in Central Asia, 94% of the urban, and 64% of the rural population had access to improved drinking water sources (Table 2). About 7.5 million in Uzbekistan, 4.8 million people in Tajikistan, and 2 million in Kyrgyzstan lack proper access to clean drinking water [22,50,51]. The World Bank points at the Kyrgyzstan's worsening health indicators are caused by poor sanitation and hygiene [52]. However, the World Health Organisation (WHO) notes that improved access to water in Central Asia has occurred since 2011 [53]. WHO [53] summarized problems with safe access to drinking water in Kazakhstan but this also holds for most of Central Asia [54]:

1. Degradation of the water supply infrastructure that is 80%–100% in some cases leading to frequent interruption of water supply and massive losses of water;

2. Unclear relationships between willingness to pay for water supply service and installment of meters in relation to perceived gains;

3. Despite measures aiming at providing public access to fresh water it is still unavailable for a majority of the rural population.

Table 2. Access to drinking water and sanitation in Central Asia (adapted from [23]).

	Access to Water for Population (%)		Coverage of Sanitation System (%)	
	Urban	Rural	Urban	Rural
Kazakhstan	78	>35	84	10
Kyrgyzstan	82	58	68	28
Tajikistan	93	49	20	5
Turkmenistan	85	42	62	2
Uzbekistan	90	71	85	40

On the positive side, Roberts *et al.* [55] found that there have been improvements in access to piped water in Central Asia between 2001 and 2010. However, the authors state that still significant gaps remain. This is particularly true for the rural and poor households. There is a need for sustained investment in basic infrastructure for piped water in the region that ensures poorer and rural populations to benefit. In Kazakhstan, 78% of the urban population and >35% of the rural population have access to a varying degree of safe drinking water [23] (Table 2). In general, urban areas are supplied rather well with both drinking water and safe sanitation. Rural areas at large distances from central water supplies and piped sewage systems face serious difficulties. In Kyrgyzstan, the same figures are 82% and 58%, respectively. The wear-and-tear of the rural water supply system is >40% that often means contamination of microbes and chemicals [23]. According to Global Water Partnership, Central Asia and Caucasus [23], a population in excess of 600,000 people in Kyrgyzstan does not have access to safe sources of drinking water and take water from irrigation canals, ditches, and rivers. In Tajikistan, 93% of the urban population and less than 49% of the rural population have access to piped drinking water. About 80% of the rural population use water that does conform to hygienic standards [23]. In Turkmenistan about 85% and 42%, in urban and rural areas, respectively, have access to piped water. The same figures for

Uzbekistan are about 90% and 71%. According to the above, it is clear that especially the rural and poor households do not have access to safe water supply. The situation is similar all over Central Asia.

4.2. Treatment of Drinking Water

Inferior drinking water quality contributes to the general health problems of the population [55,56,57,58,59,60,61,62,63,64,65]. The low level of access to high-quality drinking water is only one of the major problems in Central Asian countries. Ageing and lack of maintenance of water pipeline networks cause emergency conditions, in which people have to use water from other, mostly untreated sources of water. Other factors also play a major role such as pollution of water supply sources, discharge from industry and agriculture, and secondary pollution of drinking water by bacterial activity caused by deterioration of the anticorrosion coating of pipe surfaces [31,66].

The majority of existing drinking water treatment plants and water supply systems in Central Asia were built during the period 1950–1980. The construction philosophy was to build systems that needed low capital investment and small cost for process equipment but considerably high operational costs [23]. The main emphasis was to exploit new water supply sources in terms of pumping stations, water treatment facilities, and maximum flow capacity in main water pipes. Efficient development of water distribution systems, their zoning and rational water use, metering and administrational issues were considered to be outside of the operators' scope of activities [23]. The main problem of today is often to supply rural areas with safe water in a degraded pipe system. In many cases disinfection by chlorine is the only method used. Sometimes no disinfection is taking place. However, even chlorination will not be satisfactory due to leaks and contamination during flooding. In order to modernize the water supply system that on average is older than 35 years over vast areas such as, e.g., the Kazakhstan territory, huge investments are necessary. A possible strategy would be to also invest in locally adapted small-scale water supply and sanitation systems.

5. HEALTH IMPACTS

K. Frenken [18] investigated health effects from water-related problems in Turkmenistan, Kyrgyzstan, and Uzbekistan. The authors summarized that, although no information is available, similar problems may be present in also the other countries of Central Asia. Major factors affecting the spread of water-related diseases were identified as:

1. Use of untreated wastewater to meet water shortage;
2. Lack of infrastructure, especially related to wastewater treatment and discharge;
3. Lack of health awareness and proper handling of polluted water; and
4. Lack of regulations related to the protection of the environment and public health.

In Kyrgyzstan, 122,800 inhabitants were reported to be affected by water-related diseases (2005). In Turkmenistan, inhabitants affected by water-related diseases were estimated to be 12,295 (2004). Out of these, 7955 were intestinal infections, 22 were typhoid, and 4318 were virus hepatitis. An outbreak of malaria with 137 cases occurred in 1998. Since then malaria is reported as eliminated. In Uzbekistan, agricultural productivity is reduced due to land and water salinization. The population in these areas suffers from high levels of anaemia, together with growing levels of tuberculosis. Children are affected by liver, kidney and respiratory diseases, micronutrient deficiencies, cancer, immunological problems, and birth defects. In Karakalpakstan 40% of the rural population depend on small subsistence plots for their livings. These plots are negatively affected by water scarcity and pollution. The situation worsened during 2001–2002 in Karakalpakstan and Khorezm from a drought period and water shortage negatively impacted domestic and personal hygiene [18]. During this period the poor rural people were exposed to a higher risk for water-related diseases such as typhoid, diarrhoea, and worm infections. The government has made progress, however, still only 54% of the urban and 3% of the rural population have access to adequate sewage systems.

5.1. Non-Microbiological Contaminants

Release of contaminants from industry and agricultural areas can spread pollutants to extensive downstream areas. Raised concentrations of organochlorine pesticides and toxic metals such as arsenic have been found in human blood, milk, hair, and urine of the exposed population living in the Aral Sea Drainage Basin [67,68,69,70]. The exposed population is experiencing increasing maternal and infant mortality, delay in growth and puberty of children, liver and digestive problems, allergies, and diseases related to occurring bacteria in water [67,71,72]. In the area, only 20%–30% of the rural population have access to piped water and 25% use water from irrigation channels as their main drinking water source [73]. Also other elements have been found in the drainage basin's surface and groundwater with higher concentrations as compared to the World Health Organization (WHO) maximum recommended contents such as nitrogen species, copper, lead, chromium, and uranium [67,74,75,76,77]. The type of health effects depends on type of contaminant and exposure time [59,67,78]. Organic pollutants such as DDT, DDD, and DDE are used in controlling weeds and pests in the Aral Sea Drainage Basin and probably elsewhere in Central Asia. Törnqvist *et al.* [68] found that the surface water constituents copper, arsenic, nitrite, and DDT in Mejdurechye reservoir are above recommended maximum concentrations by WHO and represent cumulative health hazards. Even more alarming results were that groundwater was overall associated with much higher health risks as compared to surface water. Switching from surface water to groundwater, e.g., during drought periods, is therefore not encompassing better quality water.

5.2. Microbiological Elements

As mentioned above, contact between inappropriate sanitary facilities and drinking water supplies can result in serious infection risks. Matthys *et al.* [79] investigated the overall prevalence of infection with helminths and pathogenic intestinal protozoa in school children for a rural area in Tajikistan. Helminthiasis and intestinal protozoa infections are of considerable public health importance in Tajikistan and elsewhere in Central Asia [80,81]. The overall prevalence of infection with helminths and pathogenic intestinal protozoa was found to be 32% and 47%, respectively. There was pronounced spatial heterogeneity. The most common helminth species were *Hymenolepis nana* (26%), whereas the prevalence of *Ascaris lumbricoides*, hookworm, and *Enterobius vermicularis* was below 5%. The prevalence of pathogenic intestinal protozoa, namely *Giardia intestinalis* and *Entamoeba histolytica/E. dispar* was both 26%. Almost half of the households took drinking water from open water sources, such as irrigation channels, rivers, and unprotected wells. The households used sanitary facilities such as pit latrines, mostly private, and sometimes shared with neighbours. The use of public tap/standpipe as a source of drinking water supply proved to be a protective factor against *G. intestinalis* infection. Protected spring water also reduced the risk of infection with *E. histolytica/E. dispar* and *H. nana*.

Climate change is likely to have less effects on human vulnerability as compared to changes in land use and inefficient water management [82]. However, climate change is likely to increase the region's vulnerability through the water supply and thereby water-borne and vector-borne infections. The rapid economic decline after the Soviet State collapse in

1991 brought back epidemic typhus, tuberculosis, diphtheria, meningitis as well as other infectious diseases to Central Asia [83,84]. Another example is malaria through the *Plasmodium falciparum* that was almost eradicated in Soviet Union by the end of the 50ies through campaigns of insecticides, anti-malarial therapy, land reclamation, and water management. However, in the 1990s malaria was back in Uzbekistan, Kyrgyzstan, Turkmenistan, and Tajikistan [85]. Climate change resulting in increase in winter temperature and changing precipitation patterns may increase the region's vulnerability by creating more favorable conditions for vectors and parasites. In addition, infectious diseases spreading through oral-fecal mechanisms, e.g., typhoid, paratyphoid, salmonella, dysentery, amebiasis, and helminthiasis, are all associated to warm climate [82].

6. CONCLUSIONS AND DISCUSSION

The Central Asian countries share many common problems and unaddressed tasks in health-related water supply and sanitation [23]. The degradation rate of the water supply systems and sewage treatment plants are high. This leads to potentially high water loss rates and inadequate accessibility to safe water supply which is a specific problem for rural areas. Low tariffs in combination with absent metering and low collection rates for water fees mean that operation and maintenance costs for basic services of water supply and sanitation are not covered. The human resource base for the water supply and sanitation systems needs to be better trained and capacity building is needed for the governance. Large distances between remote and sparsely populated villages in rural areas means that alternative systems may be needed. Such systems may be constituted by efficient and small-scale water supply and sanitation systems.

The environmental codex and regulations need to be implemented and strengthened. Ecological expertise needs to be supplied to industrial as well as agricultural activities in order to ensure appropriate environmental monitoring of pollutants. The polluter pay principle needs to be introduced and legislated [31].

Legislative and controlling environmental authorities need be strengthened and made more efficient. There is a general lack of integration between ministries, nongovernmental organizations, and the general public. By implementing integrated water resources management all ministries for environmental protection, agriculture, water resources, health, local governments, municipal authorities, nongovernmental organizations, and representatives for industry can jointly solve water and environmental problems, create national action plans, as well as plans on the regional and local levels.

Ecological thinking needs to be stressed in education. Also, the general awareness regarding environment and detrimental effects of human and industrial waste needs to be strengthened.

REFERENCES

1. Cowan, P.J. Geographic usage of the terms Middle Asia and Central Asia. *J. Arid Environ.* **2007**, *62*, 359–363.
2. Severskiy, I.V. Water-related problems of Central Asia: Some results of the (GIWA) international water assessment program. *Ambio* **2004**, *33*, 52–62.
3. Tussupova, K.; Berndtsson, R.; Bramryd, T.; Beisenova, R. Investigating willingness to pay to improve water supply services: Application of contingent valuation method. *Water* **2015**, *7*, 3024–3039.
4. Siegfried, T.; Bernauer, T.; Guiennet, R.; Sellars, S.; Robertson, A.W.; Mankin, J.; Bauer-Gottwein, P.; Yakovlev, A. Will climate change exacerbate stress in Central Asia? *Clim. Chang.* **2012**, *112*, 881–899.
5. Dukhovny, V.A. Water economy problems of the Aral Sea basin countries: Ten years of cooperation and perspectives. *Improv. Water Econ.* **2002**, *1*, 66–70. (In Russian)
6. Kipshakbayev, N.K.; Sokolov, V.I. Water resources of the Aral Sea basin—Formation, distribution, usage. In *Water Resources of Central Asia*; ICWC Press: Tashkent, Uzbekistan, 2002; pp. 63–67. (In Russian)
7. Kamalov, T.K. Water crisis of southern Priaralje-contemporary status and perspectives. In *Water Resources of Central Asia*; ICWC Press: Tashkent, Uzbekistan, 2002; pp. 74–78. (In Russian)
8. Ramazanov, A.M. Water resources of Kazakhstan: Problems and perspectives for usage. *Improv. Water Econ.* **2002**, *1*, 10–15. (In Russian)
9. Ryabtsev, A.D.; Akhmetov, S.K. Kazakhstan water resources: Problems and prospects of using. *Hydrometeorol. Ecol.***2002**, *1*, 51–73. (In Russian)
10. Lack, T. Water and health in Europe. *BMJ* **1999**, *318*, 1678–1682.
11. Adany, R.; Villerusa, A.; Bislimovska, J.; Kulzhanov, M. Public health education in Central and Eastern Europe, and Central Asia. *Public Health Rev.* **2011**, *33*, 105–133.
12. McKee, M.; White, M.; Bojan, F.; Osrbye, T. Development of public health training in Hungary—An exercise in international co-operation. *J. Public Health Med.* **1995**, *17*, 438–444.
13. Adeyi, O.; Chellarai, G.; Goldstein, E.; Preker, A.; Ringold, D. Health status during the transition in Central and Eastern Europe: development in reverse? *Health Policy Plan.* **1997**, *12*, 132–145.
14. Oreskovic, S. New priorities for health sector reform in Central and Eastern Europe. *Croat. Med. J.* **1998**, *39*, 225–233.
15. Kulzhanov, M.; Rechel, B. Kazakhstan: Health system review. European Observatory on Health Systems and Policies. *Health Syst. Trans.* **2007**, *9*, 158.
16. Islamov, U.; Khamzayeva, A.; Maksudov, F.; Maksudova, D.; Rahimov, S.; Sakiev, B. *Water Resources Management in Central Asia: Regional and International Issues at Stake*; Serie: Asia, Número 25; Fundació CIDOB: Barcelona, Spain, 2009.
17. O'Hara, S.L. Central Asia's water resources: contemporary and future management issues. *Int. J. Water Resour. Dev.***2000**, *16*, 423–441.

18. FAO Land and Water Division, Food and Agriculture Organization of the United Nations. *Irrigation in Central Asia in Figures, AQUASTAT Survey–2012*; FAO Water Reports 39; Frenken, K., Ed.; FAO: Rome, Italy, 2013.
19. CAWaterInfo. The Aral Sea Basin. 2011. Available online: http://www.cawater-info.net/aral/index_e.htm (accessed on 6 April 2016).
20. Xu, L.; Zhou, H.; Du, L.; Yao, H.; Wang, H. Precipitation trends and variability from 1950 to 2000 in arid lands of Central Asia. *J. Arid Land* **2015**, *7*, 514–526.
21. Yin, Z.Y.; Wang, H.; Liu, X. A comparative study on precipitation climatology and interannual variability in the lower midlatitude East Asia and Central Asia. *J. Clim.* **2014**, *27*, 7830–7848.
22. International Crisis Group. *Water Pressures in Central Asia*; Europe and Central Asia Report No. 233; International Crisis Group: Brussels, Belgium, 2014; p. 35.
23. Global Water Partnership, Central Asia and Caucasus. Regional Review, Water Supply and Sanitation in the Countries of Central Asia and Southern Caucasus. Available online: http://www.gwp.org/global/activities/news/wssobzor_eng_aug2009.pdf (accessed on 23 May 2016).
24. FAO. *Review of World Water Resources by Country*; Water Report 23; FAO: Rome, Italy, 2003; p. 111.
25. UNDP. *Water Resources of Kazakhstan in New Millennium*; Review of the UNDP; UNDP: Almaty, Kazakhstan, 2004; p. 114.
26. Brown, A.; Matlock, M.D. *A Review of Water Scarcity Indices and Methodologies*; Food, Beverage & Agriculture, The Sustainability Consortium, University of Arkansas, White Paper 106; The Sustainability Consortium: Fayetteville, AR, USA, 2011; p. 19.
27. Micklin, P. The Aral Sea disaster. *Annu. Rev. Earth Planet. Sci.* **2007**, *35*, 47–72.
28. Khvorog, G.V., Ed.; *Ecological Map of the Near Aral Region*; G.V. Khvorog: Alma-Ata, Kazakhstan, 1992. (In Russian)
29. Unicef. *Progress on Sanitation and Drinking Water–2015 Update and MDG Assessment*; WHO Library Cataloguing-in-Publication Data; WHO: Geneva, Switzerland, 2015.
30. Fewtrell, L.; Kaufmann, R.B.; Kay, D.; Enanoria, W.; Haller, L.; Colford, J.M., Jr. Water, sanitation, and hygiene interventions to reduce diarrhoea in less developed countries: A systematic review and meta-analysis. *Lancet Infect. Dis.* **2005**, *5*, 42–52.
31. Esrey, S.A., Gough, J., Rapaport, D., Sawyer, R., Simpson-Hébert, M., Vargas, J., Winblad, U., Eds.; *Ecological Sanitation*; Swedish International Development Cooperation Agency: Stockholm, Sweden, 1998; p. 100.
32. Waddington, H.; Snilstveit, B. Effectiveness and sustainability of water, sanitation, and hygiene interventions in combating diarrhoea. *J. Dev. Eff.* **2009**, *1*, 295–335.
33. Mara, D.; Lane, J.; Trouba, D. Sanitation and health. *PLoS Med.* **2010**, *7*, e1000363.
34. Feachem, R.G.; Bradley, D.J.; Garelick, H.; Mara, D.D. *Sanitation and Disease: Health Aspects of Wastewater and Excreta Management*; John Wiley & Sons: Chichester, UK, 1983; p. 326.
35. Tussupova, K. *The Management System of the Drinking Water and Its Relation to Health in the Republic of Kazakhstan*; Master of Science in Public Health, Medical and Health Science Center, University of Debrecen; University of Debrecen: Debrecen, Hungary, 2007; p. 26.
36. National Environmental Health Action Plan (Kazakhstan). Available online: http://enrin.grida.no/htmls/aralsoe/nehap/kazakh.pdf (accessed on 12 February 2016).
37. WHO. Water Sanitation and Health. Available online: http://www. who.int/water_sanitation_health/diseases/burden/en/index.html (accessed on 12 February 2016).

38. Nurgalieva, Z.Z.; Malaty, H.M.; Graham, D.Y.; Almuchambetova, R.; Machmudova, A.; Kapsultanova, D.; Osato, M.S.; Hollinger, F.B.; Zhangabylov, A. Helicobacter pylori infection in Kazakhstan: Effect of water source and household hygiene. *Am. J. Trop. Med. Hyg.* **2002**, *67*, 201–206.
39. Snegireva, M.S. *Detection of Quantitative Connections between Biological Impurity of Potable Water and Morbidity of the Population of Republic of Kazakhstan by Intestinal Taints*; M.S. Snegireva: Almaty, Kazakhstan, 1999; pp. 3–28.
40. Breuer, T.; Malaty, H.M.; Graham, D.Y. The Epidemiology of *H. pylori*-Associated gastroduodenal diseases. In *The Immunobiology of H. Pylori: From Pathogenesis to Prevention*; Ernst, P.B., Michetti, P., Smith, P.D., Eds.; Lippincott-Raven: Philadelphia, PA, USA, 1997; pp. 1–14.
41. Shiotani, A.; Nurgalieva, Z.Z.; Yamaoka, Y.; Graham, D.Y. Helicobacter pylori. *Med. Clin. N. Am.* **2000**, *84*, 1125–1136.
42. Abdrachmanov, Z.N.; Pozdnyakova, A.P.; Filippenko, V.I.; Tashenova, K.O. *Indexes of Oncological Services of the Republic of Kazakhstan for 1998: Statistical Data*; Prints-S: Almaty, Kazakhstan, 1999; pp. 3–73.
43. Micklin, P.P. The water management crisis in Soviet Central Asia. In *The Carl Beck Papers in Russian and East European*; Stud. No. 905; Pittsburgh: Cent. Russ./East Eur. Stud.; University of Pittsburgh: Pittsburgh, PA, USA, 1991.
44. Chub, V.Y. *Climate Change and Its Influence on the Natural Resource Potential of the Republic of Uzbekistan*; Glavgidromet: Tashkent, Uzbekistan, 2002. (In Russian)
45. Abdulkasimov, H.P.; Alibekova, A.V.; Vakhabov, A.V. *Desertification Problems in Central Asia and Its Regional Strategic Development*; Abstracts, NATO Advanced Research Workshop; NATO: Samarkand, Uzbekistan; 11–14; June; 2003. (In Russian)
46. Reteyum, A.Y. The factors and results of desertification of central Asia. In *Desertification Problems in Central Asia and Its Regional Strategic Development*; Abstracts, NATO Advanced Research Workshop; NATO: Samarkand, Uzbekistan; 11–14; June; 2003. (In Russian)
47. Chembarisov, E.I.; Lesnik, T.V. To the preserve of surface waters of Central Asia. In *Issue of SANIGMI "Fresh Water"*; Denisov, Y.M., Ed.; SANIGMI Press Tashkent: Tashkent, Uzbekistan, 1995; pp. 64–71. (In Russian)
48. Chembarisov, E.I.; Lesnik, T.V.; Ranneva, I.V. Contemporary River's water quality of Uzbekistan. In *Water and Sustainable Development of Central Asia*; Tuzova, T.V., Ed.; Materials of the projects "Regional cooperation on the usage of water and power resources in Central Asia (1998)" and "Hydroecological problems and sustainable development of Central Asia"; American University in Kyrgyzstan: Bishkek, Kyrgyzstan, 2001; pp. 39–41. (In Russian)
49. Zharkov, V.V.; Zharkov, D.V. Hydrochemical composition of waters of the Karakum-river. *Probl. Desert Dev.* **2002**, *4*, 66–71. (In Russian)
50. Normand, O. Regional rural water supply and sanitation project Ferghana Valley. In Proceedings of the 6th International Conference on World Water Forum—Cooperative Actions for Water Security, Tashkent, Uzbekistan, 12–13 May 2011.
51. Isman, A.; Toktaliev, M. Kyrgyzstan without Water. In *Radio Free Europe/RadioAzattyk*; 13; December; 2012.
52. World Bank. *Implementation Completion and Results on a Credit in the Amount of SDR 12 Million (US$15 Million Equivalent) to the Kyrgyz Republic for a Rural Water Supply and Sanitation Project*; Report IDA-35790; World Bank: Washington, DC, USA; 15; May; 2009.
53. WHO. Improved Water Source, Rural (% of Rural Population with Access). Available online: http://data.worldbank.org/indicator/SH.H2O.SAFE.RU.ZS (accessed on 26 February 2016).
54. Syrlybayeva, B.R. Fresh Drinking Water Supply for Population of Kazakhstan. Available online: http://kisi.kz/en/categories/economy-and-energy/posts

/fresh-drinking-water-supply-for-population-of-kazakhsta372 (accessed on 12 February 2016).

55. Roberts, B.; Stickley, A.; Gasparishvili, A.; Haerpfer, C.; McKee, M. Changes in household access to water in countries of the former Soviet Union. *J. Public Health* **2012**, *34*, 352–359.

56. Milibaeva, R.T. Control of the drinking water safety in Kazakhstan. In Proceedings of the 3D International Exhibition and Conference "Su Arnasy 2007", Astana, Kazakhstan, 23–25 May 2007.

57. Bain, R.; Cronk, R.; Wright, J.; Yang, H.; Slaymaker, T.; Bartram, J. Fecal contamination of drinking-water in low- and middle-income countries: A systematic review and meta-analysis. *PLoS Med.* **2014**, *11*, e1001644.

58. White, G.F.; Bradley, D.J.; White, A.U. *Drawers of Water: Domestic Water Use in East Africa*; The University of Chicago Press: Chicago, IL, USA, 1972.

59. WHO. *Guidelines for Drinking-Water Quality*; World Health Organization: Geneva, Switzerland, 2011.

60. Esrey, S.A.; Feachem, R.G.; Hughes, J.M. Interventions for the control of diarrhoeal diseases among young children: Improving water supplies and excreta disposal facilities. *Bull. World Health Organ.* **1985**, *63*, 757–772.

61. Fewtrell, L.; Colford, J.M., Jr. Water, sanitation and hygiene in developing countries: interventions and diarrhoea—A review. *Water Sci. Technol.* **2005**, *52*, 133–142.

62. Clasen, T.; Roberts, I.; Rabie, T.; Schmidt, W.; Cairncross, S. Interventions to improve water quality for preventing diarrhoea. *Cochrane Database Syst. Rev.* **2006**, *19*, CD004794.

63. Cairncross, S.; Hunt, C.; Boisson, S.; Bostoen, K.; Curtis, V.; Fung, I.C.H.; Schmidt, W.-P. Water, sanitation and hygiene for the prevention of diarrhoea. *Int. J. Epidemiol.* **2010**, *39*, i193–i205.

64. Ligon, G.C. *Waterborne Disease Outbreaks: A Systematic Review of the Health Effects of Drinking Water System Failures*; Chapel Hill (North Carolina), University of North Carolina at Chapel Hill: Chapel Hill, NC, USA, 2012.

65. Craun, G.F.; Brunkard, J.M.; Yoder, J.S.; Roberts, V.A.; Carpenter, J.; Wade, T.; Calderon, R.L.; Roberts, J.M.; Beach, M.J.; Roy, S.L. Causes of outbreaks associated with drinking water in the United States from 1971 to 2006. *Clin. Microbiol. Rev.* **2010**, *23*, 507–528.

66. UNDP. *Water Resources of Kazakhstan in the New Millennium*; UNDP Kazakhstan, A Series of UNDP Publication in Kazakhstan, # UNDPKAZ 07; UNDP: Almaty, Kazakhstan, 2004.

67. Törnqvist, R.; Jarsjö, J.; Karimov, B. Health risks from large-scale water pollution: Trends in Central Asia. *Environ. Internsh.* **2011**, *37*, 435–442.

68. Atanyiazova, O.A.; Baumann, R.A.; Liem, A.K.D.; Mukhopadhyay, U.A.; Vogelaar, E.F.; Boersma, E.R. Levels of certain metals, organochlorine pesticides and dioxins in cord blood, maternal blood, human milk and some commonly used nutrients in the surroundings of the Aral Sea (Karakalpakstan, Republic of Uzbekistan). *Acta Paediatr.* **2001**, *90*, 801–808.

69. Erdinger, L.; Eckl, P.; Ingel, F.; Khussainova, S.; Utegenova, E.; Mann, V.; Gabrio, T. The Aral Sea disaster-human biomonitoring of Hg, As, HCB, DDE, and PCBs in children living in Aralsk and Akchi, Kazakhstan. *Int. J. Hyg. Environ. Health* **2004**, *207*, 541–547.

70. Jensen, S.; Mazhitova, Z.; Zetterström, R. Environmental pollution and child health in the Aral Sea region in Kazakhstan. *Sci. Total Environ.* **1997**, *206*, 187–193.

71. Elpiner, L.I. Public health in the Aral Sea coastal region and the dynamics of changes in the ecological situation. In *Creeping Environmental Problems and Sustainable Development in the Aral Sea Basin*; Glantz, M., Ed.; Cambridge University Press: Cambridge, UK, 1999; pp. 128–156.

72. Zetterström, R. Child health and environmental pollution in the Aral Sea region in Kazakhstan. *Acta Paediatr. Suppl.* **1999**, *88*, 49–54.
73. Small, I.; Falzon, D.; van der Meer, J.B.; Ford, N. Safe water for the Aral Sea Area: Could it get any worse? *Eur. J. Public Health* **2003**, *13*, 87–89.
74. Crosa, G.; Froebrich, J.; Nikolayenko, V.; Stefani, F.; Galli, P.; Calamari, D. Spatial and seasonal variations in the water quality of the Amu Darya River (Central Asia). *Water Res.* **2006**, *40*, 2237–2245.
75. Froebrich, J.; Ikramova, M.; Razakov, R. Improved groundwater management strategies at the Amu Darya River. In *Groundwater and Ecosystems*; Baba, A., Howard, K.W.F., Gunduz, O., Eds.; Springer: Dordrecht, The Netherlands, 2006; pp. 153–165.
76. Kawabata, Y.; Yamamoto, M.; Aparin, V.; Ko, S.; Shiraishi, K.; Nagai, M. Uranium pollution of water in the western part of Uzbekistan. *J. Radioanal. Nucl. Chem.* **2006**, *270*, 137–141.
77. Friedrich, J. Uranium contamination of the Aral Sea. *J. Mar. Syst.* **2009**, *76*, 322–335.
78. Howd, R.A.; Fan, A.M. *Risk Assessment for Chemicals in Drinking Water*; John Wiley & Sons, Inc.: Hoboken, NJ, USA, 2008.
79. Matthys, B.; Bobieva, M.; Karimova, G.; Mengliboeva, Z.; Jean-Richard, V.; Hoimnazarova, M.; Kurbonova, M.; Lohourignon, L.K.; Utzinger, J.; Wyss, K. Prevalence and risk factors of helminths and intestinal protozoa infections among children from primary schools in western Tajikistan. *Parasites Vectors* **2011**, *4*, 195.
80. United Nations Development Programme Tajikistan, Republic of Tajikistan. *Progress toward the Millennium Development Goals (Tajikistan 2003)*; United Nations Development Programme: Dushanbe, Tajikistan, 2003.
81. Steinmann, P.; Usubalieva, J.; Imanalieva, C.; Minbaeva, G.; Stefiuk, K.; Jeandron, A.; Utzinger, J. Rapid appraisal of human intestinal helminth infections among schoolchildren in Osh Oblast, Kyrgyzstan. *Acta Trop.* **2010**, *116*, 178–184.
82. Lioubimtseva, E.; Henebry, G.M. Climate and environmental change in arid Central Asia: Impacts, vulnerability, and adaptations. *J. Arid Environ.* **2009**, *73*, 963–977.
83. Elpiner, L.I. A scenario of possible effect of changes in the hydrological conditions on the medical and environmental situation: On the problem of global hydroclimatic changes. *Water Resour.* **2003**, *30*, 434–444.
84. Small, I.; Bunce, N. The Aral Sea disaster and the disaster of international assistance. *J. Intern. Aff.* **2003**, *56*, 11–13.
85. McMichael, A.J.; Kovats, R.S. Climate change and climate variability: Adaptations to reduce adverse health impacts. *Environ. Monit. Assess.* **2000**, *61*, 49–64.

CHAPTER 4

Endocrine Disruptors in Water Sources: Human Health Risks and EDs Removal from Water Through Nanofiltration

J. E. Cortés Muñoz[1], C. G. Calderón Mólgora[1], A. Martín Domínguez[1], E. E. Espino de la O[2], S. L. Gelover Santiago[1], C. L. Hernández Martínez[2] and G. E. Moeller Chávez[1]

[1] *Mexican Institute of Water Technology, Mexico*

[2] *National Water Commission, Mexico*

1. INTRODUCTION

Water is fundamental for human health and well-being as well as for stimulating diverse socioeconomic activities. Paradoxically, these very activities have contributed to the alteration and deterioration of water supply sources from a microbiological, physical and chemical standpoint, causing sanitary risks for the population. For example: since the end of the 19th century, the role of drinking water in exposing populations to pathogens, and improvements in its quality in order to prevent diarrheic illnesses, has been widely analyzed, debated and documented [1,2]; in the 20thcentury, epidemiological evidence was found of cutaneous lesions [3]and various types of cancer related to hydroarsenicism [4], as well as dental and skeletal fluorosis related to fluoride in drinking water [5].

In recent decades the problem of these possible public health risks from so-called emergent contaminants (ECs) has been factored into the problem that includes a wide range of compounds whose environmental presence and impact have been proven with the advent of new sensitive and reliable quantitative analytical tools [6]: ECs are bioactive substances synthesized and used for the household, agriculture, livestock, industry, personal care products and hygiene (PCPs), and human and veterinary medicine, including byproducts of production and degradation [7].However, beyond the concentrations and environmental persistence of ECs, their relevance lies in the fact that they continue to be released into the environment through various ways which confers ubiquity.

Chemical endocrine disruptors (EDs), which are xenobiotics (compounds exogenous or foreign to living organisms) with potential to alter hormonal regulation and normal endocrine system function, consequently affecting an intact organism, its progeny or subpopulations, are among the wide range of ECs [8]. The evidence of adverse effects on aquatic organisms at relevant environmental concentrations [9,10]is well documented as well as *in vitro*, *in vivo* and epidemiological studies that associate human exposure to these compounds with: obesity, metabolic syndrome, type II diabetes mellitus [11],estrogenic, androgenic and antiandrogenic activity or combinations thereof [12], reproductive, nervous and immune systems effects, as well as some cancers and developmental effects [13].

The presence of EDs in bodies of water is due mainly to the discharge of wastewater that impacts the quality of surface water and groundwater with compounds that are not entirely removed by conventional treatment processes [14,15], which is particularly relevant in areas like the Mezquital Valley (state of Hidalgo, Mexico).The aquifer that supplies the population is recharged with the residual waters used in agricultural irrigation. Another way is through the indirect reuse of treated wastewater for potable water source augmentation. These practices could explain why some pharmaceuticals (Phs) and personal care products (PCPs) have been detected in waters treated for human consumption [16].

The concern regarding human exposure to EDs, in this case through water consumption, is based on five points: 1) evidence of adverse effects on fish and aquatic ecosystems at relevant environmental concentrations [9,10]; 2) documented clinical cases of cancers related to hormones in industrialized nations [8], as well as prevalence of reproductive disorders in adolescents and young men in Europe [17]; 3) *in vivo* studies that show endocrine disruption through exposure to certain ambient chemicals; 4) various chemical compounds classified as EDs or with potential to act as such, have been found in surface water and groundwater [18]and, 5) evidence that suggests that conventional water treatment systems are inefficient in removal of these types of contaminants [16].

The European Union, Germany, England, USA, Australia, Canada and Japan have all installed multi-stage treatment systems that effectively reduce the concentration of EDs in drinking water. The debate has begun over the need for research and regulation, analytical methods, water sources and treated water monitoring, public health and environmental risks, water treatment processes, transformation, transport and fate in the environment of EDs.

In Mexico there are few studies related to the occurrence of EDs in water, as well as few studies that document the efficacy and efficiency of water treatment processes in the removal of ECs and EDs. Considering that the Mezquital Valley is a prime example of an aquifer affected by the reuse of wastewater, and that the occurrence of ECs has been documented in supply wells in the area, the following objectives are proposed:

a. Analyze and synthesize information regarding the presence of EDs in supply sources and treated waters for potable use, sanitary, environmental and regulatory relevance, treatment processes for removal from water.

b. Analyze the problems related to ED exposure, specifically arsenic, bisphenol A, alkylphenols and their ethoxylates through the use and consumption of water in the Mezquital Valley.

c. Analyze the technical feasibility of nanofiltration (NF) process to remove mineral and organic compounds from groundwater in the area of the study.

d. Identify investigation needs regarding the occurrence concentration, persistence, transformation and destination, action mechanisms and risk assessment of EDs in water for human consumption and treatment processes for potable use

2. ENDOCRINE DISRUPTORS IN DRINKING WATER: PUBLIC HEALTH RELEVANCE

By May 16, 2012, the Chemical Abstracts Service [19], had registered over 66.67 million organic and inorganic substances. More than 100,000 man-made chemicals are available on the market including approximately 1,500 new molecules released yearly [20,21] for manufacturing products whose primary use is for human well-being and socioeconomic development. Since the 1990s, EDs have been one of the most controversial issues, attracting the attention of the scientific community, international agencies and organizations, governments and the general public.

The U.S. Environmental Protection Agency defined an endocrine disruptor as "an exogenous agent that interferes with the production, release, transport, metabolism, binding, action, or elimination of natural hormones in the body responsible for the maintenance of homeostasis, reproduction, development, and/ or behavior" [22]; the European Commission defines it as "an exogenous substance or mixture that alters function(s) of the endocrine system and consequently causes adverse health effects in an intact organism, or its progeny, or (sub)populations" [8]. From both definitions it is clear that EDs are compounds that alter hormonal regulation or homeostasis that can cause undesirable adverse effects on health as a result of exposure to a compound whose mechanism or action is endocrine disruption.

2.1. Origin and Occurrence of Eds in Drinking Water

The nature and origin of EDs is diverse and includes groups of compounds such as: active ingredients in medicines with collateral hormonal effect, pesticides and adjuvants for their application, products to increase growth and weight gain in livestock, personal care and hygiene products, flame retardants, chemicals for use in the plastic industry and other frequently used industrial chemicals, natural and synthetic hormones, as well as products for manufacturing consumer goods and degradation by products [20,22-26].

Their impact on public health and wildlife is due to their bioactivity and ubiquity in the environment, as they are introduced unconsciously and permanently in the various environmental compartments. They can be introduced as pure substances or complex mixtures through diverse ways, especially via the flow of treated or untreated wastewaters. These compounds are not totally removed or inactivated by conventional water treatment systems or by natural processes of self-purification of the receiving bodies (water or ground), frequently reaching groundwater [26-28].

For example, in the Mezquital Valley, the aquifer that supplies the population is recharged mainly with wastewaters from Mexico City from agricultural irrigation; however, while the contamination with EDs, Phs, and other organic compounds originate from these wastewaters, they also originate from the disposal of PCPs and expired or unused medications by households, municipal and hospital wastewater, leachates from landfills and local uncontrolled garbage dumps (Figure 1, adapted from [29]), which is consistent with information published by various authors for other aquifers [29-31].

In this case, as in scenarios of direct or indirect reuse of treated wastewater as water supply source, the main concern is related to the pathogens as well as nitrates, Phs, PCPs and disinfection byproducts with potential to disrupt the endocrine system and affect human and environmental health [29].

In comparison to other chemical compounds, there is little information on the transformation and fate of EDs especially regarding biotransformation, hydrolysis and photo transformation of Phs and PCPs. Their low volatility suggests that their distribution in the environment will occur mainly via aqueous transportation and dispersion through the food chain. The polar and non-volatile nature of Phs impedes their release from water [31], without geographical or climatic borders for these synthetic substances that have been found in areas that are considered to have low pollution levels [32].

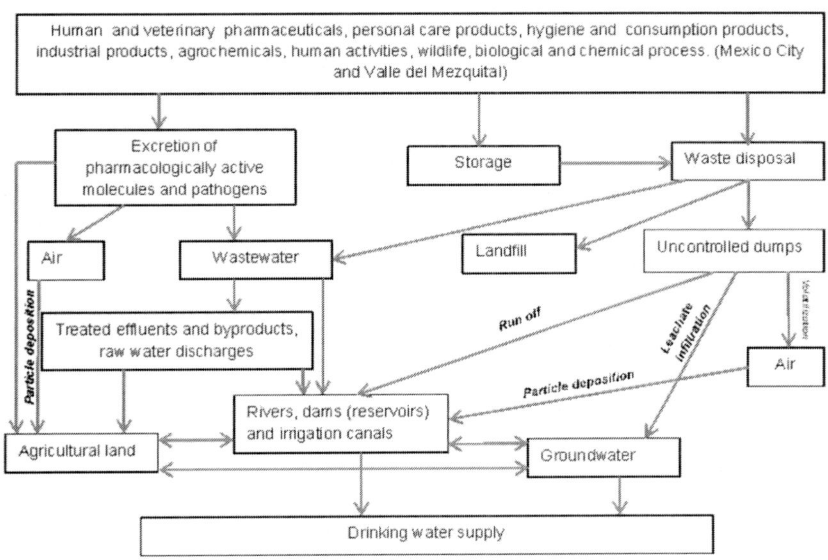

Figure 1. Conceptual model of introduction and transportation of ECs and EDs in the Mezquital Valley.

ECs, EDs and PCPs have been frequently detected in effluents and surface waters that could be present in drinking water. Ultra-trace concentrations (ng/L) of prescription and non-prescription Phs and their metabolites have been reported in samples of drinking water across the United States. These include caffeine, analgesics, anti-inflammatories (naproxen), anti-convulsants (carbamazepine, phenytoin, primidone) and anxiolytics (meprobamate), xray contrast medium (iopromide), lipid regulators (gemfibrozil), antibiotics or their metabolites, nicotine metabolites, hypotensives (atenolol), synthetic musk, a polycyclic aromatic hydrocarbon compound, a plant sterol, plastic components, an insecticide, surfactants (bisphenol-A, alkylphenols) and degradation products, a fixative used in perfumes and soaps, a flame retardant and a pesticide [16,28,33-36].

In general throughout the water cycle there is a reduction in EDs through absorption, dilution, and biodegradation [24], and yet there are still questions about their fate in the environment. Laboratory studies have shown that bioactivity is reduced over a period of hours to days due to degradation and sorption yet field studies indicate that estrogens are sufficiently mobile and persistent to impact the surface and ground waters [37], while in the ground and sediments where they can persist for long lengths of time (the half-life of clofibric acid, for example, is estimated at 21 years), they reach levels in the g/kg range [23].

Given the ubiquitous nature of EDs, all humans are exposed through different tracks: inhalation, ingestion and dermal contact. Contributing to this is their low biodegradability, air and water transportability, bio accumulation in the trophic chain and transgenerational exposure; fetuses are especially vulnerable because pregnant women accumulate EDs in adipose tissue [38].As with other environmental chemicals, their effects depend on the concentration and nature of the chemical as well as its route, frequency and intensity with which exposure occurs and in this case the phase of life at which the exposure occurs.

Due to the trace and ultra-trace concentrations in which EDs are found in drinking water, people are commonly exposed in higher quantities through medications and other sources and routes: diet, inhalation of airborne chemical substances and dermal absorption (topical medication or personal care products), which suggests the contribution by drinking water to the overall exposure and its relative importance in assessment of sanitary risks associated with these types of contaminants [39].

Likewise from a risk management viewpoint, it is important to note that the variety and chemical structure of EDs complicates their identification and quantification in water as well as the characterization of the sanitary risks associated with chronic exposure to low or environmentally relevant doses In addition little is known about the occurrence, toxicity and potential endocrine activity of the products of degradation that may result from the processes of bio and physiochemical transformation that alter the chemical structure of EDs rather than eliminate them.

2.2. Human Health Effects

2.2.1. Endocrine System

Hormones are produced by the glands that comprise the endocrine system, which is the key to communication within the human organism and control method between the nervous system and the various functions of the human body such as reproduction, immunity, energy control, metabolism, conduct, growth and development. For example:

- The Thymus gland is the source of immunologic regulatory hormones [35].
- The Hypothalamus gland releases hormones such as oxytocin; it's important in the control of reproductive endocrine processes and stimulate pituitary activity [35,40,41].
- The Pituitary gland releases steroid hormones, such as corticosteroids, androgens, and estrogens; growth hormone, oxytocin. Feedback control signals the endocrine organ (the adrenal gland or gonads), to cease production or release of the endogenous steroid or to stimulate the release of opposing hormones. This homeostatic control in response to endogenous hormones is critical for maintaining proper hormone concentrations [35,40].
- The Thyroid gland releases thyroid hormones (calcitocin and thyroxine), which are receptor nuclear of steroids, regulate metabolism, growth, development, behavior and puberty [35,40,41].
- The Adrenal glands, release corticosteroid hormones, cathecolamines to regulate metabolism and behavior [40,41].
- The Pancreas produces insulin and glucagons to regulate blood sugar levels [40].
- The Ovaries and testicles produce sex steroids such as estrogen, progesterone, testosterone (androgens and estrogens) [35,40].
- This means that all the physiological systems sensitive to hormones are vulnerable to EDs, including the brain and hypothalamic-neuroendocrine systems, cardiovascular system, mammary gland, adipose tissue, ovary and uterus in females, and testes and prostate in males [25,40].

2.2.2. Endocrine Disruption Mechanisms

Hormones circulate in the blood stream to modulate cellular and organ function through the union with complex molecular receptors and mechanisms:

- They mimic natural hormones; if exposed to relatively high doses, they join receptors within the cell and block or interfere with the ways through which hormones and receptors are synthesized or controlled [42].
- Binding and activating estrogenic and androgenic receptors N40. There are a number of estrogenic receptors in gonads, liver, brain, and sex organs; union without activation of the receptor would act like anti-estrogenics or anti-androgenics [43-45].
- Binding without activation of the receptor would act like anti-estrogenics or anti-androgenics [43-45].
- Modifying hormonal mechanisms [46], or of the number of hormonal receptors in the cell, or of the production of natural hormones, for example in the thyroid, immune or nervous systems [47].
- Selectively inhibiting DNA transcription; for example, arsenic produces a disruption in the transcription of DNA induced by the glucocorticoids mediated by receptors [47].

The mechanisms described in recent literature also include: a) the alteration of the transcriptional activity of nucleus receptors by modulating co-regulators through mediated degradation of the proteasome as well as by inhibiting histone deacetilase activity and stimulating mitogenic quinase protein activity; b) regulation of the methylation of DNA and c) modulation of lipid metabolism and adipogenesis, which possibly contribute to the current epidemic of obesity [47-50].

2.2.3. Health Effects

EDs are structurally similar to many natural hormones; some have lipophilic properties [40], act in extremely low concentrations and therefore can have effects on organisms with low dose exposures [50,51]. As a result, environmental presence in trace and ultra-trace amounts, particularly in water, may be insufficient to cause cellular death or act upon genetic material, yet could result in a source of human exposure and carry sanitary risks for more susceptible segments of the population.

The target hormones for EDs and the effects differ from one compound to another (as shown in Table 1), as well as among species and intra species;

for example, there are reports that the median Bisphenol A (BPA) level in human blood and tissues, including in human fetal blood, is higher than the level that causes adverse effects in mice [52].

The time of exposure to EDs in the organism in development is decisive in determining its character and future potential and, even when critical exposure takes place during embryonic development, the clinical manifestations may not be present until adulthood [8,43,45,46]. One compound may act on different target hormones [40] and cause different alterations. A wide and current review of the health risks is found in [22,53]. Some examples of compounds intensively used in Mexico, and thier potential effects in in humans, are presented in Table 1, as well as their normal application or use and the exposure source. It is worth mentioning that the exposure is involuntary.

2.3. NF as an Alternative To Remove ECS From Water

Between January 2001 and July 2004 the European Union conducted Project Poseidon [64]; among its objectives it proposed conducting integral studies to evaluate and improve the removal of pharmaceuticals and personal care products from residential residual waters using conventional and advanced treatments as well as with potable water. One of the conclusions of the study is that reverse osmosis, nanofiltration and ultrafiltration-powdered activated carbon are powerful processes for the removal of pharmaceuticals and personal care products, among which are found ECs and EDs (suspicious and recognized).

In spite of the conclusions from the POSEIDON Project, the question about the efficacy of NF membranes in removal of emergent contaminants persists. Many studies have been made of this topic [65-73]. The spectrum of tests covers membranes with a molecular weight cut off (MWCO) of 200, 400 and 600 Da, as well as organic compounds with different molecular weights, sizes and physiochemical characteristics.

Table 1. Example of possible sources of exposure to EDs and target hormone system.

Example	Use	Involuntary exposure source	Example of adverse effect in humans
Bisphenol A	Plasticizers	Plastic items, packaging foods, beverages,	Estrogenic [52-55], thyroid hormone, progesterone[40]

		water	, androgenic. Female breast structural anomalies, premature thelarche, cancer; pubertal timing variations, ovulatory disorders, sexual behavior, prostate cancer [56,57]. Disrupted hypothalamic estrogenic receptor distribution, altered nitric oxide syntheses signaling [57]. Affectation of human immune function [58], and disrupted behavior in children associated with early life BPA exposure, especially in girls [59].
Vinclozolin	Fungicides	Foods (fruit, vegetables, cereals), water	Estrogenic, anti-androgenic. Promotes transgenerational adult onset disease such as male infertility, kidney and prostate diseases, immune abnormalities

			and tumor development [60].
Alkylphenols	Detergents, emulsifiers, agrochemicals	Household items, water, foods (fish)	Estrogenic [40], androgenic. Environmental and health issues continue to cast uncertainty over the human risks of alkylphenols and alkylphenols ethoxylates.
Phthalates	Plasticizers	Plastic items, cosmetics, personal products care, beverages, water	Estrogenic, androgenic, thyroid hormone [40]. Abnormalities such as hypospadias, cryptorchidism, reduced anogenital distance. Oligospermia, germ cell cancer [56]. Neurodevelopment and metabolic endpoints are of concern, since studies of prenatal exposure have found associations with phthalate exposure and lowered IQs, and exposure has been implied as a risk factor for obesity, insulin

			resistance and diabetes by others [47,50,55].
Diethylstilbestrol	Contraception, hormone replace therapy	Pharmaceuticals. They have been detected in water sources, wastewater and treated effluents	Estrogenic. Abnormalities such as hypospadias, cryptorchidism, reduced anogenital distance. Female structural anomalies, breast cancer, structural anomalies, premature thelarche. Prostate cancer [56,61].
Ibuprofen, diclofenac, acetaminophen	Anti-inflammatory, analgesics		Candidates may be identified on the basis of simple assumptions regarding their use and activity: a) non estrogenic steroids may react with environmental endocrine receptors or metabolize on their way to the environment and thus form endocrine disruptors; b) many high-volume drugs released to the environment
Bezafibrate, clofibrate, gemfibrozil, fenofibrate	Lipidic regulator		
Atenolol, metoprolol, propanolol	B-blockers, antihypertensives		

			have not yet been tested for their endocrine properties, and some of these are known to interact with the human endocrine system [62].
Anabolic steroids; trenbolone acetate, melengestrol acetate	Fast growth of meat producing animals	Animal food, soil, wastewater	Androgenic. A placebo-controlled prospective study demonstrated adverse and activating mood and behavioral effects of anabolic steroids [63].

NF efficiently removes dissolved solids, organic carbon, inorganic ions and organic micro contaminants, both regulated and unregulated, and therefore its capacity is similar to that of reverse osmosis (RO); however, due to higher pressure requirements the latter has higher investment and operational costs. Furthermore the partial discrimination of calcium and bicarbonates of the NF can be advantageous since drinking water distributed through municipal networks should be saturated with calcium carbonate to avoid corrosion [74].

The mechanisms through which RO and NF reject organic compounds are mainly: a) exclusion by size or stericity; b) exclusion by repulsion of charges or ionic exclusion also known as Donnan Effect and c) the physiochemical interaction of the solvent, the solute and the membrane [75].

Size exclusion is associated with the molecular weight cut off of a membrane but this is relative to a certain type of molecule, therefore other molecules with similar molecular weight yet different physiochemical properties may not be rejected at the same rate. In the case of neutral compounds, other geometric descriptors of the molecule are far more useful such as molecular width [76]or hydrodynamic radius, especially for molecules whose radius approaches that of the membrane pores [77].

Exclusion by charge repulsion is a phenomenon in highly typified NF membranes; tests conducted with NF membranes and ultra-low pressure reverse osmosis to retain or reject 9 organic compounds (5 negative ionic charge and 4 neutral) showed that the electrically negative charged compounds were rejected with efficiencies above 91% even with molecular weights lower than the rated molecular weight cut off of the membranes, while the neutral compounds were rejected at much lower rates at levels as low as 12% for 2-naphtol [78].

The interaction between solvent, solute and membrane can lead to misleading results if enough time is not allotted for the process to normalize, since various organic contaminants tend to be adsorbed on the membrane surfaces. As a result, initially they are retained, but once the membrane surface is saturated, the true rejection of the compound can be observed [65,77-81].

NF membranes (MWCO of 200, 300 and 400 Da) and RO membranes (MWCO not reported) have a 90% reject rate of ionic and non-ionic hydrophilic organic compounds such as naproxen, diclofenac, ibuprophen, mecoprop, ketoprophen, gemfibrozil, and primidone and higher for NF membranes with MWCO of 200 Da as well as the RO membranes. Even the NF membrane with MWCO of 300 Da had reject rates of over 90% [68].

Tests conducted with a membrane with a MWCO of 300 Da (Filmtec NF200) to evaluate the removal of hormones and antibiotics with microfiltration showed that in a matrix of drinking water with a mixture of hormones and medications (i.e. hormones with sulfonamides or tetracycline) it is possible to obtain rejects of nearly all of the substances with the exception of testosterone whose reject rate was 95%. The pharmaceuticals tested were tetracyclines, (chlorotetracycline, sulfachloropiridazine, sulfamerazine, sulfametoxazole, sulfametazole) and hormones (estrogen, progesterone, testosterone and 17α-ethynilestradiol). There is also evidence that the removal efficiency of hormones and sulfonamides was lower when de-ionized water was used and a pure solute [72].

As shown in the previous examples, the results obtained in the laboratory tests on efficiency of NF membranes in the removal of ECs have been very favorable even though it has been shown that compounds with a lower molecular weight, less branched and without ionic charge, were only partially retained.

In pilot applications or full scale, the performance of NF in the control of organic micro contaminants is less documented; even so, there are two pilot systems documented [67, 69]and one municipal plant [73]. The main results of these two applications are described below.

A comparative study between ultra-low pressure reverse osmosis membranes (ULPRO) and NF showed that both types of membrane can reach efficient levels of removal of organic compounds such as pharmaceuticals, pesticides, flame retardants, plasticizers, and nitrogen, similar to that of conventional reverse osmosis, producing water with equal quality as required for indirect potable reuse of treated water. The operational conditions and results show that both types of membrane are viable for reuse projects of treated residual water in which a high quality permeate is required [67].

A NF pilot plant with a capacity of 56.4 L/min was fed with water from an advanced treatment facility for residual water. After the biological treatment, the water was micro filtered and disinfected. The pilot plant operated continuously for 1200 hours (50 days). Ten emergent contaminants were found in the source water (3 plasticizers, [TCEP, TCPP, and TDCPP], acetyl salicylic acid, naproxen, ketoprophen, diclofenac, gemfibrozil, primidone, and carbamazepine) and in all cases removal rates of 75-100% were observed, except during the first 24 hours of operation when efficiencies were lower. However, as the test progressed, an increase in removal of contaminants was observed to the degree that the organic compounds with ionic charge and the neutral ones with greater molecular weight were rejected at a rate of 90%/.

A research study was held at a municipal drinking water facility located in the north of Spain. The plant has 3 parallel trains of membrane processes, two of which are reverse osmosis (486 m^3/h each one) and one NF (360 m^3/h).The facility is fed from groundwater wells, directly influenced by infiltration from Besós River. During the study, five measurement campaigns, aimed at 12 pharmaceutical compounds, were conducted. The results obtained indicate that NF was capable of removing hydrochloratizide, ketoprofen, diclofenac, sotalol, sulfamethoxazole, metopropolol, propifenazone and carbamazepine with efficiency levels above 90%.

Compounds with negative charges like gemfibrozil and mefenamic acid had a low removal rate both in the NF (50% and 30% average), as well in RO (50% and 70% average). Meanwhile, acetaminophen, which has a molecular weight cut off of 151.16 Da., had a removal rate of 44.8% through NF and 73% through reverse osmosis. The authors conclude that the NF and reverse osmosis membranes applied in full-scale are very efficient in removal of almost all pharmaceutical residuals found in water.

Due to advantages NF has over RO, such as lower energy consumption and selective ion discrimination, as well as the capability to remove almost all pharmaceutical residuals found in water it was decided to test a NF process for water treatment at Mezquital Valley.

3. ENDOCRINE DISRUPTORS IN GROUNDWATER SOURCES OF MEZQUITAL VALLEY

3.1. Description of The Study Site

The Mezquital Valley is located in the central plateau of Mexico, 80 km. from Mexico City in the South central part of the state of Hidalgo. In the municipalities located within the limits of the aquifer of Mezquital Valley (Figure 2),there is an estimated population of 377,951 inhabitants, which represents 14.8% of the total population of the state [82].The principal economic activity in the region is agriculture, which is distributed in two irrigation districts (03 Tula and 100 Alfajayucan). In the area there are also a petroleum refinery plant, a thermoelectric plant, textile factories and a cement plant, considerable commercial activity, craft production, and the tourist industry is based on spas and hiking in the mountains [83].

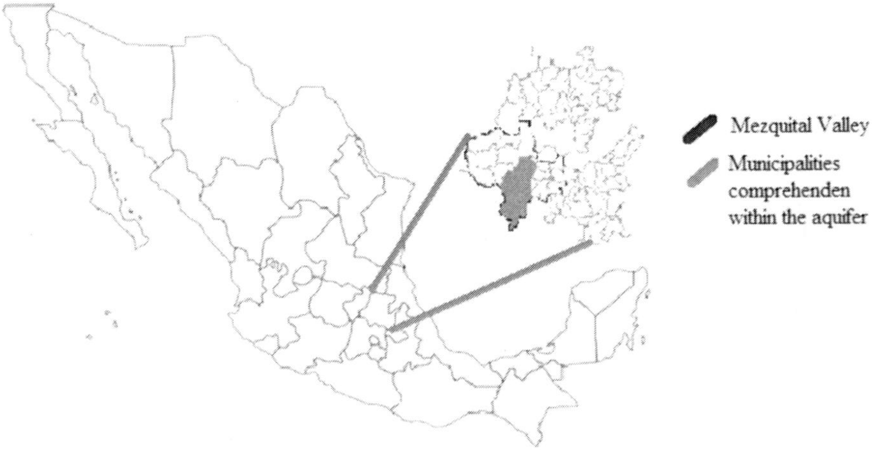

Figure 2. Location of the Mezquital Valley aquifer in Hidalgo State.

The climate is semi-arid, with a high incidence of sunlight and warm temperatures [83]. [83].Average annual precipitation is 450 mm, while evaporation is 2,100 mm; rain is limited to the months of May through October [84].

Within the Mezquital Valley are the irrigation districts 03 Tula and 100 Alfajayucan, representing one of the largest irrigation schemes of residual water in the world. It receives between 50 and 60 m^3/s of water (80% residual and 20% pluvial) from Mexico City through the Gran Canal del

Desagüe, the Interceptor Poniente, and the Emisor Central (Figure 3)[85,86].

Around 75% of this water [87]is used without formal treatment for irrigation of approximately 85,000 hectares of saline soils that lead to high levels of irrigation (1.5 to 2.2 m) [86].Approximately 10,000 hectares receive crude residual waters directly, 35,000 hectares receive mixed water (80% residual and 20% from reservoirs and pluvial sources) and 25,000 hectares receive self-purified residual water from the Requena, Endho, Rojo Gomez and Vicente Aguirre reservoirs [88].

The Emisor Central and Emisor Poniente join the Tula River to feed the Endho reservoir (approximately 200 million $m^{3)}$ El Gran Canal empties into the Salado river and reaches the Tula river beyond the dam [88] (Figure 3). 81% of the main canals and 52% of the lateral canals are not improved or coated, permitting the infiltration and artificial recharge of the aquifer. The water outlets from the aquifer are through springs that lead to the Tula river, extraction through wells, and the rest discharges toward the north and northwest, both superficially (drained by the Salado river) as well as subterranean [85]. The estimated outflow from the springs, which that spread out within the municipalities in the region, range from 100 to 600 L/s, and supply the population [86].

In the wastewaters, wells and springs in the Valle, sampling has shown traces of volatile and non-volatile organic compounds, phenolic compounds, and some PAHs and PCB's in concentrations in the order of picograms/L. Likewise, the presence of benzylbuthylphthalate, diethylhexyl phthalate, nonylphenol, salicylic acid, carbamazepine, ibuprophen and naproxen, indicate contamination from Phs and PCPs that include recognize and suspicious EDs compounds [87,89,90].

3.2. Sampling Sites

For the first monitoring campaign, 19 supply sources were selected (Figure 4) located within the polygon that outlines the aquifer of the Mezquital Valley, and distributed throughout the irrigation zone with untreated wastewater, mixed waters (wastewater and fluvial reservoir water) and at a site to the north of the region to be used as a control site. The 11 sampling sites of wastewater include the water that flows into the Valle via the Emisor Central, residual waters from irrigation canals and one point in the Xotho canal that carries agricultural runoff waters as a point of comparison (Figure 5).

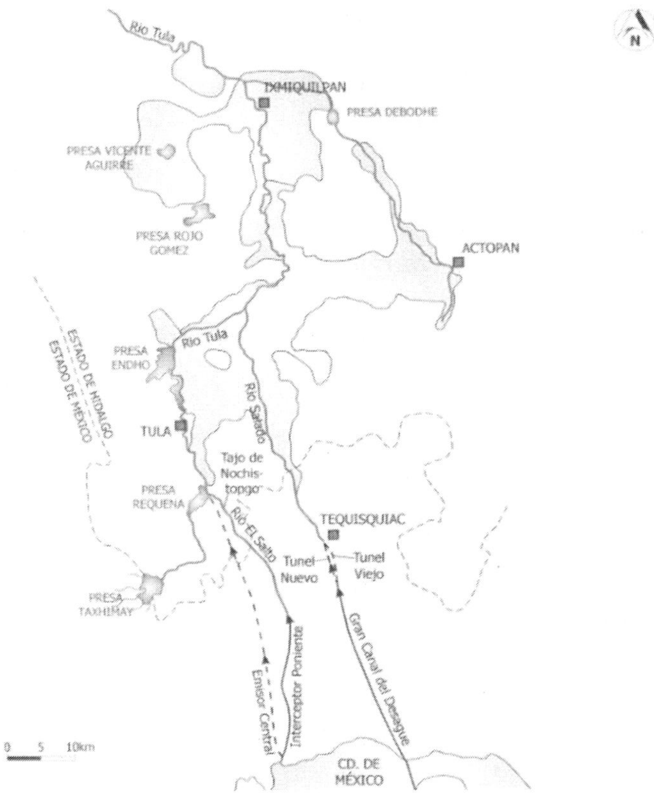

Figure 3.Sanitary sewers from Mexico City to the Mezquital Valley [85].

A second sampling was later conducted on eight supply sources to test for unregulated organic compounds, phenolic compounds, estrogenic hormones, and a screening for pharmaceuticals, personal care products and progestogenic hormones; a second more specific and selective sampling was conducted of the supply sources in accordance with the following criteria:

- Concentrations of arsenic, fluorides, mercury or lead above maximum allowed limits as established by the Mexican normative for potable waters.
- Abundance (qualitative) of unregulated organic compounds (volatile, semi-volatile and persistent)

2 "Tezoquipa" Well
3 "El Tablón" Well
4 "Dendhó" Well
5 "San Antonio" Well
6 "El Refugio" Well
7 " Ajacuba" Well
8 "Noria Tetepango" Well
9 "Doxey" Well
10 "Tlaxcoapan" Well
11 "Teltipan" Well
12 "El Puedhe" Well
13 "Atengo" Well
14 "Xochitlan" Well
15 "Progreso" Spring"
16 "Progreso" Well
17 "Cerro Colorado" Spring
18 "Fitzhi" Well
19 "Chilcuautla" Well
20 "Pedro Ma. Anaya" Well
23 "El Llano" Well

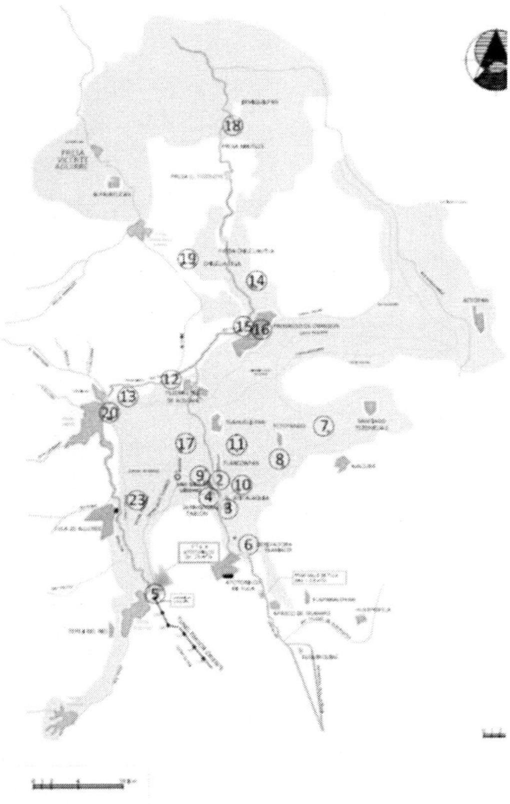

Figure 4. Supply sources.

The sources selected were: 002 (Tezoquipa), 006 (El Refugio), 013 (Atengo), 014 (Xochitlan), 015 (Progreso spring), 017 (Cerro Colorado spring), 018 (Fitzhi) y 019 (Chilcuautla)

3.3. Sampling

The procedure for gathering water samples to analyze compounds at the trace and ultra-trace levels requires special care in order to not contaminate the samples by physical contact, perspiration or exhalation. For this purpose a cap, powder free gloves and face masks were used while handling, filling, closing, labeling and packaging the containers for storage and transport.

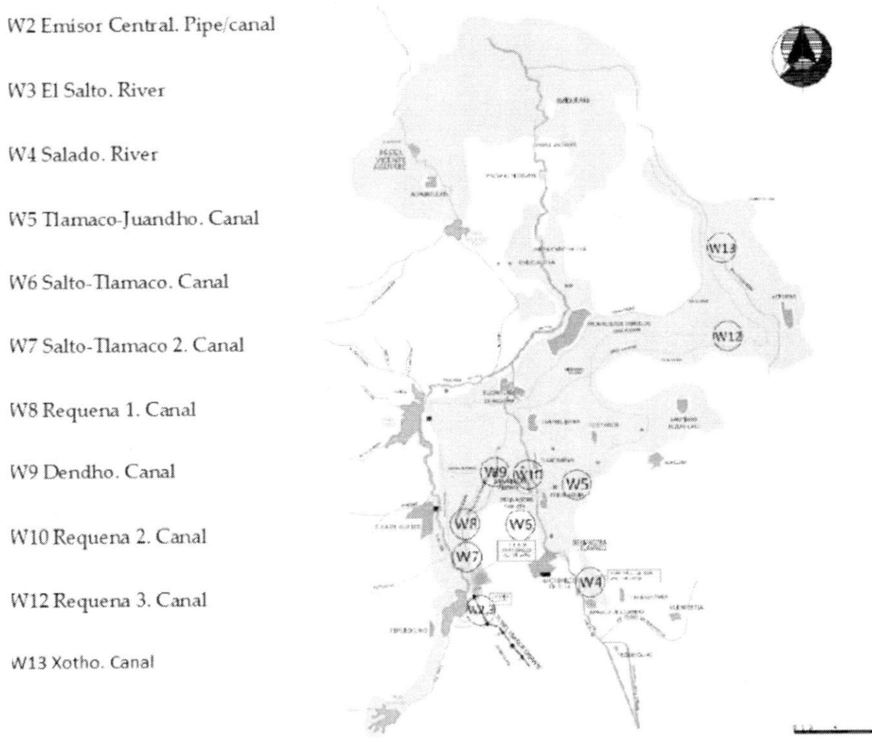

Figure 5. Wastewater sample sites.

Two people collected the samples; one was designated "dirty hands" and the other "clean hands". All operations requiring contact with the bottle and transferring the sample from the collection vessel to the bottle were conducted by "clean hands," and "dirty hands" was responsible for all activities that did not involve direct contact with the sample [91].

3.4. Analytical Methods

The samples were analyzed in laboratories accredited by the Entidad Mexicana de Acreditamiento, and the analytical methods utilized are based on the Mexican Standards or on international methods or those applied in other countries and were implemented, standardized and or validated in the selected test labs for sample analysis. For the analysis of volatile compounds, the samples were gathered in vials of 40 ml, and were kept at a temperature of 4°C during transport to the lab and until their analysis.

The samples were submitted for analysis through a methodology of USEPA Method 5030B [92],using a gas Chromatograph Varian 3800 coupled to a mass spectrometer Saturn 2200 equipped with a capillary column Sample Concentrator Teckmar/Dohrmann Model 3100 and auto sampler Varian Archon for 40 ml. vials.

The analysis of semi-volatile compounds was conducted with US-EPA Method 8270D [93]. Extractions were made in basic, neutral and acidic medium. The fractions obtained were analyzed with gas chromatography-mass spectrometry in total ion monitoring mode. For the analysis of noniphenols, method ASTM-D 7065-06 was applied. Dichloromethane was used to extract the samples. The extracts were concentrated and dried with the use of anhydrous sodium sulphate.

The detection of PPCPs was conducted by AXYS Laboratories in British Columbia Canada. The samples were analyzed following the AXYS method MLA-075: Analytical Procedures for the Analysis of Pharmaceutical and Personal Care Products in Solid, Aqueous and Tissue Samples by LC-MS/MS[91]. This method is suitable for the determination of a suite of pharmaceutical and personal care compounds in aqueous, solid and tissue samples. The analysis requires extraction at two different pH conditions: at pH 10 for analysis of fourteen analytes; and at pH 2.0 for the analysis of the other analytes. Prior to extraction and/or clean-up, samples are adjusted to the required pH and spiked with surrogates. A total of 119 analytes can be identified and quantified. Analysis of the sample extract is performed on a high performance liquid chromatography coupled to a triple quadrupole mass spectrometer. The LC/MS/MS is run in MRM (Multiple Reaction Monitoring) mode and quantification is performed by recording the peak areas of the applicable parent ion/daughter ion transitions. Some analytes are analyzed in the ESI positive mode and some are analyzed in the ESI negative mode.

3.5. Results

3.5.1. Fluoride and heavy metals: suspected endocrine disrupters

The Mexican standard for drinking water quality [94], among other chemical parameters, limits the concentration of heavy metals, fluoride, organochlorated pesticides, total trihalomethanes and volatile organic compounds. In this framework, only four contaminants with potential to cause endocrine disruption exceeded the established maximum contaminant level: Arsenic (As), Fluorides (F), mercury (Hg) and lead (Pb) [95,96].

The greatest challenge came up in the wells in the south region of the Mezquital Valley (002 through 012 and 23), that are located in the entry zone of the residual waters. In the central and northern zone of the valley high concentrations of fluorides were found (Table 2), which is consistent with data reported in a previous study [85].

Table 2. Supply sources and parameters of the Official Norm for Water for Human Consumption and Use that are not met in the study zone.

Parameter (Max allowable limit)	Supply Source (concentration µg/L or mg/L)
Arsenic (25 µg/L)	Wells: 003 (26), 004 (35.1) y006 (38.7)
Fluorides (1.5 mg/L)	Wells: 004 (1.72), 006 (1.52), 009 (1.62), 012 (1.59); Springs:015 (2.37), 017 (1.64); Wells: 016 (3.32), 018 (3.04), 019 (2.98) 023 (3.77)
Mercury (1 µg/L)	Wells:002 (2), 003(1.1), 006 (2.6), 007 (2.1) y 010 (2.6)
Lead (10 µg/L)	Wells:002 (28), 003(30) y 004 (39)

Chronic exposure to As in drinking water has been strongly associated with increased risks of multiple cancers, heart disease, diabetes mellitus II and reproductive and developmental problems in humans [65].Recent studies suggest that increased human health risks [66],at levels as low as 5-10 ppb, could be mediated, at least partially, through the capacity to alter steroid receptors [97-99].There is evidence that high exposures to F^- are associated with decreased thyroid function [100,101], increased activity of the calcitonine and parathyroid, secondary hyperparathyroidism, tolerance to glucose and possible effects over time on reaching sexual maturity [96]. Lead (Pb) and mercury (Hg), can interfere with hormone neurotransmitters and other growth factors, and in both cases any exposure can be considered dangerous to the developing organism [57,102].

3.6. Unregulated Organic Compounds and Estrogens

In the first monitoring campaign, 80 volatile and semi-volatile organic compounds (VSOCs) were identified at a qualitative level. The south zone of the valley showed greater diversity of organic contamination and in the north zone, where the irrigation water is mixed (wastewater and fluvial reservoir water), the water supply sources also showed organic contamination (Table 3 and Table 4), including naphthalene in well 019.

Water sources in the central zone (wells007, 012, 013 and015) contained the least diversity of organic compounds, nevertheless, no clear trend can be observed; therefore we assume that human activities within the Valley represent an important contribution to the organic loading (Table 3).

Table 3. Number of volatile, semi-volatile and fixed organic compounds detected in water supply sources

Supply source	Number of organic compound	Supply source	Number of organic compound
002 "Tezoquipa" Well	22	003 "El Tablón" Well	17
004 "Dendhó" Well	17	005 "San Antonio" Well	26
006 "El Refugio" Well	26	007 " Ajacuba" Well	18
008 "Noria Tetepango" Well	20	009 "Doxey" Well	17
010 "Tlaxcoapan" Well	17	011 "Teltipan" Well	22
012 "El Puedhe" Well	11	013 "Atengo" Well	17
014 "Xochitlan" Well	22	015 "Progreso" Spring"	8
016 "Progreso"	24	017 "Cerro Colorado" Spring	22
018 "Fitzhi" Well	29	019 "Chilcuautla" Well	31
020 "Pedro Ma. Anaya" Well	20	023 "El Llano" Well	17

Table 4. Number of volatile, semi-volatile and fixed organic compounds detected in wastewater

Sampling site	Number of organic compound	Sampling site	Number of organic compound
AR "002 Emisor Central" Pipe/canal	56	AR 003 "El Salto" River	37
AR 004 "Salado" River	42	AR 005 "Tlamaco Juandhó" Canal	42
AR 006 "Salto-Tlamaco 1" Canal	56	AR 007 "Salto-Tlamaco 2" Canal	39
AR 008 "Principal Requena 1" Canal	38	AR 009 "Principal Dendhó" Canal	39
AR 010 "Principal Requena 2" Canal	35	AR 012"Principal Requena 3" Canal	32
AR 013			

In the second monitoring campaign conducted during the rainy season all eight supply sources showed VSOCs (173 different ones). Qualitatively, the greatest contamination was observed mainly in the central zone; the source with the highest contamination was 015 with 69 VSOCs followed by 013 with 59 VSOCs; 017, 42; 006 and 014 with 41; 019, 38; 018, 36; and the source least contaminated was Tezoquipa (002) where 32 VSOCs were found.

From both groups of compounds, the following are notorious for their presence in residual waters and supply sources:

- Pesticides: 1,4-dichlorobencene, 1,2-dicloropropane, 3-chloro-2-methyl-1-propene, 1,1-dichloro-1-nitroethane, 1,2,3,4-tetrahydronaphtalene.
- Flavorings and essences: 3-methyl-2-pentanone, 2-methylpentanoic acid, 2-pentanone, 5-methyl-2-hexanone, 2,3-dichloro-2-methylbutane, 3-methyl-2-butanone, 2-propenilesther acetic acid and methanotiol
- Cosmetic formulations, PCPs and industrial uses: ethylbencene, dietilphthalate, 1,2,4-trimethylbenzene, methyl-isobutylcetone, eucaliptol, tetradecane, nonadecane.

Estrogenic hormones such as estrone, 17-β-estradiol, 17-β-estradiol acetate and 17-α-ethinylestradiol, were present in both residual waters and supply sources in ultra-trace amounts (ng/L). The highest concentrations were found in the residual water from Mexico City and in the Xotho canal (Table 5 and Table 6). The concentration of hormones in AR00 –Emisor Central and AR003-El Salto and AR004-Salado rivers, AR013-Xotho canal and in 019-Chilcuautla well is notable.

In the second monitoring campaign six of the eight supply sources showed contamination from Phs and PCPs (Table 8), in ultra-trace concentrations: erythromycin, sulfamethoxazole (antibiotics), carbamazepine (antiepileptic), methylprednisolone (steroid anti-inflammatory used in cases of allergy), DEET (PCP), caffeine and benzoilecgonine (metabolite of cocaine). Chlofibric acid was found in 002-Tezoquipa (55.5 ng/L) and 003-El Tablón (36.67 ng/L), and AR003-El Salto (41.32 ng/L). The same river also showed traces of gemfibrozil (63.2 ng/L).

Table 5. Total estrogens in water supply sources (ND: no detected).

Supply source	Concentration (ng/L)	Supply source	Concentration (ng/L)
002 "Tezoquipa" Well	3.021	003 "El Tablón" Well	0.524
004 "Dendhó" Well	1.228	005 "San Antonio" Well	5.633
006 "El Refugio" Well	1.320	007 " Ajacuba" Well	2.105
008 "Noria Tetepango" Well	ND	009 "Doxey" Well	ND
010 "Tlaxcoapan" Well	0.134	011 "Teltipan" Well	3.105
012 "El Puedhe" Well	3.080	013 "Atengo" Well	ND
014 "Xochitlan" Well	0.158	015 "Progreso" Spring"	ND
016 "Progreso"	1.176	017 "Cerro Colorado" Spring	5.660
018 "Fitzhi" Well	2.110	019 "Chilcuautla" Well	11.426

Table 6. Total estrogens in wastewater (ND: not detected).

Sampling site	Concentration (ng/L)	Sampling site	Concentration (ng/L)
W “002 Emisor Central” Pipe/canal	86.032	W 003 “El Salto” River	68.725
W 004 “Salado” River	ND	W 005 “Tlamaco Juandhó” Canal	ND
W 006 “Salto-Tlamaco 1” Canal	76.314	W 007 “Salto-Tlamaco 2” Canal	39.13
W 008 “Principal Requena 1” Canal	11.74	W 009 “Principal Dendhó” Canal	114.599
W 010 “Principal Requena 2” Canal	60.067	W 012“Principal Requena 3” Canal	33.855
W 013 “Xotho” Canal	12.141		

Table 7. Pharmaceuticals and personal care products (concentration in ng/L; ND: not detected).

Source	Caffeine	Benzoyl-ecgonine	DEET	Methyl-prednisolone	Carba-mazepine	Erithromycin-H_2O	Sulfa-methoxasole
002	ND	ND	0.948	12.9	ND	ND	7.25
006	ND	ND	1.64	ND	ND	0.325	ND
013	24.3	4.93	2.01	ND	11.5	1.08	15.9
014	22.9	ND	1.01	ND	ND	0.31	16.7
017	ND	0.318	0.923	5.2	5.24	1.35	10.4
019	ND	ND	1.14	6.59	ND	1.05	22.8

3.7. Bisphenol A, Alkylphenols and Alkylphenol-Ethoxylates

As previously noted, bisphenol A and alkylphenols, including their ethoxylates, are relevant from a public health standpoint for their potential to alter endocrine processes in humans and aquatic and terrestrial organisms of various taxonomy. They are found in a variety of products of frequent use and are used as an adjuvant in the application of pesticides, which explains why, in the monitoring campaign that took place during the dry season, the highest concentrations were found in residual waters of AR02-Emisor Central and in AR03-El Salto y AR04-Salado rivers, as well as in the irrigation canals in the south zone. Furthermore, the concentrations of these contaminants in the supply sources were not homogenous and there was no obvious trend that suggested that wastewater from Mexico City was the only source of contamination. This is why we concluded that there was an influence from the interior of the Valley (Table 8. and Table 9).

Table 8. Total alkylphenols and etoxilates in water supply sources (ND: not detected).

Supply source	Concentration (μg/L)	Supply source	Concentration (μg/L)
002 "Tezoquipa" Well	ND	003 "El Tablón" Well	ND
004 "Dendhó" Well	ND	005 "San Antonio" Well	2.29
006 "El Refugio" Well	ND	007 " Ajacuba" Well	4.23
008 "Noria Tetepango" Well	2.10	009 "Doxey" Well	6.07
010 "Tlaxcoapan" Well	3.7	011 "Teltipan" Well	2.51
012 "El Puedhe" Well	0.18	013 "Atengo" Well	ND
014 "Xochitlan" Well	2.0	015 "Progreso" Spring"	ND
016 "Progreso"	ND	017 "Cerro Colorado" Spring	ND
018 "Fitzhi" Well	0.18	019 "Chilcuautla" Well	1.12
020 "Pedro Ma. Anaya" Well	ND	023 "El Llano" Well	ND

Table 9. Total alkylphenols and etoxilates in wastewater.

Sampling site	**Concentration (μg/L)**	**Sampling site**	**Concentration (μg/L)**
W "002 Emisor Central"	41.28	W 003 "El Salto" River	64.57

Sampling site	**Concentration (µg/L)**	**Sampling site**	**Concentration (µg/L)**
Pipe/canal			
WW 004 "Salado" River	17.41	W 005 "Tlamaco Juandhó" Canal	40.73
W 006 "Salto-Tlamaco 1" Canal	19.03	W 007 "Salto-Tlamaco 2" Canal	24.29
W 008 "Principal Requena 1" Canal	43.82	W 009 "Principal Dendhó" Canal	94.87
W 010 "Principal Requena 2" Canal	23.84	W 012"Principal Requena 3" Canal	25.9
W013 "Xotho" Canal	1.52		

3.8. Sanitary Risks to The Mezquital Valley

Fifty-nine illnesses were identified as the primary causes of general mortality. Diabetes mellitus type II is at the top of the list, and ischemic illnesses of the heart, cerebral vascular ailments, diverse types of cancer (mainly breast, prostate, liver and leukemia) are among the primary causes of death. These diseases are multi-causal and related in great part to diet and life style as well as genetic and epigenetic factors; however environmental exposure to organic and inorganic contaminants can also contribute to the risk of contracting these ailments, which is why it is necessary to conduct studies that allows discovery or estimation of the contribution of water to these diseases.

For both samplings, hazard of exposure to 173 ECs was qualitatively identified, of which 35 are of high priority for their carcinogenic potential or systemic toxicity and among these, for their potential as EDs (acknowledged and suspicious [103-106]) and the frequency with which they were found of note are:

- Cosmetics, PCP and hygienic products formulations: ethylbenzene, octadecanoic acid (stearic acid), octyl-dimethyl-paramino benzoate (Escalol 507) and DEET
- Pharmaceuticals: carbamazepine, methylprednisolone, erythromicine
- Pesticides: 1,4-diclorobenzene
- Industrial uses: 1,2-dichloropropane, benzene, chloroform, n-propylbenzene, tetrahydrofurane, tetrachloroethylene, 1,1,2-trichloroethane
- Plasticizers: di-isobuthylphthalate, dipropylphthalate
- Estrogens: estrone, 17-β-estradiol, 17-β-estradiol acetate and 17-α-ethinylestradiol

By using a methodological approach of health risk assessment and analyzing the results of a survey completed by [107] from 1000 permanent residents in the Mezquital Valley, this study estimated the exposure to contaminants identified as hazardous through the use and consumption of water. The ingested dose via this route was compared to values of toxicity available on The Risk Assessment Information System data bases [108]. The hazard coefficient was obtained for arsenic, fluoride, bisphenol A, nonilphenol, naphthalene, and esters of ftalic acid or ftalates. The hazard coefficient or hazard index (HI) were estimated based on the volume of water consumed, concentration of the analyte in the water, intensity, frequency and magnitude of the oral exposure, body weight and the reference dose for critical effect in the target organ. A HI of 1.0 or less indicates that the exposure poses no health risks.

Hazard Index greater than 1.0 were identified for arsenic, mercury and fluoride, which means there are health risks associated with such substances in drinking water (Table 10). When such indexes are calculated for municipalities, greater hazards are identified. For example: Arsenic HI in Atitalaquia was 7.87 (± 5.08), in Progreso 2.29 (± 1.13) and in Tlaxcoapan 5.23 (± 2.47); Fluorie HI in Atitalaquia was 3.31 (± 2.14), in Atotonilco de Tula 2.98 (± 1.65), in Chilcuatla 4.68 (± 2.65), in Ixmiquilpan 5.40 (± 2.81), in Progreso 4.26 (± 2.33) and in Tlaxcoapan 3.03 (± 1.43).

Table 10. Average Hazard Index from consumption of water with EDs (recognized and suspicious) within Mezquital Valley.

Analyte	Media HI	Standard deviation	Min HI	Max HI
Arsenic	2.48E+00	4.07E+00	8.22E-03	2.41E+01
Mercury	3.26E-01	4.39E-01	0.00E+00	2.61E+00
Fluoride	3.95E+00	3.28E+00	2.74E-01	2.10E+01
Bisphenol A	7.18E-04	2.12E-03	0.00E+00	1.86E-02
4-Nonylphenol	2.81E-03	3.93E-03	0.00E+00	2.04E-02
Bis-2-(ethylhexyl)-phthalate	9.28E-03	3.51E-02	0.00E+00	3.07E-01
Diethylphthalate	5.78E-06	1.39E-05	0.00E+00	1.09E-04
Dibuthylphthalate	8.94E-04	3.04E-03	0.00E+00	2.68E-02

Even though at present it is not possible to associate health risks with chronic ingestion of organic EDs, there is uncertainty about exposure to small doses and the effects in the middle and long terms, besides the concomitant exposure.

3.9. Water Treatment Test

Pilot scale testing was carried out to make sure NF was effective at removing certain minerals, organic matter and ECs. Two supply sources within the valley were selected for this purpose.

The first phase of testing was conducted in the Cerro Colorado spring for 82 consecutive days. This source supplies water to 4 municipalities within the Valley (Mixquiahuala, Progreso, Tepacatepec and Tezontepec). Water characteristics are shown in Table 11. Due to these mineral characteristics of the water (hardness and silica content) up to 72% water recovery can be obtained.

Table 11. Cerro Colorado Spring water quality.

Parameter	Concentration	Unit	Parameter	Concentration	Unit
Cl^-	148	mg/L	Na	195.14	mg/L
F^-	1.64	mg/L	Hardness	237.1	mg $CaCO_3$/L
$HCO_3{}^{2-}$	271.53	mg $CaCO_3$/L	SDT	1,147	mg/L
SO_4	169.96	mg/L	pH	7.16	
SiO_2	63.1	mg/L			

Pilot scale NF tests were performed with a two stage skid; a 4:2 pressure vessel array with each pressure vessel holding 3 standard elements (NE4040-90 CSM), respectively, was utilized. Through brine recirculation, this design can simulate a two stage full-scale skid, employing six (8" X 40") elements per vessel.

In Figure 6 the pilot installation is shown: The whole treatment train includes sand filtration, a 5 m^3filtered water storage tank, 10 μm cartridge filtration, antifouling agent addition and NF.

The second phase of testing was conducted in the Tezoquipa well, located in Atitalaquia, Hgo. It lasted 36 days. In the first monitoring campaign, we found that the concentration of dissolved solids in this well was among the lowest in the Valley. However, heavy metals were found in the water, which is why it was selected for the second phase of the water treatment test.

3.9.1. Rejection of Mineral and Organic Compounds

To assess the performance of the NF pilot, six parameters were measured in feed water, permeate and brine: Conductivity, $SO_4{}^{2-}$, F^-, Hardness, $HCO_3{}^-$ and TOC. Figures 7, 8 and 9 show the rejection of those parameters monitored during water treatment tests.

Mineral content rejection, at the beginning of water treatment testing, was 80% (measuring conductivity). This level of efficiency was about 10% below the rejection calculated with the CSM4PRO design software (corresponding to membranes NE4040-90 of CSM). However, as the test progressed, the salt rejection rose to 93%, which was very close to the calculations made by CSM4PRO.

Figure 6. Pilot unit including multimedia filtration and NF.

HCO_3^- removal was 92% on average (± 0.9981) while hardness was rejected at a higher rate of 97.46% on average, with a minimum of 95.12 and a maximum of 98.95%.

Sulphate removal was 98.78% on average, with a maximum of 99.57% and a minimum of 95.75%. These results are consistent with the characteristics of the membrane, since it is a negatively charged hydrophobic membrane specialized in the removal of polyvalent anions.

Fluoride rejection oscillated between 80% and 96.77% (Figure 8). The variability in the results could be explained by changes in feed water temperature. Filtered water coming from the spring was stored in the $5m^3$ tank. The tank was exposed directly to sunlight, causing a rise in water temperature. While samples and measurements were meant to be taken at 13:00 hours, this was not always possible and there was no control over ambient temperatures.

Organic matter, measured as TOC, was rejected at a rate of 82% on average (Figure 9). Nevertheless, its rejection rate varied from 41% to 100%. It is worth mentioning that in two of the four measurements in which removal efficiency was less than 45%, the concentration of TOC in the raw water was lower than 2 ppm; even so the variability is considerable.

With respect to pharmaceuticals and personal care products, laboratory data indicates that the membranes were capable of rejecting all of the sulfamethoxazole% (Table 12).

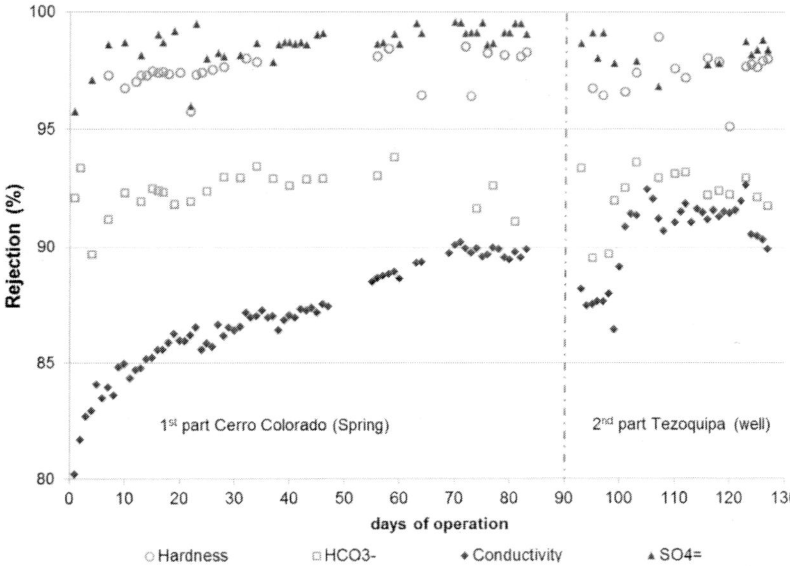

Figure 7. Major ions rejection observed in water tests.

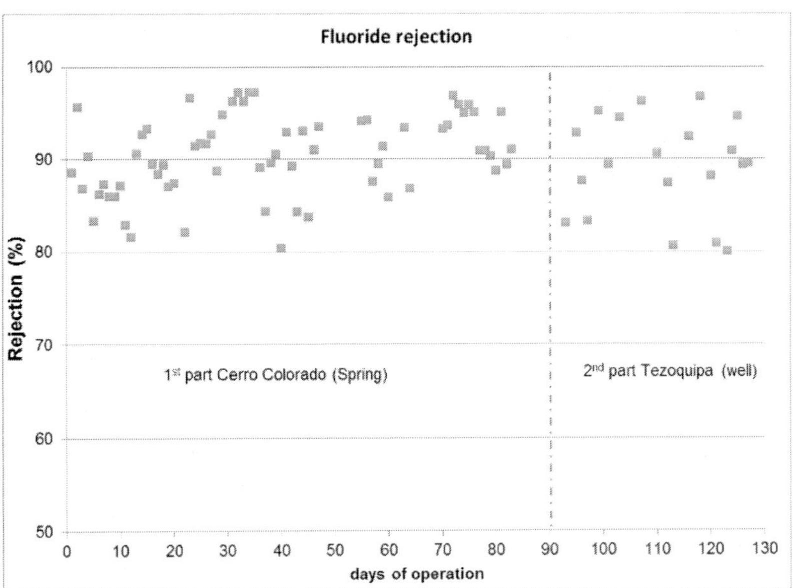

Figure 8. Fluoride removal efficency observed in water treatment test.

Erythromycin-H_2O and cocaine were found in the brine but not in the feed water or permeate. Given the low concentrations reported (fractions of nanograms), it is plausible that the raw water contained these compounds at non-detectable levels and that they were rejected with great efficiency, causing these substances to reach concentrations at a detectable and quantifiable level.

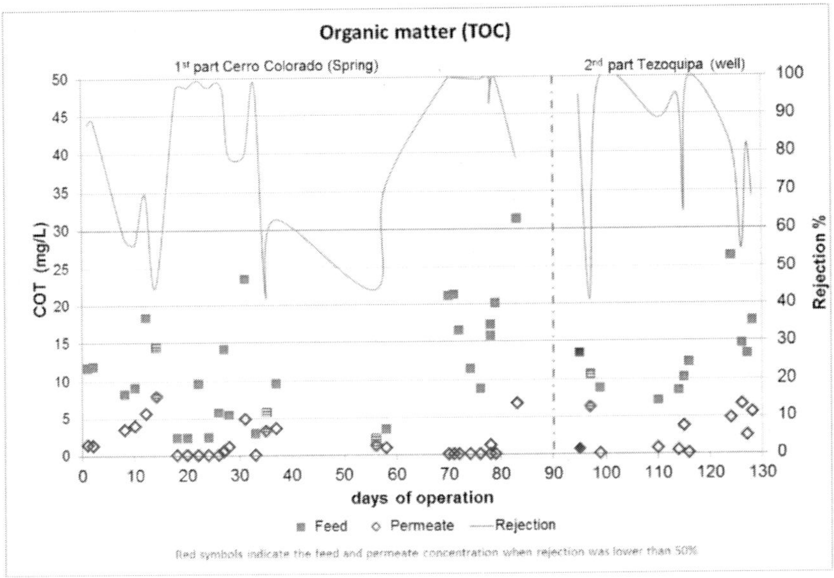

Figure 9.NF performance for organic matter removal.

Table 12. Pharmaceuticals and PCPs in water treatment test.

Site	Compound	Concentration (ng/L)
Raw water	Sulfamethoxazole DEET Metylprednizolone	7.25 0.948 12.9
Permeate	DEET Metylprednizolone	2.01 11
Brine	Caffeine Eritromicina-H2O Sulfamethoxazole Cocaine DEET	14.5 0.654 18.8 0.172 0.848

The NF membranes do not appear to be effective in the removal of DEET and are inefficient in the removal of methylprednisolone. The molecular size of DEET is smaller than the Molecular Weight Cut-Off of the membrane. Also, DEET is a neutral molecule, so negatively charged membranes won't exert electrostatic repulsion. In the case of methylprednisolone, it is of note that it was not found in the brine. This, according to the mass balance, is not possible. In other words, if the concentration in the feed and the permeate streams are the same, the same should be true for the brine.

Overall, groundwater quality within the Valley does not meet the Mexican Drinking Water Quality Standard [94], from a total dissolved solids standpoint [94]. In some sources hardness, sulphate, fluoride and heavy metals like lead, mercury and arsenic are present at concentrations exceeding the MCL. NF was able to reject all these substances at such a rate that the permeate met the Mexican Drinking Water Quality Standard. From the mineral rejection standpoint, NF seems to be an interesting alternative to treat the Mezquital Valley ground water in order to produce drinking water.

4. CONCLUSIONS AND RESEARCH NEEDS

In the semi-arid Mezquital Valley, reusing wastewater for agricultural purposes has benefitted aquifer recharge, which has increased the availability of water destined for various human activities, such as drinking water. Nevertheless, the presence of As, F, Hg and Pb and multiple organic pollutants in 19 water sources that were sampled showed a wide variety of potential health risks, including disruptions of the endocrine system.

The Hazard Quotients (HQ) estimated for the EDs (acknowledged and suspected) in the mentioned area indicated risks associated (HC > 1.0) with As and F, while no risks were found at the time of the study for organic EDs (phtalates, nonilphenols and bisphenol A). Nevertheless, in water reuse scenarios such as this one, even with the limitations inherent to both point monitoring and considering the lack of scientific knowledge.

It is necessary to emphasize the prevention of human exposure to EDs, considering that: a) the effects of critical exposure during intrauterine, perinatal and puberty periods may not manifest until adulthood; b) the population in the Mezquital Valley is exposed to a mixture of EDs that could lead to the addition, synergy, potentiation or antagonism of effects, and c) it may be assumed that the entire population is potentially exposed to these pollutants since, in qualitative terms, all water sources presented at least one ED.

More monitoring campaigns focusing on EDs in a greater number of supply sources are necessary to quantify the EDs with greater precision and estimate exposure with higher accuracy.

Studies are required to integrate environmental aspects, life style and toxicological research with epidemiology (cohort retrospective study or control cases) that allow associations to be established between EDs exposure with ailments or specific health issues: i.e. diabetes mellitus type II, the development of various types of cancer, reproductive function and processes and neurodevelopment, which are currently of interest to various groups of researchers and the governmental agencies responsible for regulation and policies in health, water and the environment.

On the other hand, the byproducts of the degradation and transformation of a compound can also alter the hormonal system and, while a large number of researchers believe that physicochemical characteristics of the molecules determine the molecules' behavior and environmental distribution, a greater understanding of their action, transformation and environmental fate are also required to assess and minimize the health risks associated with exposure to EDs in concentrations relevant to drinking water.

Water delivered through public networks must be treated in order to comply with the Mexican Standard for water quality. A NF process will remove mineral pollutants found in the Mezquital Valley Nevertheless, from an ED standpoint, further tests are needed in order to obtain enough evidence to determine if NF is capable of producing drinking water from these sources, or another process is needed, either as a substitution (like RO) or a complement (like activated carbon).

5. ACKNOWLEDGEMENTS

Projects Coordinating Office for the Valley of Mezquital /National Water Commission.

REFERENCES

1. B Arnold, J Calford, Treating water with chlorine at point of-use to improve water quality and reduce child diarrhea in developing countries: a systematic review and meta-analysis. American Journal of Tropical Medicine and Hygiene 200776235464
2. S Cairncross, C Hun, S Boisson, K Bostoen, V Curtis, I Fung, W Schmidt, Water, sanitation and hygiene for the prevention of diarrhoea. International Journal of Epidemiology 2010i193i205.

3. R Haque, D. N Guha, S Samanta, N Ghosh, D Kalman, M Smith, S Mitra, A Santra, S Lahri, S Das, De B, Smith A. Arsenic water and skin lesions: dose-response data from West Bengal, India. Epidemiology 20031417482
4. Working Group on the Evaluation of the Carcinogenic Risks to HumansArsenic in drinking water. In: IARC Monographs 84Some drinking-water disinfectants and contaminants, including arsenic. Lyon: WHO 200439270
5. S Ayoob, A Gupta, Fluoride in drinking water: a review on the status and stress effects. Critical Reviews in Environmental Science and Technology 20063643387
6. P Bradley, Potential for biodegradation of contaminants of emerging concern in stream systems. Conference Proceedings, October 14-15, 2008Charleston Area Event Center, South Carolina, USA.
7. N Mastroianni, M López, D Barceló, Emerging contaminants in aquatic environments: state of the art and recent scientific contributions. Contributions to Science 20106219397
8. T Damstra, S Barlow, A Bergman, R Kavlock, van Der Kraak G editors. Global Assessment of the State of the Science of endocrine Disruptors. USA:WHO/IPCS; 2002
9. J Brausch, G Rand, A review of personal care products in the aquatic environment: Environmental concentrations and toxicity. Chemosphera 2011821518
10. L Santos, A Araújo, A Fachini, A Pena, E Delerue-matus, M Montenegro, Ecotoxicological aspects related to the presence of pharmaceuticals in the aquatic environment. Journal of Hazardous Materials 201017545
11. C Casals, C Desvergne, Endocrine disruptors: from endocrine to metabolic disruption. Annu. Rev. Physiol20117313562
12. A Salame, F Méndez, G Aguirre, H Serrano, Disrupción endocrina de la diferenciación sexual. ContactoS 200870439
13. T Brevini, S Bertola, C Zanetto, Effects of endocrine disruptors on developmental and reproductive functions. Current Drug Targets-Immune, Endocrine & metabolic disorders 200551
14. S Snyder, D Villeneuve, E Snyder, J Giesy, Identification and quantification of estrogen receptor agonists in wastewater effluents.Environ. Sci. Technol 200135362025
15. K Barnes, D Kolpin, E Furlong, S Zaugg, M Meyer, L Barber, A national reconnaissance of pharmaceuticals and other organic wastewater contaminants in the United States- I) Groundwater. Science of the Total Environment 2008402192200
16. P Stackelberg, , E Furlong, , M Meyer, , S Zaugg, , Henderson, , Reissman D. Persistence of pharmaceutical compounds and other organic wastewater contaminants in conventional drinking water treatment plant. Science of the Total Environment 2004;329:99-113.
17. The Prague Declaration on Endocrine Disruptionhttp://www.bio.uni-frankfurt.de/ee/ecotox/news/PragueDeclaration.pdfaccessed 15 may 2012
18. M Rahman, E Yanful, S Jasim, Endocrine disrupting compounds (EDCs) and pharmaceuticals and personal care products in the aquatic environment: implications for the drinking water industry and global environmental health. Journal of Water and Health 2009222443doi:10.2166/wh.2009.021.
19. Chemcial Abstracts Servicehttp://www.cas.org/index.htmlaccessed 16 may 2012
20. C Acerini, I Hughes, Endocrine disrupting chemicals: a new and emerging public health problem? Arch dis Child 200691863338
21. N Benachour, E Clair, R Mesnage, G Séralini, endocrine disruptors: new discoveries and possible progress of evaluation. In: Berhardt V. (ed.) Advances in Medicine and Biology. France: Nova Science Publishers; 2012157
22. U. S Epa, Special report on Environmental Endocrine Disruption: An Effects Assessment and Analysis. Office of Research and Development 1997EPA/630/R-96/012, Washington D.C.

23. J López, D Barceló, Contaminantes orgánicos emergentes en aguas continentales y aspectos relacionados con el marco normativo y planificación hidrológica en España. En: Aguas continentales. Madrid:91111
24. P Burkhardt-holm, Endocrine disruptors and water quality: a state of the art review. Water Resources Development 201026347793
25. E Diamanti, J. P Bourguignon, L Giudice, R Hauser, G Prins, A Soto, T Zoeller, A Gore, Endocrine-disrupting chemicals: an Endocrine Society Scientific statement. Endocrine Reviews 200930293342
26. J Blasco, Del Valls A. Impact of emergent contaminants in the environment: Environmental risk assessment.Hdb Env Chem. 2008Part S/1):169-88.
27. A Nikolaou, S Meric, D Fatta, Occurrence patterns of pharmaceuticals in water and wastewater environments. Anal Bioanal Chem 2007387122534
28. E Godfrey, W Woessner, J Benotti, Pharmaceuticals in on-site sewage effluent and ground water, Western Montana. Groundwater 200845326371
29. T Asano, J Cotruvo, Groundwater recharge with reclaimed municipal wastewater: health and regulatory considerations. Water Res. 200438194151Bound J, Voulvoulis N. Household disposal of pharmaceuticals as a pathway for aquatic contamination in the United Kingdom. Environ Health Perspect 2005;113:1705-11.
30. D Vethaak, J Lahr, S Schrap, A Belfroid, G Rijs, A Gerritsen, J Boer, A Bulder, G Grinwis, R Kuiper, J Legler, T Murk, W Peijnenburg, H Verhaar, P Voogt, An integrated assessment of estrogenic contamination and biological effects in the aquatic environment of The Netherlands. Chemosphere 20055951124
31. SirbuD, Curseu D, Popa M, Achimas A, Moldovan Z. Environmental risks of pharmaceuticals and personal care products in water.Tenth International Water Technology Conference, IWTC10 2006, Alexandria, Egypt. http://www.iwtc.info/2006_pdf/15-2.pdf (accessed 15may 2012).
32. C Daughton, T Ternes, Pharmaceuticals and personal care products in the environment: Agents of subtle change?Environ. Health Perspect. 1999suppl. 6):907-38.
33. R Trussell, Constituents of Emerging Concern: An Overview In: WEFTEC 2006Session 20 of Proceedings of the Water Environment Federation, October, WEFTEC 20062125October 2006, Dallas Tx, USA.http://www.environmental-expert.comaccessed 15 may 2012).
34. S Snyder, R Trenholm, E Snyder, G Bruce, R Pleus, J Hemming, Toxicological relevance of EDCs and pharmaceuticals in drinking water. USA: AWWA Research Fundation/American Water Works Association/IWA; 2008
35. Committee on Hormonally Active Agents in the EnvironmentNational Research Council. Hormonally active agents in the environment. USA: National Academies Press; 1999
36. S Snyder, P Westerhoff, Y Yoon, D Sedlak, Pharmaceuticals, personal care products, and endocrine disruptors in water: implications for the water industry. Environmental Engineering Science 200320544969
37. T Hanselman, D Graetz, A Wilkie, Manure-borne estrogens as potential environmental contaminants: a review. Environmental Science & Technology 20033724547178
38. A Rivas, A Granada, M Jiménez, N Olea, Exposición humana a disruptores endocrinos. Ecosistemas 2004133712
39. E Snyder, R Snyder, R Pleus, S Snyder, Pharmaceuticals and EDCS in the US wáter industry-an update. Journal AWWA 20059711325
40. T Schug, A Janesick, B Blumberg, J Heindel, Endocrine disrupting chemicals and disease susceptibility. Journal of Steroid Biochemistry & Molecular Biology 201112720415
41. Olujimi, O Fatoki, J Ordendaal, O Okonkwo, Endocrine disrupting chemicals (phenol and phthalates) in the South African environment: a need for more monitoring. Water SSA 201036567181

42. Australian Academy of ScienceEndocrine disruption. Are chemicals in our environment reducing male fertility and causing cáncer? Do they cause deformities and effect reproduction in wildlife? Australia:Ntional Science and Industry Forum Report; 1998
43. F Argemi, N Cianni, A Porta, Disrupción endocrina: perspectivas ambientales y salud pública. Acta Bioquím Clín Latinoam 2005393291300
44. C Chichizola, Disruptores endocrinos. Efectos en la reproducción. Revista Argentina de Endocrinología 200441278105
45. S Laws, S Yavanzhay, R Cooper, J Eldridge, Nature of the binding interaction for 50 structurally diverse chemicals with rat estrogen receptors. Toxicological Sciences 20067414656
46. SandersonThe steroid hormone biosynthesis pathway as a target for endocrine-disrupting chemicals. Toxicological Sciences 2006321
47. M Tabb, B Blumberg, New modes of action for endocrine-disrupting chemicals. Molecular Endocrinology 200420347582
48. C Chichizola, H Scaglia, C Franconi, B Ludueña, C Mastandrea, A Ghione, Disruptores endocrinos y el sistema reproductive. Revista Bioquímica y Patología Clínica 2009733923
49. C Chichizola, Disruptore endocrinos. Efectos en la reproducción. Revista Argentina de endocrinología y Metabolismo 200340317288
50. E Swedenborg, J Rüegg, S Mäkelä, I Pontgratz, Endocrine disruptive chemicals: mechanisms of action and invelovement in metabolic disorders. Journal of Molecular endocrinology 200943110
51. W Welshons, K Thayer, B Judy, J Taylor, E Curran, Vom Saal F. Large effects from small exposures. I. Mechanisms for endocrine-disruptingchemicals with estrogenic activity. Environ Health Perspect 20031119941006
52. Vom Saal FHughes C. An extensive new literature concerning low-dose effects of Bisphenol A shows the need for a new risk assessment. Environ Health Perspect 200511392633
53. A Kortenkamp, R Evans, O Martin, R Mckinlay, F Orton, E Rosivatz, Annex 1 Summary of the state of the science. In: State of the art assessment of endocrine disrupters. Final report. Anex 1. European Commission, DG Environment;2012
54. H Okeda, T Tokunaga, x Liu, S Takayanagi, A Matsushima, Y Shimohigashi, Direct evidence revealing structural elements essential for the high binding ability of Bisphenol A to human estrogen-related Receptor-γ. Environ Health Perspect 2007116328
55. D Melzer, L Harries, R Cipelli, W Henley, C Money, P Mccormack, a Young, J Guralnik, L Ferrucci, S Bandinelli, A. M Corsi, T Galloway, Bisphenol A exposure is associate in vivo estrogenic gene expression in adults. Environ Health Perspect 2011119178893
56. J Bourguignon, A Parent, Early homeostatic disturbances of human growth and maturation by endocrine disrupters. Current Opinion in Pediatrics 20102247077
57. C Frye, E Bo, G Calamadrei, L Calzà, F Dessì-fulgeri, M Fernández, L Fusani, O Kah, M Kajta, Le Page G, Venerosi A, Wojtowicz A, Panzica G. Endocrine disrupters: a review of some sources, effects, and mechanisms of action on behaviour and neuroendocrine systems. Journal of Endrocrinology 2011doi:j.13652826x.
58. E Rees, M Todd, J Beam, A Aiello, The Impact of Bisphenol A and triclosan on immune parameters in the U.S. population, NHANES 2003Environ Health Perspect. 2011;119339096
59. J Braun, A Kalkbrenner, A Calafat, C Yolton, X Ye, K Dietrich, B Lanphear, Impact of early-life Bisphenol A exposure on behavior and executive function in children. Pediatrics 2011128588381
60. C Guerrero, M Settles, B Lucker, K Skinner, Epigenetic transgenerational actions of vinclozolin on promoter regions of the sperm epigenome 2010PLoS ONE 5(9):e13100. doi:10.1371/journal.pone.0013100.
61. J Palmer, A Herbst, K Noller, D Boggs, R Troisi, L Titus, E Hatch, L Wise, C Strohsnitter, R Hoover, Urogenital abnormalities in men exposed to

diethylstilbestrol in utero: a cohort study. Environmental Health 2009doi:10.1186/1476069X-8-37
62. F Ingerslev, E Vaclavick, B Halling-sorensen, Pharmaceuticals and personal care products: A source of endocrine disruption in the environment?.Pure Appl. Chem. 2003
63. T Su, M Pagliaro, T Schmidt, D Pickar, O Wolkowitz, D Rubinow, Neuropsychiatric effects of anabolic steroids in male normal volunteers. JAMA 199326921276064
64. Assessment of Technologies for the Removal of Pharmaceuticals and Personal Care Products in Sewage and Drinking Water Facilities to Improve the Indirect Potable Water Reuse: POSEIDONhttp//wwweu-poseidon.com (accessed 2 August 2011
65. Y Yoon, P Westerhoff, S. A Snyder, E. C Wert, J Yoon, Removal of endocrine disrupting compounds and pharmaceuticals by nanofiltration and ultrafiltration membranes. Desalination 200620216
66. C Bellona, J. E Drewes, The role of membrane surface charge and solute physico-chemical properties in the rejection of organic acids by NF membranes. Journal of Membrane Science 2005
67. J. E Drewes, C Bellona, J Luna, C Hoppe, G Amy, G Filteau, G Oelker, N Lee, J Bender, R Nagel, Can nanofiltration and ultra-low pressure reverse osmosis membranes replace RO for the removal of organic micropollutants, nutrients and bulk organic carbon?- A pilot-scale investigation. In: WEFTEC 2005October- 2 November 2005, Washington, DC, USA Wef Proceedings; 2005.
68. P Xu, J. E Drewes, C Bellona, G Amy, T Kim, U Adam, M Heberer, T. Rejection of Emerging Organic Micropollutants in Nanofiltration-Reverse Osmosis Membrane Applications. Water Environment Research 20057714048
69. C Bellona, J Drewes, Viability of a low-pressure nanofilter in treating recycled water for water reuse applications: A pilot-scale study. Water Research 200741348
70. C Park, H Jang, Y Yoon, S Hong, J Jung, Y Chung, Removal characteristics of endocrine disrupting compounds and pharmaceuticals by NF membranes. In: 4th IWA International Membrane Conference, Membranes for Water and Wastewater Treatment, 1517May 2007Harrogate, UK. IWA; 2007
71. S Lee, ., N Quyet, ., E Lee, ., S Kim, ., S Lee, ., Y Jung, ., S.H Choi, ., J Cho, . Efficient removals of tris(2-chloroethyl) phosphate (TCEP) and perchlorate using NF membrane filtrations. Desalination 2008; 221 234-237.
72. I Koyuncu, O. A Arikan, M. R Wiesner, C Rice, Removal of hormones and antibiotics by nanofiltration membranes. Journal of Membrane Science 200830994
73. J Radjenovic, M Petrovic, F Ventura, D Barceló, Rejection of pharmaceuticals in nanofiltration and reverse osmosis membrane drinking water treatment. Water Research 2008423601
74. I Schäfer, A. G Fane, T. D Waite, Editors. Nanofiltration- Principles and Applications. Elsevier; 2005
75. C Bellona, J Drewes, P Xu, G Amy, Factors affecting the rejection of organic solutes during NF/RO treatment-a literature review. Water Research 2004382795
76. K. O Agenson, J. I Oh, T Urase, Influence of molecular structure on the rejection characteristics of Volatile and Semi-volatile Organic Compounds by Nanofiltration. In: IMSTEC'03 5th International Membrane Science and Technology Conference, 1014November 2003Australia. New South Wales: UNESCO Centre for Membrane Science and Technology 2003
77. Y Zhang, C Causserand, P Aimar, J. P Cravedi, Removal of bisphenol A by nanofiltration membrane in view of drinking water production. Water Research 2006403793
78. K Kimura, G Amy, J Drewes, T Herberer, T. U Kim, Y Watanabe, Rejection of organic micropollutants (disinfection by products, endocrine disrupting compounds, and pharmaceutically active compounds) by NF/RO membranes. Journal of Membrane Science 2003227113

79. Y Kiso, Y Nishimura, T Kitao, K Nishimura, Rejection properties of non-phenolic pesticides with nanofiltration membranes. Journal of Membrane Science 2000171229
80. L. D Nghiem, A. I Schäfer, M Elimelech, Removal of Natural Hormones by Nanofiltration Membranes: Measurement, Modeling, and Mechanisms. Environmental Science Technology 200438618881896
81. J. H Kim, H Kwon, S Lee, C. H Lee, Removal of endocrine disruptors using homogeneous metal catalyst combined with nanofiltration membrane. Water Science and Technology 2005
82. INEGIMéxico en cifras. Hidalgo. http://www.inegi.org.mx/sistemas /mexicocifras acceded 28 September 2010
83. Gobierno del estado de Hidalgohttp://www.hidalgo.gob.mx/acceded 27 September 2010
84. Water National CommissionBritish Geological Survey, London School of Hygiene and Tropical Medicine and University of Birmingham. Impact of Wastewater Reuse onGroundwater in the Mezquital Valley, Hidalgo State, Mexico. Final Report Technical Report WC/98/42. 1998
85. L Lesser, J Lesser, S Arellano, D González, Balance hídrico y calidad del agua subterránea en el acuífero del Valle del Mezquital, México central. Revista Mexicana de Ciencias Geológicas 201128332336
86. B Jiménez, A Chávez, Quality assessment of an aquifer recharged withwastewater for its potential use as drinking source:"El Mezquital Valley" case. Water Science and Technology 200450226976
87. T Downs, E Cifuentes, E Ruth, I. M Suffet, Effectiveness of natural treatment in a wastewater irrigarion district of the Mexico City: A synoptic field survey. Water Environment Research 2000721421
88. C Siebe, E Cifuentes, Environmental impact of wastewater irrigation in Central Mexico: An overview. Int. J. Environ. Health Res 1995516173
89. J Siemens, G Huschek, C Siebe, d Kaupenjohann, M Concentrations, and mobility of human pharmaceuticals in the world's largest wastewater irrigation system, Mexico City-Mezquital Valley. Water Research 2007
90. R Gibson, E Becerril, V Silva, B Jiménez, Determination of acidic pharmaceuticals and potential endocrine disrupting compounds in wastewaters and spring waters by selective elution and analysis by gas chromatography-mass spectrometry. Journal of Chromatography A 2007
91. AXYS Method MLA-075 Rev 03 Ver 02: AXYS Method MLA-075: Analytical procedures for the analysis of pharmaceutical and personal care products in solidaqueous and tissue samples by LC-MS/MS. Canada 2011
92. U. S-E. P. A Method, B Purge, and Trap for aqueous samples.
93. U. S-E. P. A Method, D. Semivolatile Organic Compounds by Gas Chromatography /Mass Spectrometry (GC/MS).
94. Secretaría de SaludMODIFICACION a la Norma Oficial Mexicana NOM-127SSA1-1994Salud ambiental. Agua para uso y consumo humano. Límites permisibles de calidad y tratamientos a que debe someterse el agua para su potabilización. Diario Oficial de la Federación 22 de noviembre de 2000. México.
95. S Özen, Darcan Ş. Effects of environmental endocrine disruptors on pubertal development. J Clin Res Ped Endo 2011DOI:jcrpe.3i1
96. Committee on Fluoride in Drinking WaterNational. Research Council. Fluoride in drinking water: A scientific review of EPA´s standards. USA: NAP; 2006
97. J Davey, J Bodwell, J Gosse, J Hamilton, Arsenic as an Endocrine disruptor: effects of arsenic on estrogen receptor-mediated gene expression in vivo and in cell culture. Toxicological Sciences 20079817586
98. J Bodwell, J Gosse, P Nomikos, J Hamilton, Arsenic disruption of steroid receptor gene activation: complex dose-response effects are shared by several steroid receptors. Chem Res Toxicol 20081912161929

99. F Barr, L Krohmer, J Hamilton, L Sheldon, Disruption of histone modification and CARM1 recruitment by arsenic represses transcription at glucocorticoid receptor-regulated promoters. PLoS One 2009e6766;doi:10.1371/journal.pone.0006766.

100. Q Xiang, L Chen, Y Liang, M Wu, B Chen, Fluoride and thyroid function in children in two villages in China. Journal of Toxicology and Environmental Health Sciences 200913549

101. A Susheela, M Bhatnagar, K Vig, N Mondal, excess fluoride ingestion and thyroid hormone derangements in children living in Delhi, India. Fluoride 200538298108

102. J Stein, T Schettler, D Wallinga, M Valenti, In harm's way: toxic threats to child development. J Dev Behav Pediatr 2002S13S22.

103. J Walker, H Fang, R Perkins, W Tong, QSARs for endocrine disruption priority setting Database 2: The Integrated 4-Phase Model. Quant. Struct.-Act. Relat. 200322117

104. USEPAEPA's List of 134 Endocrine Disruptors. http://www.epa.gov/endo/pubs /prioritysetting/list2facts.htmaccessed 02 June 2012

105. s De Coster, B Van Lare, Endocrine disrupting chemicals, associated disorders and mechanisms of action.www.hindawi.com/journals/jeph/aip/713696,pdfaccessed02June 2012

106. Scorecard Pollution sitehttp://scorecard.goodguide.com/index.tclaccessed 16 February 2011

107. USEPARisk Assessment Guidance for Superfound.Vol. I. Human Health Evaluation Manual (Part A). USA:EPA;1989

108. Risk Assessment Information SystemChemical Toxicity values. http://rais.ornl.gov/cgi-bin/tools/TOX_searchaccessed 16 February 2011

CHAPTER 5

Microbiological Surveillance and State of the Art Technological Strategies for the Prevention of Dialysis Water Pollution

Piergiorgio Bolasco [1,*], Antonio Contu [2], Patrizia Meloni [2], Dorio Vacca [2] and Andrea Galfrè [1]

[1]*Territorial Department of Nephrology and Dialysis, ASL, Quartu Sant'Elena, Cagliari 09045, Italy; Email:*

[2]*University Department of Public Health, Clinical and Molecular Medicine, University of Cagliari, Cagliari 09121, Italy*

ABSTRACT

Methods: The present report attempts to illustrate the positive impact on the microbiological quality of dialysis patients over a 15-year period through the progressive implementation of state-of-the-art technological strategies and the optimization of microbiological surveillance procedures in five dialysis units in Sardinia. *Results:* Following on better microbiological, quality controls of dialysis water and improvement of procedures and equipment, a drastic improvement of microbiological water quality was observed in a total of 945 samples. The main aim was to introduce the use of microbiological culture methods as recommended by the most important guidelines. The microbiological results obtained have led to a progressive refining of controls and introduction of new materials and equipment, including two-stage osmosis and piping distribution rings featuring a greater capacity to prevent biofilm adhesion. The actions undertaken have resulted in unexpected quality improvements. *Conclusions:* Dialysis water should be viewed by the nephrologist as a medicinal product exerting a demonstrable positive impact on microinflammation in dialysis patients. A synergic effort between nephrologists and microbiologists undoubtedly constitutes the most effective means of preventing dialysis infections.

Keywords: microbiological methods; quality dialysis water

1. INTRODUCTION

In Europe, patients receiving hemodialysis undergo at least three treatment sessions a week. As standard hemodialysis treatment sessions last 4–6 hours, individual patients are exposed to 15,000–20,000 L of dialysis fluid yearly. In hemodiafiltration procedures, dialysis water is administered in the form of dialysate and infusate up to 3,400–6,800 L directly following the ultrafiltration through two or three dialysis water monitor ultrafilters [1]. The increasing tendency to use on-line hemodialysis procedures implies therefore the potential presence in distribution rings of microbial, fungal, and chemical substances which should be carefully monitored, as provided for in specific guidelines [2,3]. Moreover, waters should be tested by certified laboratories using significantly different analytical procedures, particularly in the case of microbiological controls, to those applied to test human blood and secretions [3]. Numerous undesirable substances originate from polluted water or infiltrations in the drinking water supply network. Humans are the major culprits, contaminating waterworks through the use of ineffective measures to reduce bacterial and/or chemical pollutant loads originating from surface waters collected in artificial basins, or from the building of drinking water networks using obsolete or inappropriate materials. The majority of toxic contaminants derive from procedures applied in the treatment of municipal waters, thereby suggesting that these are not always safe for direct use in hemodialysis applications. An unmonitored dialysis water treatment facility, an inefficient system for the distribution of dialysis waters to monitors, or a scarce regularity in disinfection constitute the most frequent causes of harmful or fatal substances which may prove to be particularly hazardous when adopting on-line procedures lacking effective ultrafilters with administration. Thus, the main parameters to be applied in the treatment of waters destined for use in hemodialysis procedures, should be underlined.

As a premise enhancing evaluation of the evolution of the plants implemented, a synthesis of the most recent ANSI/AAMI/ISO guidelines published in 2009 and 2011 is provided [2,3,4,5]. The Authors wish to underline that dialysis procedures used in the USA do not foresee the administration to patients of such a high number of infusions as occurs in Europe, when the nephrologist prescribes high convective methods such as hemodiafiltration. Furthermore, two-stage reverse osmosis (TRO) with overnight thermal disinfection is referred to, although not described in detail, in the AAMI Guidelines, and is to date applied scarcely throughout Italy and Europe.

Water utilized in hemodialysis applications is prepared in a series of different purification processes, the most common of which include [3,4,5,6,7,8]: reverse osmosis, deionization, and carbon filtration. Reverse osmosis (RO) uses high pressure to force water across a semi-permeable membrane to form product water (permeate), thereby rejecting 90 to 99 percent of ionic contaminants and >95 percent of non-ionic contaminants. A RO membrane is an effective barrier against microbiological contaminants, bacteria, viruses, and endotoxins. Currently however, two-stage reverse osmosis is actually the most effective form compared to one-stage reverse osmosis (SRO), achieving the highest degree of contaminant removal. This procedure is particularly recommended for use in on-line hemodialytic methods with high convective component [9]. Deionization (DI) uses ion exchange resins to remove ionic contaminants from water by exchanging hydrogen ions for cationic resins and hydroxyl ions for anionic resins. Mixed bed deionizers contain both cationic and anionic resins but do not remove microbiological contaminants. Deionizer performance should be closely monitored to avoid exhaustion, and it should be taken into account that ions bind to the resin with low affinity. This phenomenon, which may lead to high contaminant concentrations in the product water as the deionizer becomes exhausted [10], has been reported to cause acute, fatal fluoride toxicity in hemodialysis patients [11]. This method is virtually obsolete in Italy. In the pre-treatment procedures a variety of filters may be utilized, both as pre-treatment for main tap water purification processes and at the end of the pre-treatment system to remove coarse particulate matter prior to purification and to protect RO membranes from fine particles washed out of carbon beds. The use of chlorination and dechlorination is not mandatory, although in a few circumstances (e.g., during the drought in Sardinia in 1991) municipal authorities have implemented provisions including water supply on alternate days, and consequent reduced pressure throughout the water supply networks, which were frequently old and leaking (personal observation); this resulted in an infiltration of polluted waters from the water table in the vicinity of tubes conveying the drinking water supply. It should moreover be underlined that in Italy the use of underground water tables in supplying water to hospitals is permitted. Good quality proportional pumps should be applied to guarantee optimal chorine concentration (0.5–1 ppm), although complete water dechlorination should be ensured upstream of osmosis membranes. To prevent inadvertent exposure of patients to chloramine as the capacity of the carbon is exhausted two carbon filters in series are necessary. Inadequate chloramine removal may occur however, and assays aimed at detecting the presence of chloramines with specific rapid water checks are mandatory [12].

In water softeners resin beads inside the tanks display a high affinity for bivalent cations (calcium and magnesium) present in the feed water, releasing two sodium ions (monovalent) for each calcium or magnesium ion captured to prevent RO membranes from fouling by calcium and magnesium salts. Softeners must be regenerated by exposing resin to strong brine with a very high sodium concentration. Dechlorinators and softeners represent a suitable "pabulum" for bacterial growth due to the impossibility of disinfecting the same with a specific product unless thorough rinsing is undertaken against the flow. The treated water exits the central water treatment plant and is distributed to hemodialysis monitor by piping ring with an optimal diameter to avoid sluggish segments; metals such as brass, aluminium, or galvanized metal should not be used, and the distribution system must be designed as a closed ring. The use of polyvinyl chloride (PVC) should be eliminated and materials such as high corrosion resistance stainless steel (INOX AISI 316L) polyethylene thermoplastic polymer (PEX), and polyvinylidenefluoride (PVDF) used to construct piping and recently the polytetrafluoroethylene (PTFE) has been introduced.

The pollutant potential of reservoir storage is frequently underestimated. Tanks should be opaque, made of plastic for foodstuffs, and not be located in a "stagnant" corner of the circuit, but should guarantee a constant flow of water. However, a fundamental aspect is constituted by the bacteriological and endotoxin controls to be implemented in line with appropriate procedures in sampling and transportation, in inoculation of the culture medium with dialysis water, and in the accurate interpretation of results, not taking into account merely mesophiles from the environment, but furthering observation to evaluate the presence of bacteria or mycetes [13]. Based on these premises therefore, a study was undertaken to assess the quality of water in five dialysis units in which over 16 years numerous changes had been implemented in methods used for disinfection and monitoring and improvements made to the equipment used in the production and distribution of ultrapure dialysis water, particularly in dialysis units adopting online convective procedures with production of high volumes of infusate, such as online hemodiafiltration techniques.

2. MATERIALS AND METHODS

Characteristics of the units: five dialysis units were investigated over a 16 year 3 month period (195 months) from May 1996 to April 2012. In total 945 samples were taken (minimum three samples for each distribution ring every three months in steady state equipment). The observation periods are illustrated in Table 1, Table 2 and Table 3.

2.1. Use of Polyvinyl Chloride (PVC) Piping: Two Centres—See Table 1

PVC piping was used in two successive periods:

- (1) Use of irregular/occasional microbiological controls, chemical disinfection and SRO.
- (2) Use of regular microbiological controls, chemical disinfection and SRO in polyamide. The use of PVC however was discontinued and it was replaced with PEX rings in one unit and AISI INOX 316L steel rings in the other.

Student's T-test was used for comparing data. Results were considered significant when p was <0.05.

Table 1. PVC piping.

Two dialysis centres	Occasional chemical disinfection (average every quarter)	Regular monthly disinfection
Reverse osmosis	SRO	TRO
Technical dialysis posts	29	29 [a]
Observation time, months	60	36
Sampling, total numbers	120	72
CFU, mL^{-1}, average, bacteria *	238.7 ± 374.3	153.9 ± 160.7
CFU, mL^{-1}, average, mycetes *	19.2 ± 34.4 [b]	5.2 ± 7.5 [c]
CFU, mL, medianbacteria	122.2	25.5
CFU, mL, minimum and maximum values bacteria	7–1,773	0–1,500

[a] Six posts switch to INOX and another 23 switch to PEX piping (Tables 2 and 3); the PVC piping ring was abandoned due to its inadequate safety; [b] *vs.* [c]: $p < 0.05$; (*) values represent averages ±1 standard deviation from (heterotrophic plate count) HPC 22 °C measurements. Average and median are both derived from bacteria and mycetes at HPC 22 °C.

2.2. Use of AISI INOX 316L Piping: Three Centres—See Table 2

We used high corrosion resistance stainless steel piping in three successive periods:

- (1) Use of AISI INOX 316L steel distribution ring with irregular/occasional microbiological controls, chemical disinfection and one stage reverse osmosis (SRO) system in polyamide;
- (2) Use of AISI INOX 316L steel distribution ring with regular microbiological controls, chemical disinfection and SRO system in polyamide;
- (3) Use of AISI INOX 316L steel distribution ring with regular microbiological controls and daily overnight pulsed thermal disinfection by means of TRO.

Table 2. AISI INOX 316L piping.

Three dialysis units	Occasional chemical disinfection (average every quarter)	Regular monthly chemical disinfection	Regular daily thermal disinfection
Reverse osmosis	SRO	SRO	TRO
Technical dialysis posts	21	27 [a,b]	27
Observation time, months	60	87	36
Sampling, total numbers	180	261	108
CFU, mL/min, average, bacteria	160.4 ± 321.1 [1]	43.1 ± 76.1 [2]	15.0 ± 24.6 [3]
CFU, mL/min, average, mycetes	35.8 ±53.9 [4]	8.1 ± 8.7 [5]	1.75 ± 0.41 [6]
CFU, mL, median bacteria	19.3	21.2	0
CFU, mL, minimum and maximum values bacteria	0–520	0–107	0–2

[a] Six piping rings of PVC (Table 1); [b] increase of technical dialysis posts due to enlargement of two dialysis centers; Significance: [1] *vs.* [2] and [3] *vs.* [1], [2]: $p < 0.03$; [4] *vs.* [5], [6]: $p < 0.02$.

2.3. Use of PEX (Polyethylene Thermoplastic Polymer) Piping: Two Centres—See Table 3

PEX piping was used in three successive periods:

- (1) Use of PEX distribution ring with irregular microbiological controls, occasional chemical disinfection and SRO system in polyamide;
- (2) Use of PEX distribution ring with regular microbiological controls, chemical disinfection and SRO system in polyamide;
- (3) Use of PEX distribution ring with regular microbiological controls and daily overnight pulsed thermal disinfection by means of TRO.

Table 3. PEX piping.

Two dialysis units	Occasional chemical disinfection (average every quarter)	Regular monthly chemical disinfection	Regular daily thermal disinfection
Reverse osmosis	mono-osmosis	mono-osmosis	bi-osmosis
Technical dialysis posts	23 [d]	23 [d]	31 [e]
Observation time, months	24	24	36
Sampling, total numbers	48	48	72
FU, mL/min, average, bacteria CFU, mL/min, average, mycetes	398.1 ± 627^{1} 18.4 ± 38.8^{4}	33.7 ± 29.9^{2} 0.37 ± 0.84^{5}	2.4 ± 2.3^{3} 0.30 ± 0.4^{6}
CFU, mL, median bacteria	101	19.1	0.5
CFU, mL, minimum and maximum values bacteria	0–1,028	0–75	0–0.9

[d] 23 piping rings of PVC (Table 1); [e] increase of technical dialysis posts due to enlargement of two dialysis centers; Significance ^{1} *vs.* ^{2}, ^{3} and ^{3} *vs.* ^{1}, $p < 0.01$; Significance ^{4} *vs.* ^{5}, 6: $p < 0.01$.

2.4. Brief Description of Characteristics of Materials and Operations

Pre-Treatment

Hard water should undergo mandatory treatment with double water softeners. Softeners are sized in grains of capacity and are invariably regenerated on a routine daily basis with concentrated sodium chloride solution (brine). This method was applied throughout the observation study.

Dechlorination represents another fundamental step. We used tanks containing approximately 400 L granular activated carbon (CAG) which had never been regenerated. Throughout the first five years of operation the carbon beds had not been replaced, although in the last seven years CAG had been changed on a yearly basis and carbon tanks subjected to daily back-washing. Softeners and dechlorinators cannot be disinfected. The product commonly applied in disinfection may damage their material beds. The amount of softeners used in the units investigated was invariably overestimated, in order to guarantee an effective softening of inflow water featuring >1,000 mS/cm.

One-stage reverse osmosis. Thin film RO membranes are made from polyamide and spiral wound around a permeate-collecting tube. This material is compatible with peracetic acid used in disinfection. Mono-osmosis techniques have been abandoned over the last three years and substituted by TRO. The new CWP WRO 131/132 ROHH Gambro® equipment with ISO 14021 (using polyamide membranes) adopted in the latter stage, comprising a completely automated system of thermal disinfection that no longer required the use of chemical disinfection products, featured an integrated system for the thermal disinfection of membranes and distribution rings. Thermal disinfection of biosmosis systems provides for an automated daily thermal disinfection of dialysis room distribution circuit using water heated to 90/95 °C, to prevent the formation of biofilms. The heated disinfection unit, integrated in the biosmosis plant, is programmed to heat water from 63 °C to 90/95 °C at a pre-specified time of day, towards the end of the dialysis session with the possibility of varying programming for each day of the week. The circulation of hot water in the dialysis room distribution circuit occurs at the end of the dialysis session and continues throughout the night until shortly before the start of the morning session when piping temperatures return to non-dangerous levels. In addition to providing correct operating parameters, the integrated computer in the TRO system records data for the last three operating sessions and elaborates these data as required, in the form of text, synoptic diagram or a graph.

Chemical disinfection of distribution ring downstream of dialysis monitors was carried out on a monthly basis and was abandoned following the introduction of thermal disinfection.

The connection between distribution ring and monitors remains a highly critical and somewhat underestimated aspect. Use of a mesh tube has been definitely prohibited. A mesh tube constitutes an ideal *pabulum* for the growth of bacteria, mycetes, and algae. Approximately 15 days later swabs were obtained from the inside of mesh tubes, yielding significant results for microbiological load >10,000 CFU/cm^2 (personal observation), and were subsequently replaced at all kidney positions by large extendable PEX springs suitable for monthly disinfection using peracetic acid and/or chlorine oxidant.

2.5. Microbiological and Endotoxin Controls

Assumption

In agreement with microbiologists, the authors opted to extend tests to a higher spectrum of micro-organisms frequently detected in previous studies, particularly as the hydro-geographic situation in Sardinia is almost exclusively characterised by the presence of surface waters collected in artificial basins. Indeed, in view of the latter and to the high temperatures registered during summer months, the Italian Department of Health has not ruled out the potential development of harmful bacteria, including cyanobacteria [14] resulting in the release of cyanotoxins. Preparations are therefore currently being made to assemble procedures for use in detecting the latter.

Sampling is carried out according to a series of procedures aimed at allowing an appropiate aliquot of water to be obtained for analysis. Depending on the test to be performed, prior to sampling the volume of water required for testing should be ascertained. Bottles and vials used to collect samples for microbiological analysis should never be washed at time of sampling. In addition to exposing recipients to possible contamination, rinsing would remove traces of sodium thiosulphate present in the tap. Water should be left to run long enough to eliminate disinfectants prior to sampling.

Taps should be cleaned and disinfected prior to sampling; both taps and the inside of collars should be cleaned and residues, dust, mucilage, detergents and disinfectants, and other substances capable of affecting the outcome of microbiological analysis should be removed. A solution of sodium hypochlorite or similar types of disinfectant should be used in cleansing: 10% solutions of commercially available sodium hypochlorite or sodium dichloroisocyanurate may be used. Due to the corrosive nature of these products, particular caution should be applied in using the solutions. Fatty deposits should be removed by rubbing with isopropyl alcohol. The disinfectant solution should be left on the taps for 2–3 minutes, and all traces of the product subsequently removed. It is fundamental that during sampling procedures contamination should be avoided and the quality of water sample maintained. In addition to the mandatory cleansing and disinfection procedures, metal taps may also be flame gouged. However, if the latter procedure is performed in a perfunctory manner, it will not produce the desired effect on potential microbial contamination. As previously, the tap should be left to run for 2–3 minutes and flow left untouched during sample collection. On collection, open the sterile bottle taking care to not touch the inside of the cap or the inside of the bottle neck which will come into direct contact with the sample, and close immediately after sampling. Particular care should be taken to not overfill the bottle in order to enhance effective lab homogenisation of the sample at the time of testing.

2.6. Transportation

During transportation and storage of samples for microbiological testing, the representative nature of samples should be ensured by preventing regrowth of micro-organisms. As far as possible alterations which may at times be inevitable in a small aliquot of water stored in a closed container, should be kept to a minimum. The sample should be stored away from the light (both ultraviolet and visible) and from high temperatures, and appropriate conditions of hygiene applied during transportation. From the time of sampling to arrival in the lab, all samples should be stored at a temperature of below 10 °C, with the optimum range of 2–8 °C being recommended. The samples must be inseminated within two hours (ISO 19458:2006). Between sampling and the analysis, the maximum time for the microbiological analysis depends on the type of analysis or better by the parameters to be searched as shown in the Table 4.

Table 4. Maximum time of analysis between sampling and analysis.

Parameter	Maximum time of analysis in hours
Total bacterial count at 22 °C or 37 °C	8–12
Escherichia coli and coliforms	12–18
Enterococci	12–18
Bacteria and spores of sulphite-reducing clostridia	48–72
Pseudomonas aeruginosa	8–12

2.7. Analytical Procedures

In the Authors' opinion, microbiological testing procedures should not only assess the presence of environmental mesophiles and mycetes, but should also address the issue of detecting specific pathogens. For each of microbiological procedures we referred to ISO 13959:2009 and (UNI EN ISO 6222:2001) [2,13] for the different bacteria in order to obtain further implemented improvements and applying the specific Italian methodologies described in the sections below [15,16,17]. We also have considered the procedures applied by other authors [18,19]. The choice of parameters to test depends in particular on the water supply system. In our case over 90% of drinking waters are surface water, resulting in poor water quality. In addition, losses of about 35–40% are observed in most of the water network. Given the precariousness of the sample, also associated with the low concentration of free chlorine residual detected (0:01 to 0:02 mg/L), and the use for which it is intended, it was considered important to search also indicators of faecal contamination in countries where surface water is used for drinking water production.

2.7.1. Total Coliforms

An aliquot (100 mL) of the sample collected should be filtered through a 47 mm cellulose ester membrane with filtration characteristics corresponding to a nominal pore size of 0.45 μm. Place the membrane over the M-Endo agar LES medium and incubate at 36 ± 1 °C for 18–24 hours. Red-coloured colonies with a metallic sheen that grow over a 24-hour period, generally accompanied by a dark brick-red colouring in the area beneath the membrane, are deemed to be coliform bacteria (presumed coliforms). Tests should be carried out on colonies aged no more than 24 hours.

Tests should then be undertaken to verify coliform status: the presence of the cytochromoxydase enzyme on all, or at least a representative number of typical colonies should be assessed and biochemical identification tests performed (UNI EN ISO 9308-1:2002) [20]. Tests should be carried out on colonies aged no more than 24 hours.

2.7.2. Enterococci

An aliquot (100 mL) of the sample collected should be filtered through a 47 mm cellulose ester membrane with filtration characteristics corresponding to a nominal pore size of 0.45 μm. Place the membrane over the Slanetz and Bartley agar medium and incubate at 36 ± 1 °C for 40–48 hours. Following incubation, colonies ranging from pink to dark red and brown (in the centre or throughout the colony) are deemed typical (presumed *Enterococci*). Tests should be undertaken to confirm enterococcus status: the presence of the catalase enzyme and the esculin hydrolysis test should be undertaken on all, or at least a representative number of typical colonies and biochemical identification tests performed. Tests should be carried out on colonies aged no more than 24 hours (UNI EN ISO 7899-2: 2003) [20].

2.7.3. Pseudomonas aeruginosa

An aliquot (250 mL) of the sample collected should be filtered through a 47 mm cellulose ester membrane with filtration characteristics corresponding to a nominal pore size of 0.45 μm. Place the membrane over the Pseudomonas agar/CN medium and incubate at 36 ± 1 °C for 40–48 hours After (2 ± 2) and (44 ± 4) hours assess the growth of typical colonies on isolation medium. *Pseudomonas aeruginosa* may develop as greenish-blue colonies that produce pyocyanin, fluorescent or reddish-brown colonies. Count all characteristic colonies (presumed to be *Pseudomonsas aeruginosa*). Tests should then be undertaken to confirm *Pseudomonsas aeruginosa* status: place under a Woods lamp and following observation on all, or at least a representative number of typical colonies, biochemical identification tests should be performed. Tests should be carried out on colonies aged no more than 24 hours (UNI EN ISO 12780:2002) [20].

2.7.4. Clostridium perfringens

An aliquot (100 mL) of the sample collected should be filtered through a 47 mm cellulose ester membrane with filtration characteristics corresponding to a nominal pore size of 0.45 μm. Place the membrane over the Tryptose Sulfite Cycloserine Agar medium, and cover the membrane completely with 5–6 mL of liquid culture medium at a temperature of 50 ± 5 °C. Leave to set and incubate at a temperature of 44 ± 1 °C for 21 ± 3 hours under anaerobic conditions. Micro-organisms belonging to the *Clostridium perfringens* species form black or grey/yellowish-brown colonies on the Tryptose Sulfite Cycloserine Agar medium, conveying a slightly darker shade above or below the filtrating membrane. Count all typical colonies and treat as presumed *Clostridium perfringens*. Isolate colonies on two parallel blood Agar plates and incubate one plate under anaerobic conditions and the other under aerobic conditions at 36 ± 1 °C for 21 ± 3 hours. Following incubation, examine plates to check the presence or absence of growth. *Clostridium perfringens* colonies typically produce areas of clear hemolysis on culture medium. Undertake confirmation tests only on colonies growing on blood Agar under anaerobic conditions, and subsequently perform biochemical identification tests. Tests should be carried out on colonies aged no more than 24 hours. Method ISS A 005B REV.00 (National Standard Method W 5 ISSU 3 dell'Health Protection Agency).

2.7.5. Mycetes

An aliquot (100 mL) of the sample collected should be filtered through a 47 mm cellulose ester membrane with filtration characteristics corresponding to a nominal pore size of 0.45 μm. Place the membrane over a Sabouraud Dextrose Agar medium and incubate at 22–25 °C for 3 ± 5 days. Recognition of fungal species is a complex issue requiring lengthy observation and proven experience. The aspect of colonies varies according to the type of substrate (natural or artificial) and colonies develop on the basis of numerous factors. To identify yeasts sample from the centre of the colony and smear on to a glass slide in a drop of sterile distilled water. View under a microscope, preferably under contrast and at 20× or 40× magnification to identify pseudomycelium and spores; if necessary, perform commercially available miniaturised biochemical identification tests (UNI EN ISO 6222:2001). Use direct observation technique to view filamentous fungi under the microscope. Carefully attach a piece of transparent sticky tape to the surface of the colony to be examined; transfer the tape with the sticky part facing downwards, onto a glass slide containing a drop of lactophenol blue solution and cover with a glass cover slip; observe at 100X to enhance distinguishing of fructiferous bodies [20,21].

2.7.6. Counting of Colonies 22 °C–37 °C

Analytical procedures undertaken to assess the number of micro-organisms at 37 °C and 22 °C are identical for both parameters, comprising the agar inclusion technique. Seed 1 mL of the sample onto the bottom of a Petri dish. Pour approximately 15 mL of isolation substrate maintained in a liquid state at a temperature of 45 ± 1 °C on to the dishes containing the inoculum. Mix thoroughly by rotating dishes backwards and forwards to facilitate a complete mixing of culture medium and sample. Leave to set and incubate one dish at 36 ± 1 °C for 40–48 hours, and the other at 22 ± 1 °C for 64–72 hours (UNI EN ISO 6222:2001) [20].

2.7.7. Counting at 22 °C (Mesophiles) (UNI EN ISO 6222:2001)

An aliquot (100 mL) of the sample collected should be filtered through a 47 mm cellulose ester membrane with filtration characteristics corresponding to a nominal pore size of 0.22–0.45 µm. Place the membrane on the agar medium and incubate at 20–23 °C for 7 days (UNI EN ISO 6222:2001)) [20].

2.8. Endotoxin Tests

Tests are carried out both on osmotic water as monitors are detached [22] and at random initial-intermediate and end points in the ring. Spectrophotometric methods are used (Endosafe®-PTS™, Charles River®) with gel-clotting at 37 °C. Acceptable limits of detection on ring detachment are <0.05, and <0.01 for online infusion liquid. A training course in use of this piece of equipment is required, the machine should be calibrated by the manufacturers on a yearly basis. Indispensable requisites for on-line hemodiafiltration are as follows: CFU: 0 colonies/mL bacteria and mycetes, endotoxins <0.03 EU/mL. This method should not be used if the requisites are lacking.

3. RESULTS

Samples were obtained as far away as possible from the most recent point of disinfection according to a rotation schedule between at least three locations situated at the beginning, the centre and the end of the ring; cultured dishes were read as provided for in standard guidelines. Significance difference was observed adopting a microbiological improvement strategies and plants as illustrated in Table 1, Table

2 and Table 3 showing the experience gained throughout the period of observation.

4. DISCUSSION

The Authors' experience in five territorial dialysis units, implemented on the basis of microbiological results obtained at several points throughout the central plant and piping ring, have led to an improvement in microbiological/endotoxinic quality of water subsequent to changes made in the use of materials and procedures. Specifically, it has been demonstrated how, in the presence of materials other than PVC, regular disinfection and frequent controls (even monthly) represented a fundamental step forward in achieving low levels of CFU [23]. The use of PVC should be avoided; it is not suited for use with thermal disinfection methods and evidence has indicated it's potentially cancerogenic status [24]. The overnight thermal disinfection achieved by means of biosmosis resulted in a dramatic decrease in microbial load [7], which appeared to be more significant following the use of PEX compared to INOX in construction of the distribution ring. In the Authors' opinion, this difference may be associated to diameter of piping; a minor diameter decreased the possibility of stagnant areas creating a less suitable habitat for biofilm growth. In fact it is fundamental that pressure along the walls, and velocity and intensity of fluid shear stress on the smaller inner surfaces, are higher than those recorded in larger-sized sections; in other words a higher water pressure and velocity enhance filling of the circuit and ensure against the formation of concealed areas where bacteria and hyphae may flourish. Moreover, frequent thermal disinfection does not rapidly prevent the formation of bacterial biofilm and hyphae along inner surfaces. Therefore, in the wake of experience gained in units operating with PVC rings, the Authors opted to abandon use of this material entirely. The application of meticulously implemented periodic thermal and chemical disinfection procedures should be further supported by additional checks and immediate correction of any "pathological" areas detected by microbiological results. Accordingly, a specialized team should be appointed to carry out controls and periodic and non-scheduled maintenance based on the finding of microbiological levels exceeding established safety limits.

Irrespective of the material used in construction of the distribution ring, the connection valves to dialysis monitors should be of stainless steel AISI 316L, in view of the potential liability of the connection and the increased risk of stagnant areas at this level: this option results in a decreased probability of microbial adhesions. In all the processes described, chemical, physical and bacteriological tests should be undertaken by a certified laboratory. The Authors maintain that standard procedures should be established in conjunction with an institution specializing in environmental water testing; labs should moreover possess a working knowledge of the equipment used and ensure that methods described in the leading guidelines in the field adhered to. A synergic action between the dialysis unit staff, the maintenance team and the lab is mandatory, with purpose-trained specialised technicians performing sampling throughout all units and providing for a correct transportation to the testing lab.

The Authors underlined how systems should be monitored continuously and the following taken into account [25]: (a) on opting for the sole chemical disinfection, apply every 15–30 days solutions capable of removing both bacterial biofilms and potential mineral scales that may protect bacteria and/or enhance their survival; (b) daily nocturnal thermal disinfection; to promote safety, an additional monthly or twice monthly disinfection of osmosis membranes and ring as described in item.

The experience gained has led the Authors to progressively undertake alterations aimed at optimizing the microbiological quality of dialysis water: weak points may be represented by an excessive inner diameter of piping, both in PVC and INOX, resulting in pipes that are disproportionate to water flow and volume; the definitive abolishing of PVC and mesh tubes connecting the ring to monitors is fundamental.

Conversely, the strong points that emerged from the present study focused largely on the use of PEX. However, very few literature reports published to date have focused on the efficacy of the latter material in preventing the adhesion of bacterial biofilms. Stainless steel INOX AISI 316 L is a long-lasting but rather costly material; the length of duration however may promote the amortization of costs. On opting to use stainless steel assembly and soldering should be undertaken by specialized technicians in order to warrant a perfect connection between the various segments. The studies undertaken with stainless steel were perfectly satisfactory; however, the Authors prefer to underline the suitability of PEX circuits which have no seams and, according to the quality of water, provide for a scheduled replacement of the circuit every 7–10 years. The importance of applying a correctly-sized piping and an antireflux gooseneck drainage spout should be underlined.

5. CONCLUSIONS

The carrying out of a regular and scrupulous microbiological surveillance enables physicians and microbiologists to identify the faulty part of equipment and/or ring piping and implement the required improvements to materials, equipment, and disinfection procedures. Indeed, this result was achieved in the territorial dialysis centres investigated in the present study, all of which featured significantly diverse hydrological and geographic differences.

Over a 16-year period the dynamic procedures set up have demonstrated how TRO should be considered a necessary investment to be faced with the aim of establishing a safer microbiological profile and producing a positive impact on microinflammation in dialysis patients, particularly in the case of increased infusion volumes during on-line hemodiafiltration procedures [26]. Likewise, daily overnight thermal disinfection procedures have proved at times to be more effective than frequent chemical disinfection.

Finally, it has become increasingly clear how staff operating in nephrology and microbiology units should possess a strong, motivated cultural synergic background on this complex subject in order to supply the best possible quality water to consumers. Indeed, dialysis water today should be considered a "medicinal product" in its own right.

REFERENCES

1. Locatelli, F.; Altieri, P.; Andrulli, S.; Bolasco, P.; Sau, G.; Pedrini, L.A.; Basile, C.; David, S.; Feriani, M.; Montagna, G.; *et al.* Hemofiltration and hemodiafiltration reduce intradialytic hypotension in ESRD. *J. Am. Soc. Nephrol.* **2010**, *21*, 1798–1807.
2. International Organization for Standardization (IOS). *Water for Haemodialysis and Related Therapies; ISO 13959*; IOS: Geneva, Switzerland, 15 April 2009.
3. International Organization for Standardization (IOS). *Dialysis Fluid Quality for Haemodialysis and Related Therapies; ISO 11663*; IOS: Geneva, Switzerland, 15 April 2009.
4. International Organization for Standardization (IOS). *Water Treatment Equipment for Haemodialysis and Related Therapies; ISO 26722*; IOS: Geneva, Switzerland, 15 April 2009.
5. Martin, K.; Laydet, E.; Canaud, B. European best practice guidelines for haemodialysis (Part 1). *Nephrol. Dial. Transplant.* **2002**, *17*, 50.
6. Alloatti, S.; Bolasco, P.; Canavese, C.; Cappelli, G.; Pedrini, L.; Pizzarelli, F.; Pontoriero, G.; Fuiano, G. Guidelines on water and solutions for dialysis. Italian Society of Nephrology. *G Ital Nefrol.* **2005**, *22*, 246–273.
7. Bolasco, P.; Scotto, P. Dialysis Fluid Purity in Hemodialysis: Normal Values and Undesirable Substances. *Proceedings of CIN 2011 6th Congress of Nephrology in Internet*, Available

online: http://trabajos.cin2011.uninet.edu/180/bolasco.pdf (accessed on 17 July 2012).

8. International Organization for Standardization (ISO). *Guidance for the Preparation and Quality Management of Fluids for Haemodialysis and Related Therapies; ISO 23500*; ISO: Geneva, Switzerland, 15 May 2011.
9. Pontoriero, G.; Pozzoni, P.; Andrulli, S.; Locatelli, F. The quality of dialysis water. *Nephrol. Dial. Transplant.* **2003**, *18*, 21–25.
10. Bland, L.A.; Arnow, P.M.; Arduino, M.J.; Bova, J.; McAllister, S.K. Potential hazards of deionization systems used for water purification in hemodialysis. *Artif. Organs.* **1996**, *20*, 2–7.
11. Arnow, P.M.; Bland, L.A.; Garcia-Houchins, S.; Fridkin, S.; Fellner, S.K. An outbreak of fatal fluoride intoxication in a long-term hemodialysis unit. *Ann. Intern. Med.* **1994**, *121*, 339–344.
12. Smith, M.P.; Marr, F.E.; Kanagasundaram, N.S. Chloramine reduction by reverse osmosis membranes. *Hemodial. Int.***2011**, *16*, 120–121.
13. Ray, J. Microbiological monitoring of dialysiswater systems—Which culture method? *J. Ren. Care.* **2007**, *33*, 66–69.
14. Azevedo, S.M.; Carmichael, W.W.; Jochimsen, E.M.; Rinehart, K.L.; Lau, S.; Shaw, G.R.; Eaglesham, G.K. Human intoxication by microcystins during renal dialysis treatment in Caruaru-Brazil. *Toxicology* **2002**, *27*, 181–182, 441–446.
15. *Reference Analytical Methods for Drinkable Water Second Law by Decree 31/2001; Chemical Reports ISTISAN07/5*; Health Ministerial: Rome, Italy, 2007.
16. *Reference Analytical Methods for Drinkable Water Second Law by Decree 31/2001; Microbiological Reports ISTISAN07/5*; Health Ministerial: Rome, Italy,, 2007.
17. *Guidelines for Environmental Infection Control in Health-Care Facilities; Recommendations of Practices Advisory Committee (HICPAC) CDC and the Healthcare Infection Control*; U.S. Department of Health and Human Services Centers for Disease Control and Prevention (CDC): Atlanta, GA, USA, 2003.
18. Arvanitidou, M.; Vayona, A.; Spanakis, N.; Tsakris, A. Occurrence and antimicrobial resistance of Gram-negative bacteria isolated in haemodialysis water and dialysate of renal units: Results of a Greek multicentre study. *J. Appl. Microbiol.* **2003**, *95*, 180–185.
19. Laurence, R.A.; Lapierre, S.T. Quality of hemodialysis water: A 7-year multicenter study. *Am. J. Kidney Dis.* **1995**, *25*, 738–750.
20. Köster, W.; Egli, T.; Ashbolt, N.; Botzenhart, K.; Burlion, N.; Endo, T.; Grimont, P.; Guillot, E.; Mabilat, C.; Newport, L.; Niemi, M.; Payment, P.; Prescott, A.; Renaud, P.; Rust, A. Analytical Methods for Microbiological Water Quality Testing. In *Assessing Microbial Safety of Drinking Water: Improving Approaches and Methods*; IWA Publishing: London, UK, 2003; pp. 237–282, Chapter 8.
21. Pires-Gonçalves, R.H.; Sartori, F.G.; Montanari, L.B.; Zaia, J.E.; Melhem, M.S.; Mendes-Giannini, M.J.; Martins, C.H. Occurrence of fungi in water used at a haemodialysis centre. *Lett. Appl. Microbiol.* **2008**, *46*, 542–547.
22. Pizzarelli, F.; Cerrai, T.; Biagini, M.; Malaguti, M.; Bargagna, R. Dialysis water treatment systems and monitoring in Italy: Results of a national survey. *J. Nephrol.* **2004**, *17*, 565–569.
23. Bambauer, R.; Schauer, M.; Jung, W.K.; Daum, V.; Vienken, J. Contamination of dialysis water and dialysate. A survey of 30 centers. *ASAIO J.* **1994**, *40*, 1012–1016.
24. Pirastu, R.; Comba, P.; Reggiani, A.; Foa, V.; Masina, A.; Maltoni, C. Mortality from liver disease among Italian vinyl chloride monomer/polyvinyl chloride manufacturers. *Am. J. Ind. Med.* **1990**, *17*, 155–161.
25. Cappelli, G.; Ravera, F.; Ricardi, M.; Ballestri, M.; Perrone, S.; Albertazzi, A. Water treatment for hemodialysis: A 2005 update. *Contrib. Nephrol.* **2005**, *149*, 42–50.
26. Canaud, B. The early years of on-line HDF: How did it all start? How did we get here? *Contrib. Nephrol.* **2011**, *175*, 93–109.

CHAPTER 6

Monitoring of Water Quality Parameters in Upper and Lower Reaches of Dudhganga Catchment, India

Mohd Iqbal[1], Haroon Sajjad[1], F.A. B hat[2] and Imran Ahmad[3]

[1] Department of Geography, Faculty of Natural Sciences, Jamia Millia Islamia, New Delhi, India

[2] Department of Geology and Geophysics, University of Kashmir, India

[3] Department of Industries and Earth Sciences, Tamil University, Thanjavur, India

1. INTRODUCTION

Water is extremely essential for survival of all living organisms. The quality of water is vital concern for mankind since it is directly linked with human welfare. In India, most of the population is dependent on surface water (damp water) as the only source of drinking water supply. The groundwater is believed to be comparatively much clean and free from pollution than surface water. But prolonged discharge of industrial effluents, domestic sewage and solid waste dump causes the groundwater to become polluted and created health problems [1]. The problems of surface water quality are much more acute in the areas which are densely populated. The rapid growth of urban areas has further affected groundwater quality due to overexploitation of resources and improper waste disposal practices. Hence, there is always a need for and concern over the protection and management of surface water and groundwater quality.

[2,3,4] Considering the above aspects of water contamination, the present study was undertaken to investigate the possible impact of the surface water quality. In Kashmir, most of the population is dependent on surface water as the only source of drinking water supply Thus, in this paper an attempt has been made to assess the physical and chemical properties of surface water of the Dudhganga stream of Dudhganga watershed, Kashmir valley, India.

Dudhganga watershed of Jammu and Kashmir (Figure 1) is located in the northern part of India between 34^0 42' to 34^0 50'N and 74^0 24' to 74^0 54' E, covers an area of 660 km^2. The area supports a varied topography exhibiting altitudinal extremes of 1610 to 4700 m above mean sea level. From southwest to northeast, the area consists of the lofty Pir-Panjal, and flat-topped karewas as foothills and plains. The Pir-Panjal mountain range covers the Kashmir valley on the south and southwest, separating it from the Jammu region. The soils in the area are generally loamy soil, karewa soil and poorly developed mountain soil [5]. Drainage of the area is quite significant with most of the drainage flowing into river *Jhelum*. Dudhganga is the important tributaries of river *Jhelum* which originates near Tatakuti Mountain. The literature survey reveals that no water quality management studies are made in this region so far. Hence the present study was planned and undertaken

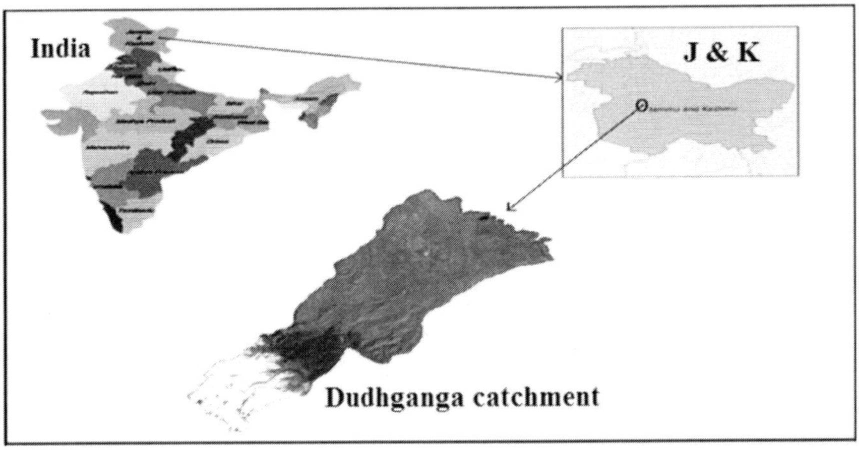

Figure 1. Geographical location of Dudhganga Catchment, India

2. METHOD AND MATERIALS

The present study was planned and undertaken at following sampling locations.

Table 1. Sampling locations

Station No.	Station Name	Latitude and Longitude in degrees	Altitude (meters)
S1(Forest Area)	Branware	33° 50' 37.3"N 74° 39' 21.3"E	2278
S2(Urban Centre)	Barzulla Baghat	34° 03' 00.0"N 74° 48' 00.6"E	1605

2.1. Preparation of Water Samples

The sample were collected from all the stations at 09.00 am to 11.00am for physico-chemical examinations, different methods of collection and handling were adopted based the standard procedures [6]. Water samples were collected carefully in a glass stoppered sterilized container (Volume approx. 1000ml) at a depth of 15cm without any air bubbles. The instruments were used of accuracy. The pH of two locations were checked on spot. The instruments used for the pH was portable pH meter. Water samples from two sampling sites were collected during May to July months (2011).

2.2. Physicochemical Analysis

Analysis was carried out for various water quality parameters such as pH, electrical conductivity (EC), dissolved oxygen (DO), free carbon dioxide, total alkalinity (TA), total hardness (TH), calcium hardness, calcium content, chloride (Cl^-), nitrate, nitrite, sulphate, phosphate, Ortho Phosphate Silicates, and TDS and ammonia using standard method [7,8].

2.3. Statistical Analysis

The simple linear correlation analysis has been carried out to find out correlation between two sampling sites (Branware and Barzulla Baghat) using SPSS.

3. RESULTS AND DISCUSSION

The average results of the physicochemical parameters for water samples are presented in (Figure 2).

3.1. PH

pH is a term used universally to express the intensity of the acid or alkaline condition of a solution. Most of the waters are slightly alkaline due to presence of carbonates and bicarbonates. The pH values of water samples varied between 7.52 to 7.1 and were found Within the limit prescribed by WHO.

Table 2. Average results of the physicochemical parameters

Parameters/Sites	Branware (Upper Reaches)	Barzulla Baghat (Lower Reaches)	Average	Standard Deviation
DO mg/L **(X1)**	12.25	5	8.625	5.127
CO2mg/L **(X2)**	3.17	12.31	7.74	6.463
Nitrite μg/L **(X3)**	0.05	18.73	9.3888	13.21
Ammonia μg/L **(X4)**	7.0	400	203.85	278.4
Phosphate μg/L **(X5)**	62.37	155	108.69	65.5
Cl^- mg/L **(X6)**	2.73	12.56	7.6488	6.946
pH **(X7)**	7.52	7.1	7.31	0.297
Total Hardness mg/L **(X8)**	85.5	280	182.75	137.5

Ca. Hardness mg/L **(X9)**	22.25	68.66	45.455	32.82
Ca. Content mg/L **(X10)**	8.64	64.53	36.585	39.52
Alkalinity mg/L **(X11)**	55.5	83.11	69.305	19.52
Conductivity ohm.cm **(X12)**	37.25	300.23	168.74	186
Nitrate mg/L **(X13)**	0.15	0.99	0.5721	0.588
Sulphate mg/L **(X14)**	1.62	7.56	4.5925	4.197
Ortho Phosphate µg/L **(X15)**	32.375	120.25	76.313	62.14
Silicates mg/L **(X16)**	1.75	4.625	3.1875	2.033
TDS mg/L **(X17)**	22.35	180.138	101.23	111.6

3.2. Electrical Conductivity

Electrical conductivity is a measure of water capacity to convey electric current. It signifies the amount of total dissolved salts [9]. EC values were in the range of 37.25ohm./cm to 300.23 ohm /cm High EC values was observed at sampling station Barzulla Baghat indicating the presence of amount of dissolved inorganic substances in ionized form.

3.3. Dissolved Oxygen

Dissolved oxygen is important parameter in water quality assessment and reflects the Physical and biological processes prevailing in the water. The DO values indicate the degree of pollution in water bodies. DO values varied from 5 to 12.25. The sampling point Barzulla Baghat showed low DO values indicating heavy contamination by organic matter.

3.4. Free Carbon Dioxide

The origin of free carbon dioxide is from air, algal respiration and organic breakdown. The Free Carbon Dioxide ranged from minimum of 3.17 mg/L to maximum of 12.31 mg/L at site1 and site 2 respectively. It indicates the presence of sewage at site 2 in the Dudhganga stream.

3.5. Total Alkalinity

Alkalinity of water is its capacity to neutralize a strong acid and it is normally due to the presence of bicarbonate, carbonate and hydroxide compound of calcium, sodium and potassium. Total alkalinity values for all the investigated samples were found within the limits prescribed by WHO.

3.6. Total Hardness

Hardness is the property of water which prevents the lather formation with soap and increases the boiling points of water [10]. Hardness of water mainly depends upon the amount of calcium or magnesium salts or both. The hardness values shown range from 85.5mg/L to 280 mg/L. All samples were found within the limits prescribed limits.

3.7. Calcium Hardness and Calcium Content

Calcium is directly related to hardness. Calcium hardness ranged between 22.25 mg/L to 68.86 mg/L and Calcium content ranged from 8.64 to 64.53 found within permissible limit, prescribed by WHO.

3.8. Chloride

The chloride concentration serves as an indicator of pollution by sewage. In the present analysis, chloride concentration was found in the range of 2.73 mg/L to 12.56 mg/L. The values are within the limits. The length of the river from Branware to Barzulla Baghat is small approximately 30 kilometers so an increase of even 5 mg/l at one station from another station may give rise to suspicions of a sewage discharge [11].

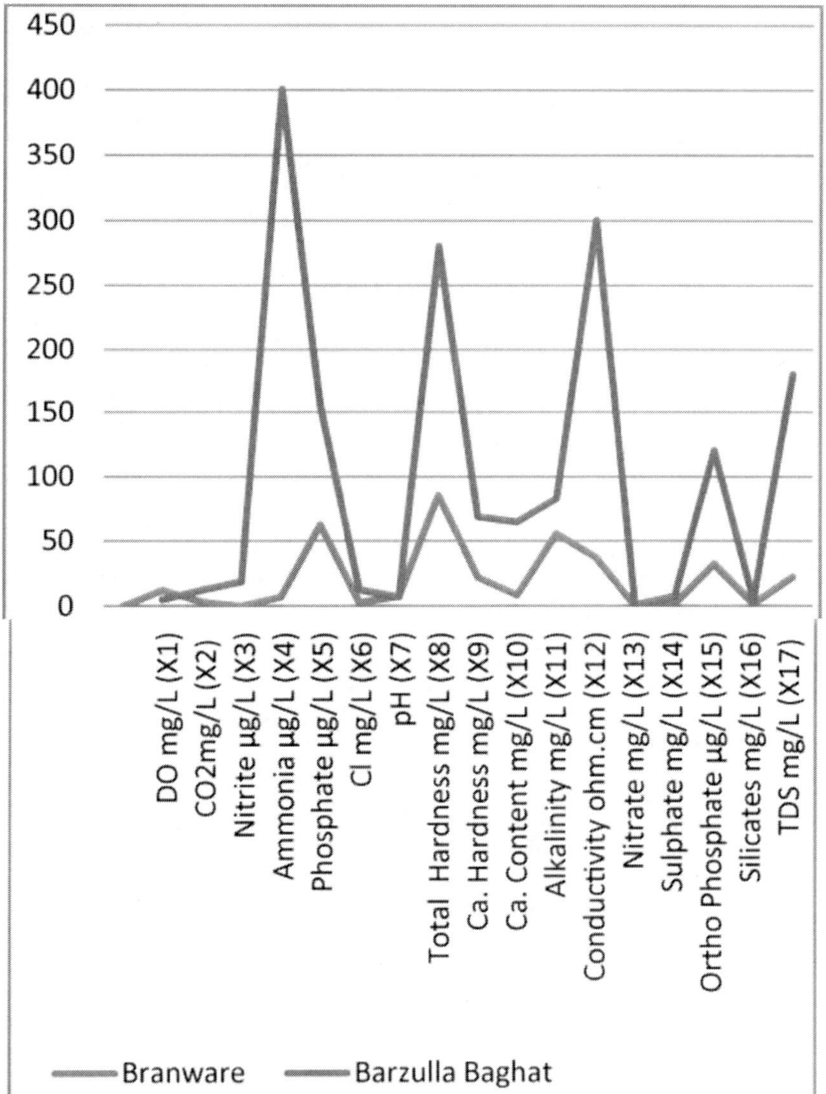

Figure 2. Water quality parameters

3.9. Nitrite

Nitrite exists normally in very low concentrations and even in waste treatment plant effluents levels are relatively low, principally because the nitrogen will tend to exist in the more reduced (ammonia; NH^3) or more oxidized (nitrate; NO^3) forms. Because nitrite is an intermediate in the oxidization of ammonia to nitrate, because such oxidation can proceed in soil, and because sewage is a rich source of ammonia nitrogen, waters which show any appreciable amounts of nitrite are regarded as being of highly questionable quality. Levels in unpolluted waters are normally low, below 0.03 mg/l Values greater than this may indicate sewage pollution [11]. In present study it ranges from 0.05µg/L to18.73µg/L.

3.10. Nitrate

Surface water contains nitrate due to leaching of nitrate with the percolating water. Surface water can also be contaminated by sewage and other wastes rich in nitrates The nitrate content in the study area varied in the range 0.15 mg/L to 0.99 mg/L and found within the prescribed limit.

3.11. Sulphate

Sulphate occurs naturally in water as a result of leaching from gypsum and other common minerals [12]. Discharge of industrial wastes and domestic sewage tends to increase its concentration. The sulphate concentration varied between 1.62 mg/L and 7.56 mg/L. and found within the prescribed limit.

3.12. Phosphate

Phosphate may occur in groundwater as a result of domestic sewage, detergents, agricultural effluents with fertilizers and industrial waste water. The phosphate content in the study area was found in the range of 62.37µg/L to 155µg/L and found within the prescribed limits.

3.13. Ammonia

Ammonia is generally present in natural waters, though in very small amounts, as a result of microbiological activity which causes the reduction of nitrogen-containing compounds. When present in levels above 0.1 mg/l N, sewage or industrial contamination may be indicated [11]. In present study ammonia value ranges from 7µg/L to 400µg/L at site1 and site 2 respectively.

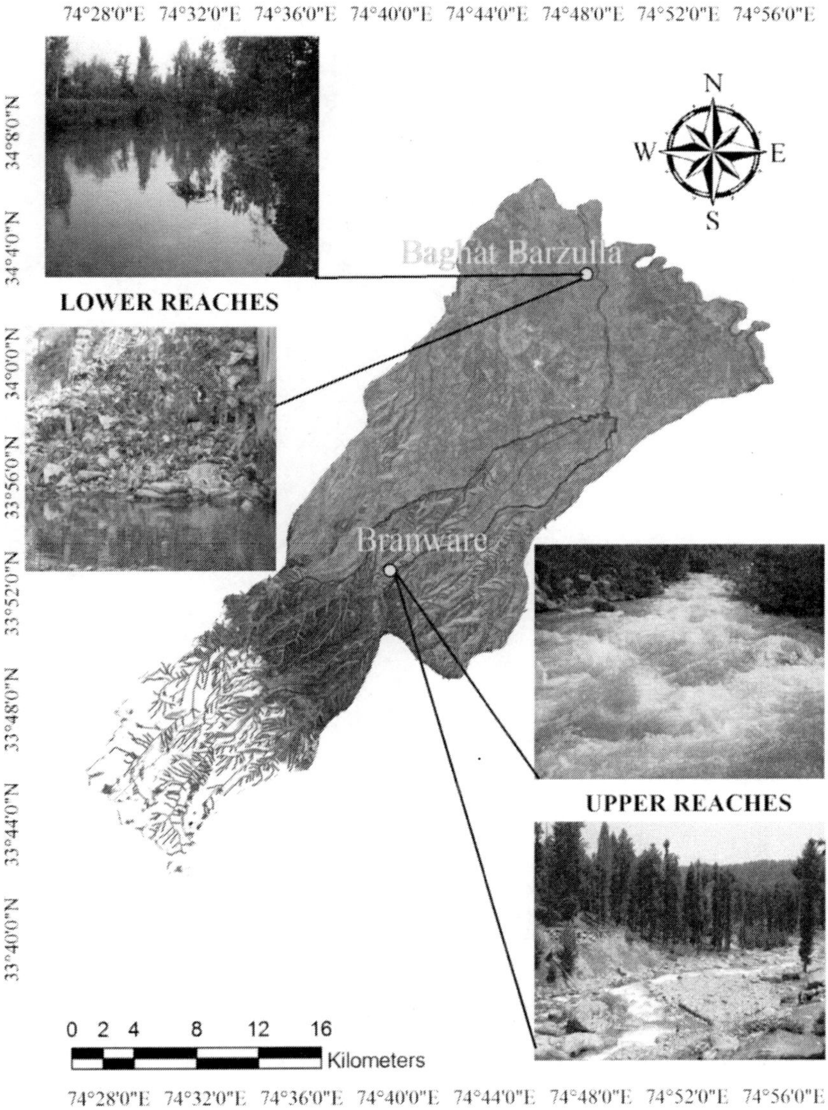

Figure 3. Shows sampling sites in upper and lower reaches of Dudhganga Catchment, India

3.14. Statistical Analysis

Interrelationship studies between two sampling sites are very helpful tools in promoting research and opening new frontiers of knowledge. The study of correlation reduces the range of uncertainty associated with decision making. The numerical values of correlation coefficient (r) for two sites are tabulated in Table 3.

Table 3. Correlation matrix of two sampling stations.

Correlations			
		BRANWARE	BARZULLA BAGHAT
BRANWARE	Pearson Correlation	1	.514*
	Sig. (2-tailed)		.035
	N	17	17
BARZULLA BAGHAT	Pearson Correlation	.514*	1
	Sig. (2-tailed)	.035	
	N	17	17

[i] - *. Correlation is significant at the 0.05 level (2-tailed).

4. CONCLUSION

Dudhganga is fresh water stream of Kashmir valley; it has a length of almost 35 km^2. All lotic water bodies have regenerating capacity then lentic water bodies because they are continuously flowing so little deviations in their water quality parameters are of great concern. Deviations were observed in water quality parameters from Site 1 to Site 2. As such no contamination of water is observed at Site S1 (Branware) because this region lies in upper catchment and having mostly forest as land cover. In upper catchment physicochemical parameters are within the water quality standards and the quality of water is good and it is fit for drinking purpose. But at last site S2 (Barzulla Baghat) which lies in the urban centre has the level of DO beyond the permissible limits and water has been contaminated by direct passage of sewage from households and it is unfit for drinking purpose

REFERENCES

1. Raja R E, Lydia Sharmila, Princy Merlin, Chritopher G, Physico-Chemical Analysis of Some Groundwater Samples of Kotputli Town Jaipur, Rajasthan, Indian J Environ Prot., 22(2), 137, (2002).
2. Patil P R, Badgujar S R. and Warke A. M. Evaluation of Ground Water Quality In Ganesh Colony Area Of Jalgaon City, Oriental J Chem., 17 (2), 283, (2001).
3. Patil V T. and Patil P R. Physicochemical Analysis of Selected Groundwater Samples of Amalner Town in Jalgaon District, Maharashtra, India, E-Journal of Chem., 7(1), 111-116, (2010).
4. Murhekar G H. Determination of Physico-Chemical parameters of Surface Water Samples in and around Akot City, Int. J. Res. Chem. Environ., 1(2), 183-187, (2011).
5. Raza M, Ahmad A, and Mohammad A. (1978). The Valley of Kashmir: A Geographical Interpretation, (New Delhi: Vikas Publishing House, Pvt. Ltd).
6. Standard Methods for the Examination of Water and Waste Water, 20th Ed., APHA, AWWA, WEF. Washington DC, (1998).
7. Standard Methods for the examination of water and waste water, American Public Health Association, 17th Ed., Washington, DC, (1989).
8. Manivaskam N., Physicochemical examination of water sewage and industrial effluent, 5th Ed. Pragati Prakashan Meerut., (2005).
9. Sudhir Dahiya and Amarjeet Kaur, physic chemical characteristics of underground water in rural areas of Tosham subdivisions, Bhiwani district, Haryana, J. Environ Poll., 6 (4), 281, (1999).
10. Trivedy R K. and Goel P K. Chemical and Biological methods for water pollution Studies, Environmental Publication, Karad. (1986).
11. Environmental Protection Agency, Parameters of Water Quality Interpretation and Standards, Johnstown Castle, Co. Wexford, Ireland. (2001).
12. Shrinivasa R. B and Venkateswaralu P, Physicochemical Analysis of Selected Groundwater Samples, Indian J Environ Prot., 20 (3), 161, (2000).

CHAPTER 7

Marine Pollution: Let us not Forget Beach Sand

Francois Galgani[1], Katrin Ellerbrake[2], Elke Fries[2]* and Chantal Goreux[3]

[2]Institute of Environmental Systems Research, University of Osnabrueck, Osnabrueck, 49076, Germany

ABSTRACT

Background

Assessing the chemical or bacterial contamination in marine waters and sediments is a very common approach to evaluate marine pollution and associated risks. However, toxicity and organic pollution of beach sands have not yet been considered, except in adjacent waters. In the present study, the toxicity and the chemical contamination of natural beach sands collected 20 m from the shoreline at two sites located on the Mediterranean Sea (Marseille and La Marana, Corsica) were studied.

Results

Up to 16.93% (net percentage) abnormal or dead larvae was observed in elutriates prepared from the urban beach sand sample (Marseille); no significant toxicity was observed in the sample collected from the reference beach in La Marana. Results of Fourier transform infrared spectroscopy analyses revealed that no microplastics were present in either of the samples. Several polycyclic aromatic hydrocarbons [PAHs] in both samples and a larger number of individual PAHs in the urban sample than in the sample collected from the reference beach were detected. In addition, the antioxidant dioctyldiphenylamine was detected in both beach sand samples, whereby a higher concentration was found in La Marana than in Marseille. Calculated PAH concentrations in elutriates were generally higher than measured ones.

Conclusions

The results of this preliminary study provide evidence of toxicity and the presence of organic trace contaminants in beach sands from France. According to our results, monitoring using a combination of biotests and chemical analyses is recommended, especially of sediments from beaches abandoned to urban and industrial areas.

Keywords: oyster larvae biotests, sediments, infrared spectroscopy, gas chromatography-mass spectrometry, microplastics, polycyclic aromatic hydrocarbons, Mediterranean Sea, bioavailability, organic contaminants

BACKGROUND

Assessing the chemical or bacterial contamination in marine waters, sediments and organisms is a very common approach to evaluate marine pollution and associated risks [1, 2, 3, 4, 5, 6, 7]. In addition, the impact of plastic debris on the chemical contamination of marine ecosystems following the sorption of organic contaminants from seawater or the release of organic plastic additives has been discussed recently [8, 9, 10]. Sorption to sediments is the main process that determines the fate of hydrophobic organic compounds in the marine environment. Organic carbon has been shown to be a significant factor in controlling the sorption process [11, 12, 13]. Although organic chemicals and heavy metals accumulate in organisms living in/on the sand or on sandy eggs [14, 15], the contamination of beaches is not normally considered, except in adjacent waters. This might be due to the lower extent of organic materials in beach sediments prone to pollution by organic contaminants. As with many other organic contaminants, some specific compounds, such as anti-UV agents, have been described in marine organisms [16], but not in beach sands, where their occurrence is also expected. Surprisingly, and to our knowledge, no information is available on the toxicity and the presence of organic trace contaminants in natural beaches.

In addition to toxicity tests, the potential for risk may be typically assessed using the proportion of dose estimated to penetrate through exposed skin [17]. However, this approach is still under development. In addition, only recommended values exist for adverse biological effects on organisms living in marine and estuarine sediments [18, 19].

From an environmental perspective, knowledge of the extent to which beaches are contaminated could lead not only to the evaluation of the levels and trends of contaminants, but also, following human contact with sand, to an assessment of the effects on public health. More fundamentally, these measurements should also specify fluxes of contaminants at the interface between coasts and seawater.

In the present study, beach sands were collected from two sites on the Mediterranean Sea in France (Marseille and La Marana, Corsica) and studied for toxicity and chemical pollution. The samples were analysed by Fourier transform infrared [FTIR] spectroscopy to assess the presence of microplastics. Polycyclic aromatic hydrocarbons [PAHs] - a class of hydrophobic organic chemicals - were analysed due to their widespread occurrence and persistence in the environment and their toxic and carcinogenic properties [20] with sufficient evidence from animal studies [21]. PAHs are widely found in surface soils, where typical concentrations range from 5 to 100 μg/kg in forests and 600 to 3,000 μg/kg in urban areas [21]. However, no guidelines on PAHs for beaches or non-industrial soils have ever been described in relation to their effects on human health.

RESULTS

Biotests

The larval development of the oyster *Crassostrea gigas* was tested for 24 h, until stage D was reached, in the presence of various concentrations of elutriates prepared from beach sands collected at both sites, Marseille and La Marana (Figure 1). After 24 h, we found up to 16.93 ± 7.14% (net percentage, two test series times three replicates each) abnormal or dead larvae in 100% elutriates prepared from the sediment sample collected in Marseille by seawater extraction. In the sample collected from the reference beach (La Marana), no significant toxicity was observed (0.87 ± 0.74%, net percentage, two test series times three replicates each).

FTIR spectroscopy analysis

FTIR spectroscopy was used to test whether microplastics occur in the beach sands. CH_2 groups were found only in the beach sand collected from La Marana. Moreover, no signal was found in this sample after all CH_2groups were removed following extraction using tetrachloromethane.

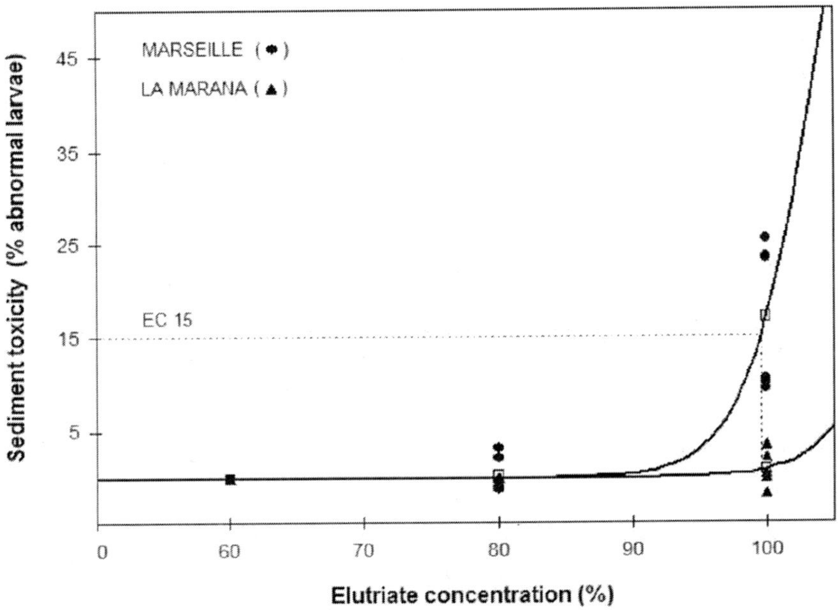

Figure 1 Effects of sand extracts from Marseille and La Marana on larval development of *C. gigas*. Toxicity is expressed as the percentage of abnormal larvae (net percentage, toxicity minus control) after 24-h development in various concentrations of beach sand extracts (elutriates). EC_{15} (99.60%) is the elutriate concentration causing 15% of abnormalities. Empty squares, mean values; filled circle, Marseille; filled triangle, La Marana.

GC-MS analysis

Concentrations of PAHs measured in the beach sand samples are shown in Table 1. The following PAHs (including several EPA-PAHs) were detected in the urban sample (Marseille): acenaphthylene [ACY], acenaphthene [ACE], phenanthrene [PHE], anthracene [ANT], fluoranthene [FLT], pyrene [PYR], benz(a)anthracene [BAA], chrysene [CHR], benzo(b)fluoranthene [BBF], benzo(k)fluoranthene [BKF], benzo(e)pyrene [BEP], benzo(a)pyrene [BAP], indeno(1,2,3-cd)pyrene [IPY], dibenz(a,h)anthracene [DBA] and benzo(ghi)perylene [BPE]. The total PAH concentration in the urban sample was 539 μg/kg dry weight [dwt]. The concentrations of individual PAHs were between 1.52 (ACE) and 101 μg/kg dwt (PHE). In the sediment sample collected from La Marana, only three individual PAHs were detected (ACE, BAP, IPY) with a much lower total of 1.56 μg/kg dwt (Table 1).

Table 1 Concentrations of PAHs measured in beach sand samples collected from Marseille and La Maran

Compound	M (μg/kg dwt)	LM (μg/kg dwt)
NAP	<LOD	<LOD
ACY	61.7	<LOD
ACE	1.52	0.77
FLN	<LOD	<LOD
PHE	101	<LOD
ANT	4.48	<LOD
FLT	61.6	<LOD
PYR	29.0	<LOD
BAA	27.5	<LOD
CHR	27.4	<LOD
BBF	16.3	<LOD
BKF	36.0	<LOD
BEP	37.4	<LOD
BAP	52.2	0.31
IPY	29.4	0.48
DBA	7.76	<LOD
BPE	45.6	<LOD
ΣPAHs	539	1.56

LOD, limits of detection; M, Marseille; LM, La Marana; SW, seawater; NAP, naphthalene; ACY, acenaphthylene; ACE, acenaphthene; FLN, fluorene; PHE, phenanthrene; ANT, anthracene; FLT, fluoranthene; PYR, pyrene; BAA, benz(a)anthracene; CHR, chrysene; BBF, benzo(b)fluoranthene; BKF, benzo(k)fluoranthene; BEP, benzo(e)pyrene; BAP, benzo(a)pyrene; IPY, indeno(1,2,3-cd)pyrene; DBA, dibenz(a,h)anthracene; BPE, benzo(ghi)perylene.

Concentrations of PAH in elutriates prepared from the sediment samples collected at both sites and in seawater are shown in Table 2. The limits of detection [LOD] are also shown. Since most PAHs in elutriates and seawater used for sediment extraction were below the limits of quantitation [LOQ] and LOD, with the exception of PYR, BAA and CHR in the elutriate prepared from the sediment sample from Marseille, the equilibrium concentrations in the elutriates were calculated. The results are shown in Table 3. The calculated PAH levels in elutriates obtained from the sediment samples in

Marseille and La Marana were between 0.07 (DBA) and 426 ng/L (ACY; sum of PAHs 677 ng/L) and between 0.01 (BAP, IPY) and 4.26 ng/L (ACY; sum of PAHs 4.28 ng/L), respectively. Since agitation was performed for 8 h, equilibrium could be assumed, and much lower measured PAH concentrations than calculated ones might be attributed to a loss of compounds during elutriate preparation.

Dioctyldiphenylamine - an antioxidant in oils, lubricants, rubber and plastics - was also identified in both samples, whereby the concentration in La Marana (>100 μg/kg dwt) was higher than that in Marseille (>0.3 μg/g dwt).

Table 2 Measured concentrations of PAHs in elutriates prepared from beach sand samples and in seawater

Compound	LOD (ng/L)	M (ng/L)	LM (ng/L)	SW (ng/L)
NAP	2.25	<LOD	2.25	<LOD
ACY	0.44	<LOD	0.44	<LOD
ACE	0.93	<LOD	0.93	<LOD
FLN	0.46	<LOD	0.46	<LOD
PHE	1.67	<LOQ	1.67	<LOQ
ANT	0.13	<LOD	0.13	<LOD
FLT	5.14	<LOQ	5.14	<LOQ
PYR	0.76	6.59	0.76	6.59
BAA	0.51	1.74	0.51	1.74
CHR	0.50	2.60	0.50	2.60
BBF	0.53	<LOD	0.53	<LOD
BKF	0.53	<LOD	0.53	<LOD
BEP	0.35	<LOD	0.35	<LOD
BAP	0.70	<LOD	0.70	<LOD
IPY	0.57	<LOD	0.57	<LOD
DBA	1.46	<LOD	1.46	<LOD
BPE	1.02	<LOD	1.02	<LOD

M, Marseille; LM, La Marana; NAP, naphthalene; ACY, acenaphthylene; ACE, acenaphthene; FLN, fluorene; PHE, phenanthrene; ANT, anthracene; FLT, fluoranthene; PYR, pyrene; BAA, benz(a)anthracene; CHR, chrysene; BBF, benzo (b)fluoranthene; BKF, benzo(k)fluoranthene; BEP, benzo(e)pyrene; BAP, benzo (a)pyrene; IPY, indeno(1,2,3-cd)pyrene; DBA, dibenz(a,h)anthracene; BPE, benzo (ghi)perylene; ΣPAHs, summation of polycyclic aromatic hydrocarbons; LOD, limits of detection; LOQ, limits of quantitation.

Table 3 Calculated equilibrium concentrations of the PAHs in elutriates prepared from beach sand samples

Compound	M (ng/L)	LM (ng/L)
NAP	-	-
ACY	426	-
ACE	8.38	4.26
FLN	-	-
PHE	170	-
ANT	6.00	-
FLT	33.2	-
PYR	14.3	-
BAA	2.09	-
CHR	3.00	-
BBF	1.16	-
BKF	2.55	-
BEP	0.71	-
BAP	2.41	0.01
IPY	0.42	0.01
DBA	0.07	-
BPE	0.56	-
ΣPAHs	677	4.28

M, Marseille; LM, La Marana; SW, seawater; NAP, naphthalene; ACY, acenaphthylene; ACE, acenaphthene; FLN, fluorene; PHE, phenanthrene; ANT, anthracene; FLT, fluoranthene; PYR, pyrene; BAA, benz(a)anthracene; CHR, chrysene; BBF, benzo(b)fluoranthene; BKF, benzo(k)fluoranthene; BEP, benzo(e) pyrene; BAP, benzo(a)pyrene; IPY, indeno(1,2,3-cd)pyrene; DBA, dibenz(a,h) anthracene; BPE, benzo(ghi)perylene; ΣPAHs, summation of polycyclic aromatic hydrocarbons

DISCUSSION

From our study on sand samples collected from two beaches, evidence of toxicity was found in a beach sand aqueous extract. The occurrence of abnormal larvae in the course of larval development in the extract from Marseille could be linked to the occurrence of contaminants in the sediments. The non-specific test is used to measure the overall impact of contaminants. Although the toxicity rate is low compared to immersed sediments from industrial areas or towns [22, 23], it is significant and in the same range as some affected areas such as coastal lagoons and bays [24]. It is difficult to draw any conclusions on the basis of these results with regard to the source and nature of the contaminants involved. However, the results suggest the presence of bioavailable toxic contaminants.

Several organic plastic additives, such as phthalates, organotins, bisphenol A and polybrominated diphenyl ethers, and chemicals sorbed to plastics, such as PAHs and polychlorinated biphenyls, have potentially toxic or endocrine effects [9, 10, 25]. Migration of such chemicals into elutriates or the direct ingestion of microplastics by oyster larvae followed by leaching into the organisms could be a possible explanation for the observed toxic effect. From our results of FTIR spectroscopy analysis, it can be assumed that CH_2 groups in the beach sands were long-chain hydrocarbons and that no plastic polymers were present in beach sands from either of the sampling locations. The absence of microplastics in the sediment samples suggests the existence of bioavailable fractions of organic contaminants in the sediments collected in Marseille.

The detection of PAHs in both sediment samples indicated that beach sands are a possible sink for such hydrophobic chemicals. The occurrence of PAHs in both samples must be due primarily to atmospheric deposition because samples were taken from areas that had not been flooded for some time. Possible PAH emission sources are car and truck traffic, industrial activities using fuel oils and the combustion of coal/coke (iron and steel industries). In addition, aerosols from the sea may also contribute to the PAH burden in beach sands, especially following stormy weather, as hydrophobic compounds accumulate at the surface, albeit to a lesser extent [26]. The sum of PAH concentrations in the urban sample (Marseille) was within the lower range found in immersed sediments from Marseille Bay, France [7] and Dar es Salaam, Tanzania [3], but higher than in coastal sediments from Togo [5]. In the study by Benlahcen et al. [6], the sum of PAHs was usually higher in Mediterranean coastal sediments than that found in the urban sample of the present study.

PAH ratios, such as ANT/(ANT + PYR) (mass 178), FLT/(FLT + PYR) (mass 202) and BAA/(BAA + CHR) (mass 228), are commonly used to characterise types of PAH sources in soils [27] and river sediments [28]. The ANT/178 ratio of 0.04 calculated for the sediment sample collected in Marseille was lower than 0.1, indicating petrogenic sources. On the other hand, the FLT/202 ratio of 0.7 suggests pyrogenic sources of PAHs, such as kerosene, grass, coal and wood combustion [28, 29]. A BAA/228 ratio of 0.5 supports this suggestion [28]. The petrogenic origin of PAHs is unlikely for the selected sampling locations because samples were collected 20 m from the shoreline where flooding had never occurred. The usefulness of ANT/178 ratio as a recipient matrix has already been called into question before for soil [13, 27].

Previous studies on marine sediments and caged mussels [7, 24] have demonstrated the different PAH contamination at both sites. Differences in the PAH burden of both samples are explained by the large number of PAH sources in the Bay of Marseille, including maritime activities, refineries, traffic and urbanisation, whereas the reference station La Marana, which has an adjacent open sea, is unaffected by local sources.

The larger number of individual PAHs and the higher sum of PAHs measured in the urban sediment sample (Marseille) compared to the sample from the reference beach (La Marana) indicate that abnormal or dead oyster larvae may result from bioavailable fractions of PAHs. Toxic effects of PAHs on oyster larvae have been reported for a total elutriate PAH concentration of 940 ng/L [30]. The total PAH concentration measured in the elutriate obtained from the sample from Marseille was much lower (Table 3). Although the toxic effect was also lower, these results might also indicate effects of other unknown chemicals.

Since the bioavailable fraction of dioctyldiphenylamine is expected to be very low according to the very high estimated octanol-water partition coefficient ($\log K_{ow}$) of 11.26 [31], it is rather unlikely that the observed toxic effect was caused by this compound. However, this non-polar chemical is not readily biodegradable, suggesting its potential to persist in the environment [31]. Hence, more data on its occurrence in sediments and its bioavailability are required. However, other contaminants that could not be detected using the GC-MS methods applied, such as metals or more polar organic compounds, may also be involved.

CONCLUSIONS

This first evidence of toxicity and the presence of organic contaminants, such as PAHs and dioctyldiphenylamine in beach sands, raise important questions that need to be addressed in the future by the scientific community. Further investigations into toxic chemicals in beach sands are recommended in combination with biotests, especially at beaches in the vicinity of urban, industrial and tourist areas. In addition, the question whether aerosols from the sea may be a source of organic pollutants in beach sand should be addressed in future research. The problem of toxicity due to organic contaminants in beach sands will be of broad interest to stakeholders, policymakers and the general public. The abundance of such toxic chemicals as PAHs in beach sand is also of concern from a health perspective since most beaches are frequented by tourists. In addition to providing new insights into marine pollution, evaluating the problem on a larger scale may also help to create a new, simple method of monitoring the state of the environment, environmental quality standards and guidelines.

METHODS

Sampling

One sand sample made of ten pooled fractions from a 10 × 10-m area was collected from the top 2 cm of an urban beach in Marseille (France) and compared with a reference beach in La Marana (Corsica, France). Both sampling locations were 20 m from the shoreline. Marseille was chosen because it is a large urban, industrial and coastal area with various pollution sources. Marseille and the surrounding area are home to more than 1.2×10^6 inhabitants. The Rhone river delta is adjacent (west) to the city. This is one of the areas affected most severely by contamination in the Mediterranean, especially by PAHs [32]. The sampling location at La Marana was selected as a background station. It is located south of the city Bastia (30×10^3 inhabitants) and has no industry, but tourist activities. Beach sand samples were placed in 100-mL brown glass bottles (Omnilab, Hannover, Germany) precleaned using methanol (HPLC grade, Merck, Darmstadt, Germany) and ultrapure water taken from an Arium 611VF water purification system (Sartorius, Goettingen, Germany). The caps were equipped with Lupolen inserts (Omnilab) wrapped in an aluminium foil. Sediment samples were kept frozen until the chemical analysis was undertaken.

Biotests

The toxicity of the beach sand samples was assessed using oyster larvae biotests [33]. Sands were agitated for 8 h after adding filtered seawater (4:1 volume/weight) and decanted for 4 h. Biotests were performed at the Institute Pasteur Lille (IPL), France by exposing oyster embryo larvae to five concentrations of elutriates prepared from both samples (20%, 40%, 60%, 80% and 100% elutriates + control). Elutriates were stored at +4°C until the chemical analysis was undertaken. Mature genitors (*C. gigas*) were obtained from the Guernsey Sea Farms hatchery. These genitors were carefully cleaned and immersed in unfiltered reference water at 18°C for 30 min before a thermal shock (28°C, 30 min). After the emission of gametes, fecundation was monitored under a microscope; then, after dilution, the larvae were transferred to Iwaki microplates (final volume 3 mL, number

300 larvae/well) and placed in culture with sand extracts at 24 ± 1°C for 24 h. After incubation and after reaching the typical stage D [33], the larvae were fixed in 4% formaldehyde and decanted. The abnormality rate, including dead larvae, was determined on the basis of a count of 100 larvae/well (two tests, three replicates per concentration). Abnormalities in controls were 6.16 ± 2.4%. The results are given as the net percentage of abnormalities (toxicity minus toxicity of control The dose/response relationship was obtained using the REGTOX: macro Excel [34].

FTIR spectroscopy analysis

Air-dried beach sand samples were analysed using FTIR spectroscopy with an infrared spectrometer and ATR MIRacle (attenuated total reflection; PIKE Technologies, Madison, WI, USA) with a diamond crystal [35]. The system enabled to analyse the surface of the sand for the presence of CH_2-related organic compounds or only polymers after tetrachloromethane (CCl_4) extraction. The depth penetration of light in the samples was 4 μm. Sensitivity tests were made with pure sand washed with demineralised water and spiked with micronized polyethylene (Total Petrochemicals, Feluy, Belgium). It was checked that 0.01% and 0.1% of the micronized polyethylene can be clearly detected.

GC-MS analysis

The beach sand samples were analysed for the presence of organic contaminants using Soxhlet extraction followed by gas chromatography-mass spectrometry [GC-MS]. All sand samples were freeze-dried. To determine the procedural blank, one aliquot (10 g) was precleaned with dichloromethane [DCM] (HPLC grade, Promochem, Wesel, Germany) for 12 h in a Soxhlet extractor. Aliquots of 10 g were spiked with 100 μL of a deuterated PAH standard solution containing acenaphthene-D_{10}, chrysene-D_{10}, perylene-D_{12} and phenanthrene-D_{12} (Dr. Ehrenstorfer GmbH, Augsburg, Germany) with a concentration of 5 ng/μL in acetone using an aluminium laboratory spoon. During mixing, contact between the spoon and the solution was strictly avoided. The samples were extracted for 12 h in the Soxhlet extractor. The solvent was evaporated to dryness and dissolved again in 100 μL DCM. Eight microlitres of a standard solution containing fluazifop-butyl (Sigma-Aldrich, Seelze, Germany) in methanol (HPLC grade, Promochem) were added to the final extracts as a recovery standard.

One microlitre of the extracts was injected into a GC injector. GC analyses were performed on a 6890N GC system coupled with a 5973 inert mass selective detector (Agilent Technologies, Santa Clara, CA, USA). An HP-5MS capillary column (5% diphenyl-95% dimethylpolysiloxane), 30 m in length

× 0.25 mm in inner diameter and 0.25 μm in film thickness (Agilent Technologies), was used for chromatographic separation. The GC programme was as follows: 80°C (held for 5 min), then at 15°C/min to 180°C followed by 5°C/min to 310°C (held for 5 min). Helium was used as the carrier gas at a flow rate of 1.5 mL/min. Identification of compounds was carried out by comparing the mass spectrum collected from the Wiley library and retention times and mass spectra to those obtained from the analyses of authentic standards (PAHs were purchased from Neochema, Bodenheim, Germany; dioctyldiphenylamine was purchased from Chemos GmbH, Regenstauf, Germany). Quantification of PAHs was performed by calculating the peak area ratios of the individual PAHs and selected deuterated PAHs corrected by a factor resulting from different recoveries. The procedural blanks for PAHs were between 0.28 (BAA) and 4.04 μg/kg (PHE). The LOD were either determined by blanks or by a peak-to-noise ratio of 3:1. The LOQ were set at three times the LOD.

PAHs were also analysed in elutriates (7 mL) using solid-phase microextraction and GC-MS. A one-aliquot sample of seawater used for sediment extraction was also analysed. Tap water was analysed as a blank of the whole procedure. Nine microlitres of the deuterated PAH standard solution with a concentration of 50 pg/μL in acetone was added to the samples as an internal standard. The method was described in detail in Kukucka et al. [36]. The LOD were either determined by blanks or by a peak-to-noise ratio of 3:1. The LOQ were set at three times the LOD.

The equilibrium concentrations of PAHs in elutriates (C_{aq}) were calculated using Equation 1:

$$C_{aq} = 1/K_d * C_{sed}, \tag{1}$$

where K_d is the sediment-water distribution coefficient and C_{sed} is the concentration of the individual PAH in the sediment.

The K_d values were obtained from the organic carbon normalised distribution coefficient (K_{oc}) according to Equation 2:

$$K_d = K_{oc} * f_{oc}, \tag{2}$$

where f_{oc} is the fraction of organic carbon [OC] in the sediment. The mean values of f_{oc} ($n = 3$) in the samples collected from Marseille and La Marana were determined at 0.05 ± 0.001% and 0.002 ± 0.0001%, respectively, using a total OC analyser (Elementar, Hanau, Germany). According to Wu and Sun [37], the factor between K_d for seawater and freshwater was 1.21 for PHE. Since no factors exist for the other PAHs, the effect of the saline solution on K_d was neglected.

The K_{oc} value was determined according to Karickhoff [38] by Equation 3:

$$\log K_{OC} = 0.989 * \log K_{OW} - 0.346, \quad (3)$$

where K_{ow} is the octanol-water partition coefficient. The values of K_{ow} were taken from Ma et al. [39].

REFERENCES

1. Claisse D: Chemical contamination of French coasts. The results of a ten years mussel watch. Mar Poll Bull 1989, 20:523-528.
2. Barbee GC, Barich J, Duncan B, Bickham JW, Matson CW, Hintze CJ, Autenrieth RL, Zhou GD, McDonald TJ, Cizmas L, Norton D, Donnelly KC: In situ biomonitoring of PAH-contaminated sediments using juvenile coho salmon (Oncorhynchus kisutch). Ecotoxicol Environ Saf 2008, 71:454-64.
3. Gaspare L, Machiwa JF, Mdachi SJ, Streck G, Brack W: Polycyclic aromatic hydrocarbon (PAH) contamination of surface sediments and oysters from the inter-tidal areas of Dar es Salaam, Tanzania. Environ Pollut 2009, 157:24-34.
4. Fleisher JM, Fleming LE, Solo-Gabriele HM, Kish JK, Sinigalliano CD, Plano L, Elmir SM, Wang JD, Withum K, Shibata T, Gidley ML, Abdelzaher A, He G, Ortega C, Zhu X, Wright M, Hollenbeck J, Backer LC: The BEACHES study: health effects and exposures from non-point source microbial contaminants in subtropical recreational marine waters. Int J Epidemiol 2010, 39:1291-1298. Galgani et al. Environmental Sciences Europe 2011, 23:40 http://www.enveurope.com /content/23/1/40 Page 6 of 7
5. Gnandi K, Musa Bandowe BA, Deheyn DD, Porrachia M, Kersten M, Wilcke W: Polycyclic aromatic hydrocarbons and trace metal contamination of coastal sediment and biota from Togo. J Environ Monit 2011, 13:2033-2041.
6. Benlahcen KT, Chaoui A, Budzinski H, Bellocq J, Garrigues Ph: Distribution and sources of polycyclic aromatic hydrocarbons in some Mediterranean coastal sediments. Mar Poll Bull 1997, 34:298-305.
7. Asia L, Mazouz S, Guiliano M, Doumenq P, Mille G: Occurrence and distribution of hydrocarbons in surface sediments from Marseille Bay (France). Mar Poll Bull 2009, 58:424-455.
8. Endo S, Takizawa R, Okuda K, Takada H, Chiba K, Kanehiro H, Ogi H, Yamashita R, Date T: Concentration of polychlorinated biphenyls (PCBs) in beached resin pellets: variability among individual particles and regional differences. Mar Poll Bull 2005, 44:1103-1114.
9. Teuten EL, Saquing JM, Knappe DRU, Barlaz MA, Jonsson S, Björn A, Rowland SJ, Thompson RC, Galloway TS, Yamashita R, Ochi D, Watanuki Y, Moore C, Viet PH, Tana TS, Prudente M, Boonyatumanond R, Zakaria MP, Akkhavong K, Ogata Y, Hirai H, Iwasa S, Mizukawa K, Hagino Y, Imamura A, Saha M, Takada H: Transport and release of chemicals from plastics to the environment and to wildlife. Phil Trans R Soc B 2009, 364:2027-2045.
10. Hirai H, Takada H, Ogata Y, Yamashita R, Mizukawa K, Saha M, Kwan C, Moore C, Gray H, Laursen D, Zettler ER, Farrington JW, Reddy CM, Peacock EE, Ward MW: Organic micropollutants in marine plastics debris from the open ocean and remote and urban beaches. Mar Poll Bull 2011, 62:1683-1692.
11. Goring CAI: Control of nitrification by 2-chloro-6-(trichloro-methyl) pyridine. Soil Science 1962, 93:211-218.

12. Goring CAI, Hamaker JW: In Organic Chemicals in the Soil Environment. Volume I. New York: Marcel Dekker Inc.; 1972.
13. Bucheli TD, Blum F, Desaules A, Gustafsson Ö: Polycyclic aromatic hydrocarbons, black carbon, and molecular markers in soils of Switzerland. Chemosphere 2004, 56:1061-1076.
14. Ungherese G, Ugolini A: Sandhopper solar orientation as a behavioural biomarker of trace metals contamination. Environ Poll 2009, 157:1360-1364.
15. Vidal M, Domínguez J, Luís A: Spatial and temporal patterns of polycyclic aromatic hydrocarbons (PAHs) in eggs of a coastal bird from northwestern Iberia after a major oil spill. Sci Total Environ 2011, 409:2668-2673.
16. Brown S: Sunscreen wipes out corals. Nature 2008, doi:10.1038/news.2008.537.
17. Knafla A, Petrovic S, Richardson M, Campbell J, Rowat C: Development and application of a skin cancer slope factor for exposures to benzo[a] pyrene in soil. Regul Toxicol Pharmacol 2011, 59:101-110.
18. Cachot J, Geffard O, Augagneur S, Lacroix S, Le Menach K, Peluhet L, Couteau J, Denier X, Devier M, Pottier D, Budzinski H: Evidence of genotoxicity related to high PAH content of sediments in the upper part of the Seine estuary (Normandy, France). Aquat Toxicol 2006, 79:257-267.
19. OSPAR: CEMP Assessment Report: 2008/2009 Assessment of Trends and Concentrations of Selected Hazardous Substances in Sediments and Biota. London 2009.
20. Douben PET: PAHs: An Ecotoxicological Perspective (Ecological & Environmental Toxicology Series) West Sussex: John Wiley & Sons; 2003.
21. WHO: Health Risks of Persistent Organic Pollutants from Long-range Transboundary Air Pollution. Copenhagen 2003.
22. Dalmazzone C, Blanchet D, Lamoureux S, Durrieu J, Dutrieux E, Camps R, Galgani F: Impact of drilling activities in warm sea: re-colonization capacities of seabed. Oil Gas Sci Technol 2004, 59:625-647.
23. Galgani F, Chiffoleau JF, Orsoni V, Costantini L, Boissery P, Calendini S, Andral B: Chemical contamination and toxicity of sediments from coastal areas of Corsica islands. Chem Ecol 2006, 22:299-312.
24. Galgani F, Senia J, Guillou JL, Laugier T, Munaron D, Andral B, Guillaume B, Coulet E, Boissery P, Brun L, Bertrandy MC: Assessment of the environmental quality of French continental Mediterranean lagoons with oyster embryo bioassay. Arch Environ Contam Toxicol 2009, 57:540-551.
25. Fries E, Zarfl C: Sorption of polycyclic aromatic hydrocarbons (PAHs) to low and high density polyethylene (PE). Environ Sci Poll Res 2011, doi: 10.1007/s11356-011-0655-5..
26. Guigue C, Tedetti M, Giorgi S, Goutx M: Occurrence and distribution of hydrocarbons in the surface microlayer and subsurface water from the urban coastal marine area off Marseilles, Northwestern Mediterranean Sea. Mar Poll Bull 2011, doi:10.1016/j.marpolbul.2011.09.013.
27. Brändli RH, Bucheli TD, Ammann S, Desaules A, Keller A, Blum F, Stahel WA: Critical evaluation of PAH source apportionment tools using data from the Swiss soil monitoring network. J Environ Monit 2008, 10:1278-1286.
28. Yunker MB, Macdonald RW, Vingarzan R, Mitchell RH, Goyette D, Sylvestre S: PAHs in the Fraser River basin: a critical appraisal of PAH ratios as indicators of PAH source and composition. Org Geochem 2002, 33:489-515.
29. Budzinski H, Jones I, Bellocq J, Piérard C, Garrigues P: Evaluation of sediment contamination by polycyclic aromatic hydrocarbons in the Gironde estuary. Mar Chem 1997, 58:85-97.
30. Geffard O, Geffard A, His E, Budzinski H: Assessment of the bioavailability and toxicity of sediment-associated polycyclic aromatic hydrocarbons and heavy metals applied to Crassostrea gigas embryos and larvae. Mar Poll Bull 2003, 46:481-490.

31. USEPA (United States Environmental Protection Agency): Screening-level Hazard Characterization for High Production Volume Chemicals: Substituted Diphenylamines. Washington, DC 2008.
32. Galgani F, Martinez-Gomez C, Giovanardi F, Romanelli G, Calvo M, Caixap J, Cento A, Scarpato A, Benbrahim S, Sammari C, Deudero S, Boulahdid M, Andral B: Assessment of polycyclic aromatic hydrocarbon concentrations in mussels (Mytilus galloprovincialis) from the North West basin of the Mediterranean Sea (the Mylilos project). Environ Monit Assess 2011, 172:301-317.
33. His E, Seaman M, Beiras R: A simplification of the bivalve embryogenesis larval development bioassay method for water quality assessment. Water Res 1997, 31:351-355. REGTOX. [http://www.normalesup.org/~vindimian/].
34. Coleman PP: Practical Sampling Techniques for Infrared Analysis Boca Raton: CRC Press; 1993.
35. Kukučka P, Lammel G, Dvorská A, Klánová J, Möller A, Fries E: Contamination of Antarctic snow by polycyclic aromatic hydrocarbons dominated by combustion sources in the polar region. Environ Chem 2010, 7:504-513.
36. Wu W, Sun H: Sorption-desorption of phenanthrene - effect of nanopores, solute concentration, and salinity. Chemosphere 2010, 81:961-967.
37. Karickhoff SW: Semi-empirical estimation of sorption of hydrophobic pollutants on natural sediments and soils. Chemosphere 1981, 8:833-846.
38. Ma Y-G, Lei YD, Xiao H, Wania F, Wang W-H: Critical review and recommended values for the physical-chemical property data of 15 polycyclic aromatic hydrocarbons at 25°C. J Chem Eng Data 2010, 55:819-825.

CHAPTER 8

Mercury in Marine and Oceanic Waters—A Review

Barbara Gworek, Olga Bemowska-Kałabun, Marta Kijeńska, Justyna Wrzosek-Jakubowsk

Institute of Environmental Protection–National Research Institute,

Warsaw, Poland

ABSTRACT

Mercury contamination in water has been an issue to the environment and human health. In this article, mercury in marine and oceanic waters has been reviewed. In the aquatic environment, mercury occurs in many forms, which depend on the oxidation-reduction conditions. These forms have been briefly described in this article. Mercury concentrations in marine waters in the different parts of the world have been presented. In the relevant literature, two models describing the fate and behavior of mercury in saltwater reservoirs have been presented, a conceptual model which treats all the oceans as one ocean and the "ocean margin" model, providing that the ocean margins manifested themselves as the convergence of continents and oceans, covering such geological features, such as estuaries, inland seas, and the continental shelf. These two conceptual models have been summarized in the text. The mercury content in benthic sediments usually reflects is level in the water reservoir, particularly in reservoirs situated in contaminated areas (mines, metallurgical plants, chemically protected crops). The concentrations of mercury and its compounds determined in the sediments in surface waters in the different parts of the world have been presented. Due to the fact that the pollution caused by mercury is a serious threat for the marine environment, the short paragraph about mercury bioaccumulation in aquatic organisms has been included. The cited data demonstrated a large scatter of mercury contents both between the fish species and the water areas. Mathematical models, valuable tools which provide information about the possible responses of ecosystems, developed to simulate mercury emissions, both at a small scale, for local water reservoirs, and at a global scale, as well as to model mercury bioaccumulation in the chain web of aquatic systems have been described.

Keywords Mercury Marine waters Benthic sediments Pollution Environment

1 INTRODUCTION

Mercury is an element the emissions of which from its natural sources exceed its anthropogenic emissions. Therefore, it is difficult to estimate mercury emission levels from its natural sources. This is related, e.g., to the absence of sufficient data on this issue and the complexity of processes affecting natural mercury emissions (to mention only geological processes which are characterized by enormous spatial and temporal variabilities). In general, natural mercury emissions into the environment can be divided into original emissions, e.g., volcanic emissions, and secondary reemissions of natural origin. Natural mercury emissions can also be divided into emissions from oceans which are estimated at 2680 Mg/year and those of terrestrial origin, estimated at 1850 Mg/year [Mason 2009; Pirrone et al. 2010; Pirrone and Cinnirella 2012) with volcanic emissions to the atmosphere estimated as 94–112 Mg/year (Nriagu and Becker 2003; Nriagu and Becker 2004). Natural mercury emission processes mainly include

- Degassing from mercury deposits;
- Degassing from aquatic and terrestrial systems (through the reduction of Hg^{2+} to Hg^0);
- Geological activity—volcanic and geothermal processes (underwater exhalations from geothermal vents);
- Biomass burning—e.g., fires of forests and steppes;
- Erosion of mercury-containing minerals; and
- Plant growth.

Mercury emissions into the air from sources of natural origin are an important element of global mercury flows, as it is the largest element of the global mercury cycle. Many research teams have estimated these emissions, using different numerical models in addition to direct measurements. Their balance is hampered by the share of reemissions from anthropogenic sources which is difficult to estimate (Mason 2009; Pirrone et al. 2010; Pirrone and Cinnirella 2012).

In the aquatic environment, mercury occurs in many forms which depend on the oxidation-reduction conditions. The forms $HgCl_4^{2-}$ and $HgOH^-$ dominate in the good oxidation conditions, whereas sulfur-related forms (HgS^{2-} and CH_3HgS^-) prevail in the reduction conditions.

In the intermediate conditions, the alkyl forms of mercury, MeHgCl and EtHgCl, can most often be found (Kabata-Pendias and Mukherjee 2007; Kabata-Pendias 2011; Tyler 1992). Soluble forms of mercury, such as $[HgOH]^+$, $[HgCl]^+$, $[HgCl_2]$, $[HgCl_3]^-$, $[HgCl]^{2-}$, and $[HgS_2]^{2-}$, can often be encountered. Higher concentrations of Cl^- ions, which form stable complexes with mercury, such as $HgCl_3^-$, $HgCl_2^-$, $HgCl_4^{2-}$, or $HgBrCl^-$, lead to increased dissolution of solid phases of mercury (Grassia and Nettib 2000).

Moreover, mercury can occur in soluble non-ionic organic compounds and other organic and inorganic compounds. Examples of sparingly soluble forms of mercury include CH_3Hg^+ or $Hg(CN)_2$. There are many paths of mercury transport into the aquatic environment. Inorganic forms, such as Hg(II) and methylmercury (MeHg), can be directly introduced into reservoirs through (wet and dry) depositions from the atmosphere (Lina et al. 2006). On the other hand, Hg(II) and MeHg are transported into water reservoirs through surface runoff as well as through leaching from the upper levels of a soil profile to groundwater and, subsequently, to surface waters (Fitzgerald et al. 2007; Morel et al. 1998; US EPA 1997).

Wet and dry atmospheric depositions are the most frequent paths of transport of Hg (THg) into the surface waters of the Arctic and the Antarctic. It is estimated that the total annual quantity of mercury penetrating the atmosphere from natural and anthropogenic mercury emission sources is 5000–6500 t (Berg et al. 2006; Bone et al. 2007; Bookman et al. 2008; Camargo 1993; Cheng and Hu 2010; Jiang et al. 2006; Gbor et al. 2007; Gray and Hines 2006; Lamborg et al. 2002; Mason et al. 1994; Streets et al. 2009), with the emissions from anthropogenic sources estimated at about 2200 t Hg/year (Bergan et al. 1999; Gbor et al. 2007; Pacyna et al. 2006). China is the country with the highest mercury emissions (Wang et al. 2016). Every year, thousands of tons of mercury from air deposition are transported into aquatic ecosystems. Mercury-containing flowing and standing surface waters, such as rivers, streams, and estuaries, are the main paths of mercury transport into marine ecosystems.

In the aquatic environment, mercury undergoes many different chemical and biochemical processes which condition its speciation and transport between the solid and aqueous phases (Fitzgerald et al. 2007; Morel et al. 1998). In the aquatic environment (water, sediments, aquatic fauna, and flora), most mercury occurs in organic and inorganic forms of divalent mercury and Hg(0), as a form of mercury dissolved in the aqueous phase (Ullrich et al. 2001). Mercury adsorption and desorption processes in the aquatic environment play a dominating role in the distribution of different forms of mercury in the particular elements in the aquatic environment. In waters, these processes are also responsible for the course of mercury transport, transformation, and uptake by living organisms, and they also condition the toxicity of this element.

In the aquatic environment, mercury can be adsorbed on sediment particles, thus constituting a substantial mercury resource. In the sediments in water reservoirs, both as a result of chemical reactions and under the impact of biological factors, e.g., those related to the activity of microorganisms, methylmercury, $CH3Hg^+$, and dimethylmercury, $(CH_3)_2Hg$, emerge (Benoit et al. 2003; St. Louis et al. 1994).

MeHg is the most common form of organic mercury in the environment. Methylmercury is a neurodevelopmental toxicant (Obi et al. 2015), and it is also the most toxic form of mercury (Henriques et al. 2015). MeHg and dioxin-like compounds are considered as the most important toxic compounds in the case of large-scale consumers of seafood (Szefer 2013).

Methylation is a result of abiotic and biotic processes, which are affected by such factors as pH, temperature, the presence of sulfates, and the availability of organic carbon.

The impact of mercury in the environment on human health was found for the first time in relation to the Minamata disease in the 1950s, which caused mass-scale poisoning by methylmercury. It had accumulated in aquatic organisms which were subsequently eaten by humans. A similar case of poisoning by mercury accumulated in fish also took place in Sweden (Zaib et al. 2015).

Apart from spectacular cases of poisoning, the presence of mercury in the environment also affects the human population in a more concealed manner. Every year, Trasande et al. (Trasande et al. 2005) found mercury concentrations exceeding 5.8 μg/L—a level related to IQ loss—in blood samples taken from 316,588–637,233 children. Humans are mainly exposed to methylmercury as a result of their consumption of oceanic fish (Drevnick et al. 2015).

One of the key stages of the biogeochemical cycling of mercury in the environment is its biomobilization, which is mediated by microorganisms. Microorganisms mediate in the following four types of mercury transformation: in the reduction of Hg(II) to Hg(0), in the degradation of $CH_3Hg(I)$ and other organic mercury compounds, in the methylation of Hg(II), and in the oxidation of Hg(0) to Hg(II). The transformation processes listed above also unfold in the environment without any involvement of microorganisms. Table 1 shows a summary list of known mercury transformation mechanisms.

Table 1 Mechanisms of mercury transformation processes

Mechanisms	Reduction of Hg(II)	Degradation/demethylation of $CH_3Hg(I)$	Methylation	Oxidation of Hg(0)
Biotically	Enzymatically—mercury reductase	Enzymatically—organomercurial lyase (OL)	Transfer of methyl groups by corrinoid coenzyme (bacterial methyltransferase)	Hydroperoxidase (e.g., catalase)
	Indirectly—reduced metabolites	Oxidative demethylation—anaerobic bacteria	Disturbances in methionine synthesis pathway in fungi	–
Abiotically	Free radicals related to humic substances	Photodegradation	Photochemically—induced by humic and fulvic acids	In the atmosphere via H_2O_2 under low pH conditions
	Disproportionation of Hg(I)	–	Transformation of $CH_3Hg(I)$ to $(CH_3)Hg$ in the presence of H_2S	–

Source: Barkay (2000)

Biological mercury transformations, along with the processes of its chemical transformations, provide the basis for mercury cycling in the biosphere; furthermore, the role played by the biological transformations themselves is not fully known and probably depends on mercury concentrations and the conditions occurring in the environment.

The reduction of Hg^{2+} and the degradation of organic mercury compounds are processes constituting natural detoxification mechanisms, which unfold in bacteria and enable their growth and development in the presence of these toxic compounds. Bacteria that are resilient to the harmful impact of Hg^{+2} are capable of producing mercury, which catalyses the reaction with the involvement of nicotinamide adenine dinucleotide phosphate (NADPH). Due to its low water solubility and high vapor pressure, Hg^0 which emerges as a result quickly escapes into the atmosphere. Some bacteria which are resilient to Hg^{2+} due to the presence of organomercurial lyase (OL) also demonstrate resistance to the toxic impact of organic mercury compounds (Walsh et al. 1988). The resistance to the harmful impact of organic forms of mercury is related to the action of both enzymes, lyase and reductase, while all the bacteria which are not vulnerable to the impact of organic mercury compounds are at the same time resilient to the toxic effect of Hg^{+2} (a wide range of resilience). However, a large part of bacteria do not show the capacity to produce organomercurial lyase; as a result, they are vulnerable to the presence of this form of mercury in the environment (a narrow range of resilience).

Whereas the reactions of the reduction of Hg^{2+} to Hg^0 may unfold without any involvement of microorganisms, in contrast, the process of the degradation of organic forms of mercury without any involvement of microorganisms unfolds extremely slowly and is of no greater significance in the processes of mercury cycling in nature (Barkay et al. 1992). The chemical reactions in which Hg^0 escapes into the atmosphere include the reduction of Hg^{2+} to $2Hg^+$, with the subsequent disproportionation to

Hg^0 and Hg^{2+} (Baltisberger et al. 1979), and the reduction through the interaction of Hg^{2+} with free-radical electrons of humic acids (Alberts et al. 1974).

As shown in Table 1, abiotic sources of methylmercury include transformations of Hg^{2+} as a result of its reactions with fulvic and humic acids (Nagase 1984) and the transfer of methyl groups from the soluble fraction of organic matter to Hg^{2+}. These reactions are initiated phototochemically in the presence of sulfur (Akagi et al. 1974).

The intensity of methylation grows as the temperature and the content of Hg(II) in the environment increase, while the optimum level is reached at pH of less than 5.

The methylation reaction is also stimulated by the presence of other metals which play the role of catalysts. Photochemical methylation takes place in the presence of chemical donors of methyl groups, such as methanol or ethanol; moreover, the greatest efficiency is achieved in the presence of acetic acid. The anaerobic conditions and the presence of colored compounds as photosensitizers also enhance the efficiency of this process (Barkay 2000).

2 MERCURY IN MARINE AND OCEANIC WATERS

In the biogeochemical mercury cycling, a substantial part of it reaches seas and oceans and its important source is its atmospheric deposition (Bindler 2003; Wang et al. 2004).

In oceanic waters, mercury mainly occurs in the forms of Hg^0, Hg^{2+}, MeHg, and diMeHg and in colloidal form (Morel et al. 1998). In marine waters, mercury forms compounds with chlorine ($HgCl_3^-$ and $HgCl_4^{2-}$) to a greater extent than oxides, as is the case in freshwaters (Mason and Fitzgerald 1993). It has been demonstrated that in saltwaters, Hg^{2+} forms complexes with halides, such as those of chlorine, and these complexes do not undergo the reduction and methylation processes (Gardfeldt et al. 2003; Whalin et al. 2007) as quickly as the other Hg^{2+}compounds do. Dissolved gaseous mercury (DGM) arises as a result of the transformations of both Hg^0 and Me_2Hg (Lamborg et al. 1999; Mason et al. 1995), with the form Hg^0 dominating in the upper layer of the ocean (Table 2) (Gardfeldt et al. 2003; Laurier et al. 2003).

Table 2 Mercury concentrations in marine waters in the different parts of the world, in ng/L

Survey site	THg[a] Seawaters	HgCR[a]	DGM[a]	diMeHg[a]	HgR[a]	MeHg[a]	HgC[a]	Source
Minamata Bay (Japan)	1,600–3,600	–	–	–	–	–	–	Hosohara et al. (1961)
Atlantic Ocean	400–1,600	–	–	–	–	–	–	Aidin'yan and Belavskaya (1963)
Red Sea	700–2,000	–	–	–	–	–	–	Aidin'yan (1962)
Pacific Ocean, Ramapo Deep	80–150	–	–	–	–	–	–	Hamaguchi et al. (1961)
Baltic Sea	0.6 ± 0.2	–	–	–	–	~20 % THg	–	Pempkowiak et al. (1998)
Gulf of Gdańsk	27?–630	–	–	–	–	–	–	BMEPC (1987)
Gulf of Finland	10–140	–	–	–	–	–	–	
Bothnian Bay	2–40	–	–	–	–	–	–	
Gulf of Riga	10–40	–	–	–	–	–	–	
Matsalu Bay	5–130	–	–	–	–	–	–	
North Sea	0.56 ± 0.24	–	–	–	–	–	–	Coquery and Cossa (1995)
North Sea	0.5–200	–	–	–	–	–	–	Schmidt (1991)
Mediterranean Sea (Tunisia)	990–27,060	–	–	–	–	–	–	Nasfi (1995)
Atlantic Ocean (N–E)	2.7	–	–	–	–	–	–	Cossad et al. (1996)
Sepetiba Bay (Brazil)	–	0.12–0.67	0.032–0.092	–	0.11–0.36	<0.02–0.55	–	Marins et al. (2000)
Sepetiba Bay (Brazil)	–	–	<2.6	–	0.2–32.2	0–246	–	Marins et al. (2000)
South China Sea	0.8–2.3	–	–	–	–	0.05–0.22	–	Fu et al. (2010)
Andaman Sea (Thailand)	<10–540	–	–	–	–	–	–	Thongra-Ra and Parkpian (2002)
Yellow Sea	1.34–5.5	–	–	–	–	–	–	Ci et al. (2011)
Wali Estuary	39–430	–	–	–	–	0.046–0.28	–	Wang et al. (2009a)
Pacific Ocean (N)	–	–	–	–	–	173 ± 118[b]	–	Sunderland et al. (2009)
Pacific Ocean (equatorial part)	–	–	–	<10–670[b]	–	<50–500[b]	–	Mason and Fitzgerald (1990, 1993)
Atlantic Ocean (N)	–	–	–	81 ± 68[b]	–	ND[c]	–	Mason et al. 1998
Atlantic Ocean (S and equatorial part)	–	–	–	<10–110[b]	–	25–200[b]	–	Mason and Sullivan (1999)
Mediterranean Sea	–	–	–	3 ± 3[b]	–	280 ± 120[b]	–	Kotnik (2007)
	–	–	–	3 ± 2[b]	–	200 ± 26[b]	–	Horvat et al. (2003)
	–	–	–	<20–290[b]	–	<150[b]	–	Cossa et al. (1997)

– no data or not applicable

[a] *THg* total Hg, *HgCR* total dissolved Hg, *DGM* dissolved gaseous Hg, *HgR* reactive Hg, *diMeHg* dimethylmercury, *MeHg* monomethylmercury, *HgC* Hg in elemental form

[b] Value given in fM

[c] ND below the detection limit

The impact of anthropogenic sources on the contents of mercury and its forms can be seen to a greater extent in the surface layers of the oceans (i.e., in the upper layer of the ocean waters down to a depth of about of 100 m) (Strode et al. 2007). In the near-surface layer, i.e., in the layer between 100 and 1000 m, which is the water-mixing layer, the displacement of mercury of anthropogenic origin is hampered. It depends on many factors, e.g., the local and regional movements of water masses, on the production formation and degradation, as well as the accumulation of intermediate substances (Mason and Sheu 2002).

The particular oceans are different in their total mercury concentrations, with the average concentration of about 1.5 picomoles (pM) (Lamborg et al. 2002). Higher concentrations were recorded in the Mediterranean Sea—2.5 pM (Cossa et al. 1997), and the northern part of the Atlantic—2.0 pM (Mason et al. 1998), whereas a lower concentration was found in the Antarctic Ocean—0.8 pM (Laurier et al. 2004; Sunderland and Mason 2007). Simulations carried out using mathematical models have indicated that the mercury concentrations in most oceans are not balanced with its atmospheric deposit and that they will also systematically grow over the next several dozen years (Sunderland and Mason 2007).

The mercury exchange with the ocean water currents unfolds most quickly in the Atlantic Ocean (17.99–24 Sv), while in the other intervals, it has similar values, falling within the range of 6 to 12.4 Sv.

It is believed that the mercury exchange at the interface between the ocean surface and the atmosphere unfolds relatively quickly. Just as in the freshwater systems, Hg transformation processes in saltwaters are quite complex, too. Hg^{2+} reaches the ocean through its dry and wet depositions, while Hg^0 does so through its dry deposition; in addition, Hg2+ subsequently undergoes biological or photochemical reduction to Hg^0 and absorption on the solid particles of the suspended sediment or methylation (Strode et al. 2007).

It is difficult to determine the intensity of the methylation process in saltwaters, since old data are not reliable given the analytical detection limits of measurement equipment; in consequence, difficulties emerge in respect of the MeHg measurements in these waters. The average levels of MeHg contents determined in oceanic waters fell in the range of 2–35 % (Sunderland et al. 2009). In saltwater systems, mercury methylation may unfold in sediments in the continental shelf areas (Hammerschmidt and Fitzgerald 2006), in estuaries (Gill et al. 1999; Heyes et al. 2006), in a seawater column (Monperrus et al. 2007), or in hydrothermal vents situated in deep ocean layers (Kraepiel et al. 2003).

In the relevant literature, the following two models describing the fate and behavior of mercury in saltwater reservoirs have been presented: a conceptual model which treats all the oceans as one ocean, based on the data and modeling presented by Rolfus and Fitzgerald (1995), and the "ocean margin" model developed by Cossa et al. (1996), providing that the ocean margins manifested themselves as the convergence of continents and oceans, covering such geological features, such as estuaries, inland seas, and the continental shelf. However, it should be borne in mind that these models are very general and only reflect basic knowledge of the mercury paths and transformations in the marine environment.

2.1 The Fate of Mercury in Saltwaters (a Conceptual Model of One Ocean)

Rolfus and Fitzgerald (1995) created a simple ocean model, designed to help determine the fate of mercury in saltwaters and its accumulation in fish and mostly based on the data which these authors had collected earlier. The model provided that the ocean consisted of the following three compartments: the coastal zone, the upwelling zone, and the open ocean. The open ocean zone represents more than 90 % of the total surface area of the oceans, but it is characterized by low fish productivity. The coastal and upwelling zones represent, respectively, about 10 and 0.1 % of the total surface area and generate almost all the fish production (in the model, each such area accounts for almost 50 % of the total fish production). The mercury sources adopted in the model include atmospheric deposition from the atmosphere, the flux from river systems, and the flux from deep ocean waters. On the basis of the dependence observed in the North Atlantic between the increased concentration of reactive mercury in the surface ocean waters and the predicted rate of atmospheric deposition, it is also assumed that its point of entry is the mixed layer of the waters. From this layer, reactive mercury is transported in the form of complexes with suspended particles to the regions and layers of the ocean where its methylation occurs (these areas are naturally poorer in oxygen) and mercury is released from the complexes as they fall into lower regions of the reservoir. This model also assumes that monoMeHg forms in regions which are poor in oxygen, under the thermocline of the open ocean zone and the upwelling zone, from where it is subsequently transported to the mixed layer, to a depth of less than 100 m, where it is incorporated into the lower levels of the trophic chain (Mason and Fitzgerald 1990, 1993, 1996). After its transport from the mixed layer, in the subthermocline water layer, most of reactive mercury is methylated to dimethylmercury (diMeHg), although monomethylmercury can also form directly from reactive mercury. In marine waters, dimethylmercury is unstable and most of it quickly decomposes to form MeHg (Mason and Fitzgerald 1996). Some of MeHg is then converted to Hg^0, which is transported to the surface zones of the ocean. In this layer, as a result of its supersaturation, elemental Hg can be released into the atmosphere. This is the main mechanism of removal of mercury from saltwaters, reaching a level of 1 % per day in the open ocean zone. It is presumed that the transport of reactive mercury from the mixed layer of the ocean is a factor which determines the rate of methylmercury formation (Mason and Fitzgerald 1996).

The reduction processes are related to both abiotic and biotic factors of the marine environment (Mason et al. 1995); specifically, abiotic factors are responsible for about 10–30 % of the reduction. It was demonstrated that the abiotic reduction might unfold under the impact of sunlight. In an experimental aquatic system, they demonstrated that in the conditions of the combination of fulvic and humic acids with a synthetic light source, the reduction of dissolved divalent forms of Hg occurred. In turn, Weber (1993) suggested that the process of abiotic mercury reduction was mediated by methyltin compounds (e.g., $MeSnCl_3$, Me_2SnCl_2, Me_3SnCl) and humic acids. However, for the most part, the reduction of reactive mercury results from the action of microorganisms, e.g., bacteria and cyanobacteria in the mixed layer of the ocean (Mason et al. 1995).

The model described here also assumed that mercury methylation in the coastal compartments (including the estuary regions) occurred in both the sediments and in the water column, in the vicinity of the oxycline (Rolfus and Fitzgerald 1995), from where methylmercury was transported to the mixed layer and incorporated into the lower trophic levels of the marine food web. The total deposition of mercury to marine reservoirs was estimated at 10 Mmol in a year, while the input from rivers and their estuaries was estimated at about 10 % of this value (about 1 Mmol/year); in turn, the input to the upwelling layer from deeper and colder waters was estimated at about 0.5 Mmol/year. Several authors, including Rolfus and Fitzgerald (1995), found that the enhanced mercury deposition resulting from higher anthropogenic emissions caused greater bioaccumulation in the food chain and, primarily, higher mercury contents in marine fish.

2.2 The Fate of Mercury in Saltwaters (the *Ocean Margin* Model)

Estuaries and coastal regions are particularly vulnerable to anthropogenic mercury contamination, since contaminants are both transported to them in river waters and discharged directly to the oceanic waters. In these regions, there is a higher level of reactive forms of mercury and particle-bound mercury released into the atmosphere from local sources of anthropogenic origin. It is well known that reactive Hg deposits faster than elemental Hg and that higher concentrations of oxidants may occur in the atmospheric layer over the coastal waters. On the basis of research results, it has been found that in certain mercury-contaminated estuaries, the main source of their contamination is the direct discharge of mercury into water rather than its atmospheric deposition. On the basis of data collected for the regions mentioned above, Cossa et al. (1996) developed a mass balance model for mercury for these areas and the quantity of THg estimated on the basis of the model was about 3.3 Mmol. In this model, it was assumed that the mercury flux from the river systems to the oceanic margins represented the highest share of the total mercury input to the coastal waters and that annually, about 4.8 Mmol of mercury reached oceanic waters in this way. Cossa et al. (1996) also noted that the concentrations of total mercury were very variable and that, moreover, the highest concentration was found in the waters of rivers flowing through urbanized and industrial areas. More than 90 % of mercury are transported in particle-bound form. Moreover, a substantial part of this form of mercury seems to be unreactive and is deposited in sediments. The authors estimated that the dry deposition of mercury in coastal waters amounted to about 2 Mmol/year and found that a substantial part of mercury was chemically reactive and participated in reactions unfolding in the marine environment.

Atmospheric deposition in remote ocean areas and then its transport via upwelling to these regions (2.5–3.5 Mmol/year) were also considered to be a significant source of mercury in the waters of marine coastal regions.

The impact of anthropogenic sources on the contents of mercury and its forms is more conspicuous in the surface layer of the oceans (i.e., in the upper layer of the ocean waters down to a depth of about of 100 m) (Strode et al. 2007). In the near-surface layer, i.e., in the layer between 100 and 1000 m, which is the water-mixing layer, the displacement of mercury of anthropogenic origin is hampered. It depends on many factors, e.g., the local and regional movements of water masses, on the formation and degradation, and the accumulation of intermediate substances (Mason and Sheu 2002).

It is believed that the process of mercury exchange at the interface between the ocean surface and the atmosphere unfolds relatively quickly. Just as in freshwater systems, Hg transformation processes in saltwaters are also quite complex. Hg^{2+} reaches the ocean through its dry and wet depositions, while Hg^0 does so through its dry deposition; in addition, Hg^{2+}subsequently undergoes biological or photochemical reduction to Hg^0 and absorption on the solid particles of the suspended sediment or methylation (Strode et al. 2007).

It is difficult to determine the intensity of the methylation process in saltwaters, since old data are not reliable given the analytical detection limits of measurement equipment; in consequence, difficulties emerge in respect of the MeHg measurements in these waters. The average levels of MeHg contents determined in oceanic waters fell in the range of 2–35 % (Sunderland et al. 2009).

In saltwater systems, mercury methylation may unfold in sediments in the continental shelf areas (Hammerschmidt and Fitzgerald 2006), in estuaries (Gill et al. 1999; Heyes et al. 2006), in a seawater column (Monperrus et al. 2007), or in hydrothermal vents situated in deep ocean layers (Kraepiel et al. 2003).

The mercury flux from sediments to the waters in coastal waters was regarded as its less significant source.

Taking into account the assumptions listed above and enumerating them in approximate order of significance, it is found that the flux of total mercury in the area of coastal waters includes, *inter alia*,

- The sedimentation of particles, with mercury adsorbed on their surface originating from surface flow in riverine and deep regions;

- The transport from the open ocean waters; and
- The emissions into the atmosphere which are, on a global scale, usually offset by atmospheric deposition, with a regional imbalance resulting from the domination of deposition processes over releases in the northern latitudes and a converse situation in the southern latitudes.

The model in question was also used to describe the fate and behavior of methylmercury in oceanic coastal waters (Cossa et al. 1996). It follows from the assumptions of the model that more than half of MeHg present in oceanic coastal waters originates from the waters of the bottom of the ocean, whereas the other part consists of the product of methylation of reactive mercury in coastal waters and MeHg from other sources. Mercuric ions in the anoxic layers of sediments in marine water reservoirs are transformed to monomethylmercury, primarily through microbial transformations driven by sulfate-reducing bacteria, such as *Desulfovibrio desulfuricans* (Compeau and Bartha 1985), and the organic matter content has a significant impact on the course of the process (Choi and Bartha 1994). Research has demonstrated that the bioavailability of MeHg bound to sediment particles is substantially greater than the bioavailability of divalent inorganic mercury bound to sediment particles. Dissolved species of both methylmercury and divalent inorganic mercury demonstrate greater bioavailability than particle-bound forms (Choi and Bartha 1994).

3 MERCURY IN BENTHIC SEDIMENTS

Both in soil and benthic sediments, mercury is mainly related to the presence of organic matter. In sediments, methylmercury which has formed through methylation represents not more than 1.5 % of the total quantity of mercury; this quantity is determined as an equilibrium level between the formation and removal processes. In each aquatic environment, a certain state of equilibrium is established between the reactions of methylation and demethylation, depending on the stability of the particular forms of mercury. Dimethylmercury is unstable in sediments but retains relatively high stability under the impact of such factors as its high sulfate content, salinity, anaerobic conditions, or continuous access to methane (Weber et al. 1998).

When comparing the results of the measurements of mercury contents in samples taken from sediments, it should be taken into consideration that slight changes in the conditions of the storage of the collected samples (temperature, redox conditions, and oxygenation) have a very large effect on changes in the contents of the particular forms of mercury (Horvat and Gibicar 2005).

The mercury content in benthic sediments usually reflects its level in the water reservoir, particularly in reservoirs situated in contaminated areas (mines, metallurgical plants, chemically protected crops). Table 3 shows the concentrations of mercury and its compounds determined in the sediments in surface waters in the different parts of the world. There are substantial differences in mercury concentrations among the sediments originating from different water reservoirs and different regions (Table 3).

Table 3 Mercury concentrations in marine sediments, in mg/kg d.s.

Table 3 Mercury concentrations in marine sediments, in mg/kg d.s.

Location	HgT total mercury	MeHg	Source
Marine sediments			
Baltic Sea	2–340	–	Pempkowiak et al. (1998)
Baltic Sea Proper	100 ± 50	–	Borg and Jonsson (1996)
Baltic Sea (Aland Sea)	180 ± 60	–	
Baltic Sea (Bothnian Sea)	100 ± 30	–	
Bothnian Bay	400 ± 240	–	
Gulf of Puck	0.74–5.7	–	Falandysz et al. (1993)
Gulf of Gdańsk	3.5–160	–	
Gulf of Gdańsk	0.25	–	Szumiło-Pilarska et al. (2016)
Gulf of Puck	2.8–180	–	Boszke et al. (2002)
Denmark Strait	60–220	–	Brzezińska et al. (1984)
Baltic Sea (Bosex Area)	140–190	–	
South Baltic Sea	30 ± 10	–	
Baltic Sea Proper	20–360	–	
Gulf of Gdańsk	310 ± 310	–	
Gulf of Riga	30–790	–	Ojaver (1995)
Mediterranean Sea (Israel)	10–900	–	Herut et al. (1994)
Mediterranean Sea (Italy)	100–5,330	–	Barghigiani and Ristori (1995)
South China Sea (Malaysia)	20–127	0.01–0.053	Kannan and Falandysz (1998)
Victoria Harbour, Hong Kong	47–855	0.1–1.5	Shi et al. (2007)
East China Sea (China)	<0.0005–0.0798	–	Shi et al. (2005)
	0.0041–0.0476	–	Fang and Chen (2010)
	0.042–0.072	–	Fang et al. (2004)
Coastal sediments			
Bay of Fundy, (USA/Canada)	25–514[a]	0.5–7.38	Sunderland et al. (2006)
Hainan Coast	0.02–0.1	–	Qiu et al. (2011)
Southeast China Coast	0.0023–0.9036	–	Ding et al. (2009)
Andaman Sea (Thailand)	0.047–2.135	–	Thongraa-Ra and Parkpian (2002)
Bohai Sea	0.8–25	–	Wang et al. (2009b)

[a] pmol/g d.s.

It is believed that natural THg concentrations fall within the interval from 10 to 200 ng/g d.s. (Fergusson 1990), while the quotient of the THg concentration and methylmercury can be used as a marker to describe the state of contamination of a reservoir (Kannan and Falandysz 1998); moreover, in the uncontaminated regions, this indicator takes values of less than 1. In the benthic sediments of the Gulf of Gdańsk, the determined average concentration of methylmercury was 0.65 ng/g d.s., representing about 0.5 % THg, while the indicator in question fell within the interval from 0.02 to 2.27. The direction of mercury transformations in the anoxic conditions of the lower layers of the water column and benthic sediments is determined by a high affinity of mercuric ions with sulfides.

In the anoxic conditions, the dominant forms of mercury include HgS, HgS_2H_2, HgS_2H^-, and HgS_2^{2-}, as well as CH_3HgS (the most important form of mercury in these conditions). Mercuric sulfide HgS, which is sparingly soluble in the aquatic environment, is mainly deposited in benthic sediments and determines the solubility of Hg(II) in waters with low oxygen content or without oxygen; moreover, it is believed that the deposition of HgS in sediments is correlated with the contents of organic matter and iron oxide ions (Boszke et al. 2002). The solubility of HgS increases with the growing content of sulfide ions, which enable the formation of soluble complexes with mercury. This is of large importance in the anoxic areas of aquatic ecosystems, where there are high concentrations of dissolved mercury. Although mercuric sulfide HgS is strongly bound to sediments, part of it may occur in dissolved form as a result of microbial transformations or under the impact of the aerobic conditions (Gagnon et al. 1997). The redeposition of mercury from benthic sediments to the water column occurs as a result of diffusion or the activity of benthic organisms. Moreover, in the latter case, it is greater by a factor of 2 to 10 than the redeposition as a result of diffusion (Rutgers van der Loeff et al. 1984). It is also believed that the existence of oxidizing conditions in the surface layer of sediments constitutes a geochemical barrier to the diffusion of methylmercury from the near-surface layers of sediments with oxidizing conditions to the water column. The mercury input from deeper sediment layers to the near-surface layers of sediments as a result of diffusion was estimated at 3 % of the total mercury input (Gagnon et al. 1997).

Research carried out in the water-sediment system demonstrated that Hg(II) underwent strong sorption on clay particles and biotic elements, which was much weaker than that on sand particles. It was demonstrated that it also underwent strong complexation on organic matter. Following their absorption on solid elements, Hg(II) and methylmercury settle and accumulate on the surface of the sediment layer, which undergoes dynamic transformations on the boundary of the water column, resuspension and bioturbation. The content of methylmercury in water and sediments is a function of its transformations in the methylation and demethylation processes. In the aquatic environment, methylmercury undergoes degradation under the impact of two microbial transformations and under the impact of light (Barkay et al. 2003).

Equilibrium research demonstrated that the in situ formation of mercury was often its main source in water reservoirs and benthic sediments (Benoit et al. 2003; Gilmour et al. 1998). Despite the contamination of aquatic ecosystems with mercury from anthropogenic sources, its direct increase can be observed in aquatic organisms rather than in these ecosystems. This primarily results from the presence of regulatory mechanisms, such as the activity of methylating microorganisms and the contents of organic matter and sulfur (Grieb et al. 1990). The quantity of accumulated mercury in aquatic organisms is affected by such factors as the level in the trophic chain (the higher level in the trophic chain it is, the more mercury is stored in it), the age of organisms, the microbial activity, the mercury concentration in the sediment layer, the content of the soluble fraction of organic matter, salinity, pH, and the redox potential (Kamman et al. 2005). The benthic sediments of oceans are regarded as the ultimate "recipient," which cumulates highly insoluble forms of HgS (Whalin et al. 2007; Wiener et al. 2003). The first stage related to mercury bioaccumulation in aquatic ecosystems is the transformation from its inorganic form to methylmercury, which occurs in both the aquatic and sedimentation phases. Although the mercury bioaccumulation process prevail over its demethylation, the mechanism of methylmercury synthesis has not been completely identified (Horvat 1996).

Cores of sediments from distant, uncontaminated areas are used to study mercury deposition in historically remote times. The overall background THg level demonstrated by Fuji (1976) is about 50 μg/kg in river sediments, 100–300 μg/kg in lake sediments, and 50–80 μg/kg in marine sediments. The MeHg content in the pool of total mercury (THg) reaches about 10 % (US EPA 1997).

Mercury is a strongly dispersed element. Its contents in sedimentary and magmatic rocks are similar, i.e., about 10 μg/kg. Higher mercury contents can be found in rocks which are rich in organic matter, i.e., in clay rocks. Natural mercury contents in sediments do not exceed 0.05 mg/kg. Due to the poor solubility of mercuric carbonates, phosphates, and sulfides and also due to the formation of bonds and connections with organic compounds, mercury undergoes ionic adsorption by organic and inorganic substances. Mercury adsorption is conditioned by many factors, depending, e.g., on the chemical form of mercury, the quantities and chemical properties of organic and inorganic colloids in sediments, the type of cations in sorption complexes, and the redox potential or the reaction of sediments. An increase in the mercury content in surface sediments is usually related to human activities in a given area, but it is rarely affected by geological factors, such as the presence of underlying mercury-containing rocks. The major mercury sources in sediments are the same as the mercury sources in soils (e.g., atmospheric deposition, pesticides, the application of sewage sludge and municipal waste for soil fertilization, coal combustion, metallurgy, and so-called artisanal and small-scale gold mining (ASGM)) and surface waters from which Hg migrates to sediments (Lacerda and Salomons 1999).

4 MERCURY BIOACCUMULATION IN AQUATIC ORGANISMS

As a result of the intensive fixation of mercury by bottom sediments, there is the risk of its accumulation in aquatic organisms (Wilken and Hintelmann 1991). Surveys on different species of seawater and freshwater fish have indicated that mercury concentrations in their tissues grow as their body mass and age increase (Squadrone et al. 2013).

The cycle of mercury transformations in the aqueous environment is a complex one. Biomethylation, the products of which can easily move between the different elements of the environment, may play an important role in the transfer of mercury. Inorganic compounds of mercury (II) in the aqueous environment undergo biochemical transformations effected by microorganisms in the anaerobic conditions, as a result of which highly toxic methylmercury compounds emerge. Both metallic mercury and the mercury contained in different compounds undergo biomethylation in hardly soluble sediments on the bottom of water reservoirs (Boening 2000; He et al. 2005; Squadrone et al. 2013; Stein et al. 1996; Wang et al. 2004; Winfrey and Rudd 1990).

It is important to investigate the methylmercury absorption and accumulation pathways in the food chain. For this purpose, research is carried out on mercury concentrations in animal tissue and determines the position which these organisms occupy in the food chain—by using the stable nitrogen isotope ^{15}N (the content of which in tissues grows with each successive trophic level occupied by the consumer) (Post 2002; Roach et al. 2013). Short-chain alkyl mercury compounds, including those of metylmercury, demonstrate strong bioaccumulation properties. As a lithophilous compound, methylmercury is also capable of easily penetrating through mucous membranes.

A percentage share of mercury in the form of methylmercury in organisms grows for a higher trophic level. Mostly because of their small body size, short lifetime, and low position in the food chain, aquatic invertebrates accumulate smaller amounts of methylmercury in their organisms compared with those in fish. Wong et al. (1997) demonstrated that an average mercury concentration in invertebrates fell within the range from 0.07 to 0.18 μg/g. Surveys to determine the mercury contents in aquatic organisms have also been carried out on crayfish. They are considered to be scavengers which feed on the lowest trophic levels; as a result of this, the methylmercury contents in their bodies are slow compared with its contents in organisms situated at higher trophic levels.

The pollution caused by mercury is a serious threat for the marine environment. This is both a hygienic and ecotoxic problem. When this element occurs at levels exceeding its natural level in the seas and oceans, it poses a large threat for aquatic organisms.

Mercury concentrations detected in fish tissues vary, depending, e.g., on the fish species or the sea area from which a given fish originates. This is important when consumers choose a product. It is important to compare the place of origin of the fish contained in a fish product with the list of sea areas which are generally recognized to be highly vulnerable to mercury pollution.

Mercury contents in the organisms of sea fish depend not only on their species. The surveys carried out by Monteiro et al. (1996) on eight North Atlantic fish species demonstrated that mercury concentrations in the fish varied depending on the depth where they lived. The average mercury concentrations in the fish species covered by the surveys varied between 57 and 377 ppb in fresh tissue and were four times higher in mesopelagic species (depth of more than 300 m) than those in epipelagic (depth of less than 200 m). The fate of organic Hg species in marine ecosystems is dependent also upon surface water temperature, nutrient supply, and on the abundance of phytoplankton and its species composition (Szefer 2002).

In addition to the depth where sea fish feed, the mercury contents in their bodies are also affected by the fact whether they live in open waters or those situated closer to the coast. The coastal regions of seas and oceans are particularly vulnerable to anthropogenic mercury contamination, since both contaminants from rivers and those discharged directly into the oceans are transported to those areas. An additional source of mercury in these sea areas is the mercury exchange at the interface between the ocean surface and the atmosphere. Table 4 lists the data on mercury contents in fish coming from different sea areas in the world. These data demonstrate a large scatter of mercury contents both between the fish species and the water areas.

Table 4 Mercury contents in various sea fish from different sea areas in the world

tbcolw100ptChemical properties	Elemental Hg		Inorganic Hg^{2+}		MeHg[a]	
Air/water partition coefficient K_{AW}	0.317	Estimated by Clapeyron for Henry's law constant at 25 °C	NA[a]	–	9.00 × 10[illegible]	Recommended literature value
Fraction absorbed on the skin	NR[b]	–	0	EA (2009)	0.1	EA (2009)
Diffusion coefficient in the air, m^2/s	6.34 × 10[illegible]	Estimated by the method according to Heinsohn and Cimbala (2003)	NA	–	8.61×10^{-6}	Estimated by the FSG method
Diffusion coefficient in water, m^2/s	2 × 10[illegible]	Estimated by the method according to Hayduk and Laudie (1974)	NA	–	1.7	Recommended literature value
log Kow—octanol/water partition coefficient (log), dimensionless	0.62	Recommended literature value	NA	–	1.9	Estimated using a non-hydrophobic relation with the octanol/water partition coefficient log Kow
log Koc/w—organic carbon/water partition coefficient (log), cm^3/g	4.16[d]	–	NA	–	251.1	Recommended literature value
Relative molar mass, g/mol	200.59	Recommended literature value	NA	–	NA	–
Soil/water partition coefficient, cm^3/g	NA	–	500	Recommended literature value	1.13 (25 °C)	Recommended literature value
Vapor pressure, Pa	0.07028	Recommended literature value	NA	–	100 (21 °C)	Recommended literature value
Water solubility, mg/L	0.056 (25 °C)	Recommended literature value	74,000 (20 °C)	Recommended literature value	0.5	EA (2009)
Soil-to-dust transport factor	0.5	EA 2009	0.5	EA 2009	1	EA (2009)

[a] *NA* not applicable; the models usually do not require these values for organic/inorganic compounds

[b] *NR* not relevant

[c] The experimentally determined value of K_{AW} for the Cl^- ionic concentration of 0.2×10^{-3} mol

[d] The value estimated from the soil/water partition coefficient for the inorganic components of mercury with the organic carbon content of 0.0348 (equivalent to 6 % of SOM)

5 MATHEMATICAL MODELS

The management strategies intended to control the emissions of mercury of anthropogenic origin require the identification of the response of ecosystems to changes in the atmospheric deposition of this element. Mathematical models are valuable tools which provide information about the possible responses of ecosystems, since the measured data themselves are often insufficient and cannot be extrapolated in order to investigate the environmental impacts in other systems (Knightes et al. 2009).

Environmental models are an important decision-making tool for authorities. The spatial and temporal scales linking the tasks related to the control of the environment and the protection of its quality usually do not enable a comprehensive approach to the problem, and thus, they do not make it possible to understand the relation between economic activity and environmental quality. Moreover, environmental models facilitate the understanding of the key research needs and priorities of the data collection in the future (NRC 2007).

There are many models which are developed to simulate mercury emissions, both at a small scale, for local water reservoirs, and at a global scale, as well as to model mercury bioaccumulation in the chain web of aquatic systems. These models are used, e.g., to investigate the time scale required for the bioaccumulation of a specific mercury content level in the tissues of predators, in response to the differentiated sources of this element in the reservoir. These models are based on data on the mercury concentrations in entire aquatic ecosystems for which simulations are carried out and in the hydrodynamic processes (surface runoff, seepage, leaching, evaporation/evapotranspiration, etc.).

As the appropriate model is designed and developed, its important component are the data on which the model will be based, in particular the data on the dynamics of the processes unfolding in sediments, in the water column, and at the interface between them (Knightes et al. 2009). Table 5 shows the selected parameters used to develop the mathematical models and programs for mercury cycling in the environment. Table 6 presents a summary comparison of certain of the available models used to simulate the fate and behavior of mercury in the aquatic environment.

Table 5 The parameters used in the models of mercury recycling in the environment

tbcolw100ptChemical properties	Elemental Hg		Inorganic Hg^{2+}		$MeHg^{+}$	
Air/water partition coefficient K_{AW}	0.117	Estimated by Clapeyron for Henry's law constant at 25 °C	NA[a]	–	9.00×10^{-6c}	Recommended literature value
Fraction absorbed on the skin	NR[b]	–	0	EA (2009)	0.1	EA (2009)
Diffusion coefficient in the air, m^2/s	6.34×10^{-6}	Estimated by the method according to Heinsohn and Cimbala (2003)	NA	–	8.61×10^{-6}	Estimated by the FSG method
Diffusion coefficient in water, m^2/s	2×10^{-9}	Estimated by the method according to Hayduk and Laudie (1974)	NA	–	1.7	Recommended literature value
log Kow—octanol/water partition coefficient (log), dimensionless	0.62	Recommendal literature value	NA	–	1.9	Estimated using a non-hydrophobic relation with the octanol/water partition coefficient log Kow
log Koc/w—organic carbon/water partition coefficient (log), cm^3/g	4.16[d]	–	NA	–	251.1	Recommended literature value
Relative molar mass, g/mol	200.59	Recommendal literature value	NA	–	NA	–
Soil/water partition coefficient, cm^3/g	NA	–	500	Recommended literature value	1.13 (25 °C)	Recommended literature value
Vapor pressure, Pa	0.07028	Recommendal literature value	NA	–	100 (21 °C)	Recommended literature value
Water solubility, mg/L	0.056 (25 °C)	Recommendal literature value	74,000 (20 °C)	Recommended literature value	0.5	EA (2009)
Soil-to-dust transport factor	0.5	EA 2009	0.5	EA 2009	1	EA (2009)

[a] *NA* not applicable; the models usually do not require these values for organic/inorganic compounds

[b] *NR* not relevant

[c] The experimentally determined value of K_{AW} for the Cl^- ionic concentration of 0.2×10^{-3} mol

[d] The value estimated from the soil/water partition coefficient for the inorganic components of mercury with the organic carbon content of 0.0348 (equivalent to 6 % of SOM)

Table 6 Models simulating mercury transport and transformation in aquatic system

Model	Properties	Survey site	Investigated media	Assessed forms of mercury
Coastal system				
ECoS (the Estuarine Contaminant Simulator model)	Estimation of contamination with Hg	Ria de Aveiro Coastal Lagoon (Portugal)	Suspended sediments, stagnant water	THg
2D STATRIM (the 2D STAtionary TRIeste gulf Mercur, a model with two submodels, 2D MIKE21MT, a model of transport in sediments, and PCFLOW2D-HD, a hydrodynamic model)	Simulations of Hg transport and transformation	The Gulf of Trieste	Suspended sediments, stagnant water, plankton	Hg^{2+}, Hg^{0}, MeHg, THg
Modified PCFLOW 3D (a hydrodynamic model containing a module for transport in sediments)	Simulations of Hg transport and transformation	The Gulf of Trieste	Suspended sediments, stagnant water	Hg^{2+}, Hg^{0}, MeHg, THg
Other				
EMMMA Environmental Mercury Mapping, Modeling, and Analysis	Determination of the characteristics of contamination with Hg, simulations of Hg transport and transformation, simulations of bioconcentration in fish	–	Suspended sediments, water, fish	Hg^{2+}, Hg^{0}, MeHg, THg

Source: Wang et al. (2004)

6 CONCLUSION

The surveys on the mercury contents in oceanic and marine waters carried out in the second half of the twentieth century showed that its levels were comparable to those that now occur in the waters of the global ocean. This phenomenon relates to the fact that the mercury level in the global ocean has not yet reached the phase of a dynamic equilibrium with its level in the atmosphere. It follows from the following three major premises:

- The average residence time of mercury in oceanic waters is 20–30 years, whereas that in the atmosphere varies between 0.8 and 2 years; thus, the mercury discharged into the ocean is removed from there much more slowly than the mercury emitted into the atmosphere;
- Oceanic waters concentrate much larger masses of mercury of natural (geogenic) origin rather than those of anthropogenic origin, which are close to the masses originating from the preindustrial period; and
- A large part of the mercury which is vertically transported from the oceanic depths to the surface layers is transported back to the mixed layer, which is more active biologically.

The premises listed above indicate that an increase in the mercury concentration level in oceanic waters will be very slow and may take hundreds of years, even if an increase in the mercury concentration level in the atmospheric air is not observed (Selin et al. 2010; Mason et al. 2012). In 1990–1996, there was a significant reduction in the quantities of mercury and other dangerous substances discharged into oceanic waters. Since 1997, the quantities of mercury discharged into the oceans have remained at a relatively constant level.

REFERENCES

1. Afonso, C., Lourenço, H. M., Nunes, M. L., & Castro, M. (2007). Contaminant metals in black scabbard fish (*Aphanopus carbo*) caught off Madeira and the Azores. *Food Chemistry, 101*, 120–125.
2. Aidin'yan, N. (1962). Content of mercury in some natural waters. *Chemical Abstracts, 57*, 16336.
3. Aidin'yan, N., & Belavskaya, G. A. (1963). The problem of supergene migration of mercury. *Chemical Abstracts, 59*, 8471–1963.
4. Akagi, H., Takabatake, E., & Fujita, Y. (1974). Photochemical methylation of inorganic mercury in the presence of solid sulfur. *Chemistry Letters, 3*, 761–764.
5. Alberts, J., Schindler, J., Miller, R., & Nutter, D. (1974). Elemental mercury evolution mediated by humic acid. *Science, 184*, 895–897.

6. Anan, Y., Kunito, T., Tanabe, S., Mitrofanov, I., & Aubrey, D. G. (2005). Trace element accumulation in fishes collected from coastal waters of the Caspian Sea. *Marine Pollution Bulletin, 51*, 882–888.
7. Baltisberger, R., Hildebrand, D., Grieble, D., & Ballintine, T. (1979). A study of the disproportionation of mercury (I) induced by gas sparging in acidic aqueous solutions for cold vapor atomic absorption spectrometry. *Analytica Chimica Acta, 111*, 111–122.
8. Barghigiani, C., & Ristori, T. (1995). Preliminary results on the role of rivers in total Hg concentrations in marine sediments and benthic organisms of coastal area of Italy. *Water, Air, and Soil Pollution, 80*, 1017–1020.
9. Barkay, T. (2000). Mercury cycle. In J. Lederberg (Ed.), *Encylopedia of microbiology* (pp. 171–181). San Diego: Academic Press.
10. Barkay, T., Turner, R., Saouter, E., & Horn, J. (1992). Mercury biotransformations and their potential for remediation of mercury contamination. *Biodegradation, 3*, 147–159.
11. Barkay, T., Miller, S. M., & Summers, A. O. (2003). Bacterial mercury resistance from atoms to ecosystems. *FEMS Microbiology Review, 27*, 355–384.
12. Benoit, J. M., Gilmour, C. C., Heyes, A., Mason, R. P., & Miller, C. L. (2003). Geochemical and biological controls over methylmercury production and degradation in aquatic systems. In Y. Cai & O. C. Braids (Eds.), *Biochemistry of environmentally important trace elements* (pp. 262e–297e). Washington, DC: American Chemical Society.
13. Berg, T., Fjeld, E., & Steinnes, E. (2006). Atmospheric mercury in Norway: contributions from different sources. *Science of the Total Environment, 368*, 3–9.
14. Bergan, T., Gallardo, L., & Rodhe, H. (1999). Mercury in the global troposphere: a three-dimensional model study. *Atmospheric Environment, 33*, 1575–1585.
15. Bilandžić, N., Dokić, M., & Sedak, M. (2011). Metal content determination in four fish species from the Adriatic Sea. *Food Chemistry, 124*, 1005–1010.
16. Bindler, G. R. (2003). Estimating the natural background atmospheric deposition rate of mercury utilizing ombotrophic bogs in southern Sweden. *Environmental Science and Technology, 37*, 40–46.
17. BMEPC—Baltic Marine Environment Protection Commission. (1987). *Progress reports on cadmium, mercury, cooper and zinc.* Baltic Sea Environment Proceedings, 24. Helsinki Commission.
18. Boening, D. W. (2000). Ecological effects, transport, and fate of mercury: a general review. *Chemosphere, 40*, 1335–1351.
19. Bone, S. E., Charette, M. A., Lamborg, C. H., & Gonneea, M. E. (2007). Has submarine groundwater discharge been overlooked as a source of mercury to coastal waters. *Environmental Science and Technology, 41*, 3090–3095.
20. Bookman, R., Driscoll, C. T., Engstrom, D. R., & Effler, S. W. (2008). Local to regional emission sources affecting mercury fluxes to New York lakes. *Atmospheric Environment, 42*, 6088–6097.
21. Borg, H., & Jonsson, P. (1996). Large-scale metal distribution in Baltic Sea sediments. *Marine Pollution Bulletin, 32*, 8–21.
22. Boszke, L., Głosińska, G., & Siepak, J. (2002). Some aspects of speciation of mercury in a water environment. *Polish Journal of Environmental Studies, 11*(4), 285–298.
23. Brzezińska, A., Trzosińska, A., Żmijewska, W., & Wodkiewicz, L. (1984). Trace metals in suspended matter and surficial bottom sediments from the Southern Baltic. *Oceanologia, 18*, 79–94.
24. Camargo, J. A. (1993). Which source of mercury pollution. *Nature, 365*, 302.
25. Cheng, H., & Hu, Y. (2010). China needs to control mercury emissions from municipal solid waste (MSW) incineration. *Environmental Science and Technology, 44*, 7994–7995.
26. Choi, S. C., & Bartha, R. (1994). Environmental factors affecting mercury methylation in estuarine sediments. *Bulletin of Environmental Contamination and Toxicology, 53*, 805–812.

27. Ci, Z. J., Zhang, X. S., Wang, Z. W., & Niu, Z. C. (2011). Phase speciation of mercury (Hg) in coastal water of the Yellow Sea, China. *Marine Chemistry, 126*, 250–255.
28. Compeau, G. C., & Bartha, R. (1985). Sulfate reducing bacteria: principal methylators of mercury in anoxic estuarine sediment. *Applied and Environmental Microbiology, 50*, 489–502.
29. Coquery, M., & Cossa, D. (1995). Mercury speciation in surface waters of the North Sea. *Netherlands Journal of Sea Research, 34*, 245–257.
30. Cossa, D., Coquery, M., Gobeil, C., & Martin, J. (1996). Mercury fluxes at the ocean margins. In R. Baeyens, W. Ebinghaus, & O. Vasiliev (Eds.), *Global and regional mercury cycles: sources, fluxes and mass balances* (pp. 229–248). The Netherlands: Kluwer Academic Publishers.
31. Cossa, D., Martin, J. M., Takayanagi, K., & Sanjuan, J. (1997). The distribution and cycling of mercury species in the western Mediterranean. *Deep-Sea Research Part II: Topical Studies in Oceanography, 44*, 721–740.
32. Ding, Z. H., Liu, J. L., Li, L. Q., Lin, H. N., Wu, H., & Hu, Z. Z. (2009). Distribution and speciation of mercury in surficial sediments from main mangrove wetlands in China. *Marine Pollution Bulletin, 58*, 1319–1325.
33. Drevnick, P. E., Lamborg, C. H., & Horgan, M. J. (2015). Increase in mercury in Pacific yellowfin tuna. *Environmental Toxicology and Chemistry, 34*(4), 931–934.
34. EA Environment Agency. (2009). *Updated technical background to the CLEA model. Science Report SC050021/SR3.* Bristol: Environment Agency.
35. Falandysz, J., Dubrawski, R., & Bolalek, J. (1993). Zawartosc rtęci w osadach dennych z Zatoki Puckiej i Gdanskiej (Mercury content in the benthic sediments in the Gulfs of Puck and Gdańsk—in Polish). *Bromatologia i Chemia Toksykologiczna, 26*, 29–32.
36. Fang, T., & Chen, R. (2010). Mercury contamination and accumulation in sediments of the East China Sea. *Journal of Environmental Sciences (China), 22*, 1164–1170.
37. Fang, J., Wang, K. X., Tang, J. L., Wang, Y. M., Ren, S. J., Wu, H. Y., & Wang, J. (2004). Copper, lead, zinc, cadmium, mercury, and arsenic in marine products of commerce from Zhejiang coastal area, China, May 1998. *Bulletin of Environmental Contamination and Toxicology, 73*, 583–590.
38. Fergusson, J. E. (1990). *The heavy elements—chemistry, environmental impact and health effects.* Oxford: Pergamon Press.
39. Fitzgerald, W. F., Lamborg, C. H., & Hammerschmidt, C. R. (2007). Marine biogeochemical cycling of mercury. *Chemical Reviews, 107*, 641–662.
40. Fuji, M. (1976). Mercury. In S. Kitamura, M. Kondo, Y. Takizawa, M. Fujii, & M. Fujiki (Eds.), *Mercury distribution in lithosphere and atmosphere*. Tokyo: Kodansha Scientific.
41. Fu, X. W., Feng, X. B., Zhang, G., Xu, W. H., Li, X. D., Yao, H., Liang, P., Li, J., Sommar, J., Yin, R. S., & Liu, N. (2010). Mercury in the marine boundary layer and seawater of the South China Sea: concentrations, sea/air flux, and implication for land outflow. *Journal of Geophysical Research-Atmospheres, 115*, 1–11.
42. Gagnon, C., Pelletier, E., & Mucci, A. (1997). Behavior of anthropogenic mercury in coastal marine sediments. *Marine Chemistry, 59*, 159–176.
43. Gardfeldt, K., Sommar, J., Ferrara, R., Ceccarini, C., Lanzillotta, E., Munthe, J., Wangberg, I., Lindqvist, O., Pirrone, N., Sprovieri, F., Pesenti, E., & Stromberg, D. (2003). Evasion of mercury from coastal and open waters of the Atlantic Ocean and the Mediterranean Sea. *Atmospheric Environment, 37*, S73–S84.
44. Gbor, P. K., Wen, D., Meng, F., Yang, F., & Sloan, J. J. (2007). Modeling of mercury emission, transport and deposition in North America. *Atmospheric Environment, 41*, 1135–1149.
45. Gill, G. A., Bloom, N. S., Cappellino, S., Driscoll, C. T., Dobbs, C., et al. (1999). Sediment-water fluxes of mercury in Lavaca Bay, Texas. *Environmental Science and Technology, 33*, 663–669.
46. Gilmour, C. C., Reidel, G. S., Ederington, M. C., Bell, J. T., Benoit, J. M., Gill, G. A., & Stordall, M. C. (1998). Methylmercury concentrations and production rates across a trophic gradient in the northern Everglades. *Biogeochemistry, 40*, 327–345.

47. Grassia, S., & Nettib, R. (2000). Sea water intrusion and mercury pollution of some coastal aquifers in the province of Grosseto (southern Tuscany, Italy). *Journal of Hydrology, 237*, 198–211.
48. Gray, J. E., & Hines, M. E. (2006). Mercury: distribution, transport, and geochemical and microbial transformations from natural and anthropogenic sources. *Applied Geochemistry, 21*, 1819–1820.
49. Grieb, T. M., Driscoll, C. T., Gloss, S. P., Schofield, C. L., Bowie, G. L., & Porcella, D. B. (1990). Factors affecting mercury accumulation in fish in the Upper Michigan Peninsula. *Environmental Toxicology and Chemistry, 9*, 919–930.
50. Hamaguchi, H., Kuroda, R., & Hosohara. (1961). Photometric determination of traces of mercury in seawater. *Chemistry Abstracts, 55*, 15222.
51. Hammerschmidt, C. R., & Fitzgerald, W. F. (2006). Methylmercury cycling in sediments on the continental shelf of southern New England. *Geochimica et Cosmochimica Acta, 70*, 918–930.
52. Hayduk, W., & Laudie, H. (1974). Prediction of diffusion coefficients for non-electrolysis in dilute aqueous solutions. *Journal of the American Institute of Chemical Engineers, 20*, 611–615.
53. He, Z. L., Yang, X. E., & Stoffella, P. J. (2005). Trace elements in agroecosystems and impacts on the environment. *Journal of Trace Elements in Medicine and Biology, 19*, 125–140.
54. Heinsohn, R. J., & Cimbala, J. M. (2003). *Indoor air quality engineering: environmental health and control of indoor air pollutants*. New York: Marcel Dekker.
55. Henriques, B., Rocha, L. S., Lopes, C. B., Figueira, P., Monteiro, R. J. R., Duarte, A. C., Pardal, M. A., & Pereira, E. (2015). Study on bioaccumulation and biosorption of mercury by living marine macroalgae: prospecting for a new remediation biotechnology applied to saline waters. *Chemical Engineering Journal, 281*, 759–770.
56. Herut, B., Hornung, H., & Kress, N. (1994). Mercury, lead, copper, zinc and iron in shallow sediments of Haifa Bay, Israel. *Fresenius Environmental Bulletin, 3*, 147–151.
57. Heyes, A., Mason, R. P., Kim, E. H., & Sunderland, E. (2006). Mercury methylation in estuaries: insights from using measuring rates using stable mercury isotopes. *Marine Chemistry, 102*, 134–147.
58. Horvat, M. (1996). *Global and regional mercury cycles: sources, fluxes and mass balances*. Netherlands: Kluwer Academic Publishers.
59. Horvat, M., & Gibicar, D. (2005). Speciation of mercury: environment, food, clinical, and occupational health. In R. Cornelis, J. Caruso, H. Crews, & K. Heumann (Eds.), *Handbook of elemental speciation II—species in the environment, food, medicine and occupational health* (pp. 281–305). Chichester: John Wiley & Sons Ltd.
60. Horvat, M., Kotnik, J., Logar, M., Fajon, V., Zvonaric, T., & Pirrone, N. (2003). Speciation of mercury in surface and deep-sea waters in the Mediterranean Sea. *Atmospheric Environment, 37*(S1), 93–108.
61. Hosohara, K., Kozuma, H., Kawasaki, K.-H., & Tsuruta, T. (1961). Total mercury content in sea water. *Chemical Abstracts, 56*, 5766.
62. Jiang, G.-B., Shi, J.-B., & Feng, X.-B. (2006). Mercury pollution in China. *Environmental Science and Technology, 40*, 3672–3678.
63. Julshamn, K., Grøsvik, B. E., Nedreaas, K., & Maage, A. (2006). Mercury concentration in fillets of Greenland halibut (*Reinhardtius hippoglossoides*) caught in the Barents Sea in January 2006. *Science of the Total Environment, 372*, 345–349.
64. Kabata-Pendias, A. (2011). *Trace elements in soils and plants* (4th ed.). Boca Raton: CRC Press - Taylor & Francis Group.
65. Kabata-Pendias, A., & Mukherjee, A. B. (2007). *Trace elements from soil to human*. New York: Springer-Verlag Berlin Heidelberg.
66. Kamman, N., Chalmers, A., Clair, T., Major, A., Moore, R., Norton, S., & Shanley, J. (2005). Factors influencing mercury in freshwater surface sediments of northeastern North America. *Ecotoxicology, 14*, 101–111.
67. Kannan, K., & Falandysz, J. (1998). Speciation and concentrations of mercury in certain coastal marine sediments. *Water, Air, and Soil Pollution, 103*, 129–136.

68. Knightes, C. D., Sunderland, E. M., Barber, M. C., Johnston, J. M., & Ambrose, R. B. (2009). Application of ecosystem-scale fate and bioaccumulation models to predict fish mercury response times to changes in atmospheric deposition. *Environmental Toxicology and Chemistry, 28*(4), 881–893.
69. Kojadinovic, J., Potier, M., LE Corre, M., Cosson, R. P., & Bustamante, P. (2007). Bioaccumulation of trace elements in pelagic fish from the Western Indian Ocean. *Environmental Pollution, 146*, 548–566.
70. Kotnik, J. (2007). Mercury speciation in surface and deep waters of the Mediterranean Sea. *Marine Chemistry, 107*, 13–30.
71. Kraepiel, A. M. L., Keller, K., Chin, H. B., Malcolm, E. G., & Morel, F. M. M. (2003). Sources and variations of mercury in tuna. *Environmental Science and Technology, 37*, 5551–5558.
72. Lacerda, L., Salomons, W. (1999). Mercury contamination from New World gold and silver mine tailings. In: *Mercury contaminated sites*. Springer-Verlag Berlin, 131–142.
73. Lamborg CH, Rolfhus KR, Fitzgerald WF, Kim G (1999) The atmospheric cycling and air-sea exchange of mercury species in the South and equatorial Atlantic Ocean. *Deep-Sea Research Part II: Topical Studies in Oceanography 46*(5), 957–977
74. Lamborg, C. H., Fitzgerald, W. F., Damman, A. W. H., Benoit, J. M., Balcom, P. H., & Engstrom, D. R. (2002). Modern and historic atmospheric mercury fluxes in both hemispheres: global and regional mercury cycling implications. *Global Biogeochemical Cycles, 16*, 1104–1114.
75. Laurier, F. J. G., Mason, R. P., Whalin, L., & Kato, S. (2003). Reactive gaseous mercury formation in the North Pacific Ocean's marine boundary layer: a potential role of halogen chemistry. *Journal of Geophysical Research-Atmospheres, 108*(17), ACH 3-1–ACH 3-12.
76. Laurier, F. J. G., Mason, R. P., Gill, G. A., & Whalin, L. (2004). Mercury distributions in the North Pacific Ocean—20 years of observations. *Marine Chemistry, 90*, 3–19.
77. Lina, C.-J., Pongprueksaa, P., Lindberg, S. E., Pehkoned, S. O., Byune, D., & Jang, C. (2006). Scientific uncertainties in atmospheric mercury models I: model science evaluation. *Atmospheric Environment, 40*, 2911–2928.
78. Marins, R. V., Lacerda, L. D., & Paraquetti, H. H. M. (2000). Dissolved reactive mercury speciation in water profiles from Sepetiba Bay, Brazil. In J. O. Nriagu (Ed.), *11th Annual International Conference on Heavy Metals in the Environment*. Ann Arbor: University of Michigan, School of Public Health.
79. Mason, R. P. (2009). Mercury emissions from natural processes and their importance in the global mercury cycle. In N. I. Pirrone & R. Mason (Eds.), *Mercury fate and transports in the global atmosphere* (pp. 173–191). New York: Springer.
80. Mason, R. P., & Fitzgerald, W. F. (1990). Alkylmercury species in the equatorial Pacific. *Nature, 347*, 457–459.
81. Mason, R. P., & Fitzgerald, W. F. (1993). The distribution and biogeochemical cycling of mercury in the equatorial Pacific Ocean. *Deep-Sea Res Part I, Oceanography Research Papers, 40*, 1897–1924.
82. Mason, R., & Fitzgerald, W. (1996). Sources, sinks and biogeochemical cycling of mercury in the ocean. In W. Baeyens, R. Ebinghaus, & O. Vasiliev (Eds.), *Global and regional mercury cycles: sources, fluxes and mass balances* (pp. 249–272). the Netherlands: Kluwer Academic Publishers.
83. Mason, R. P., & Sheu, G. R. (2002). Role of the ocean in the global mercury cycle. *Global Biogeochemical Cycles, 16*(4), 40-1–40-14.
84. Mason, R. P., & Sullivan, K. A. (1999). The distribution and speciation of mercury in the south and equatorial Atlantic. *Deep Sea Researches, Part II, 46*, 937–956.
85. Mason, R. P., Fitzgerald, W. F., & Morell, F. M. M. (1994). The biogeochemical cycling of mercury: anthropogenic influences. *Geochimica et Cosmochimica Acta, 58*(15), 3194–3198.
86. Mason, R., Morel, F., & Egmond, H. (1995). The role of microorganisms in elemental mercury formation in natural waters. *Water, Air, and Soil Pollution, 80*, 775–787.

87. Mason, R. P., Rolfhus, K., & Fitzgerald, W. F. (1998). Mercury in the North Atlantic. *Marine Chemistry, 61*, 37–53.
88. Mason, R. P., Choi, A. L., Fitzgerald, W. F., Hammerschmidt, C. R., Lamborg, C. H., Soerensen, A. L., & Sunderland, E. M. (2012). Mercury biogeochemical cycling in the ocean and policy implications. *Environmental Research, 119*, 101–117.
89. Meador, J. P., Ernest, D. W., & Kagley, A. N. (2005). A comparison of the non-essential elements cadmium, mercury, and lead found in fish and sediment from Alaska and California. *Science of the Total Environment, 339*, 189–205.
90. Monperrus, M., Tessier, E., Point, D., Vidimova, K., Amouroux, D., et al. (2007). The biogeochemistry of mercury at the sediment–water interface in the Thau Lagoon. 2. Evaluation of mercury methylation potential in both surface sediment and the water column. *Estuarine, Coastal and Shelf Science, 72*, 485–496.
91. Monteiro, L. R., Costa, V., Furness, R. W., & Santos, R. S. (1996). Mercury concentrations in prey fish indicate enhanced bioaccumulation in mesopelagic environments. *Marine Ecology Progress Series, 141*, 21–25.
92. Morel, F., Kraepiel, A. M. L., & Amyot, M. (1998). The chemical cycle and bioaccumulation of mercury. *Annual Review of Ecological Systems, 29*, 543–566.
93. National Research Council (NRC) 2007. Models in environmental regulatory decision making. National Academies Press, Washington.
94. Nagase, H. (1984). Mercury methylation by compounds in humic material. *Science of the Total Environment, 32*, 147–156.
95. Nasfi, F. H. (1995). Total mercury content of sea water on the Tunisian shore. *Fresenius Environmental Bulletin, 4*, 161–168.
96. Nriagu, J., & Becker, C. (2003). Volcanic emissions of mercury to the atmosphere: global and regional inventories. *Science of the Total Environment, 304*, 3–12.
97. Nriagu, J., & Becker, C. (2004). Volcanic emissions of mercury to the atmosphere: global and regional inventories. *Science of the Total Environment, 327*(1–3), 331–333.
98. Obi, E., Okafor, C., Igwebe, A., Ebenebe, J., Johnson Afonne, O., Ifediata, F., Orisakwe, O. E., Nriagu, J. O., & Basu, N. (2015). Elevated prenatal methylmercury exposure in Nigeria: evidence from maternal and cord blood. *Chemosphere, 119*, 485–489.
99. OJAVER, E. *(1995). Mercury content in the ecosystem of the Gulf of Riga. (w:) Ecosystem of the Gulf of Riga between 1920 and 1990*, 80–103, Tallinn.
100. Pacyna, E. G., Pacyna, J. M., Steenhuisen, F., & Wilson, S. (2006). Global anthropogenic mercury emission inventory for 2000. *Atmospheric Environment, 40*, 4048–4063.
101. Pempkowiak, J., Cossa, D., Sikora, A., & Sanjuan, J. (1998). Mercury in water and sediments of the southern Baltic Sea. *Sciences of the Total Environment, 213*, 18192.
102. Pirrone N., Cinnirella S. (2012). Mercury fluxes and emissions from natural processes. In: Pirrone N. (ed.) *Atmospheric pathways, transport and fate of mercury in the global environment, UNEP mercury air transport and fate research partnership area*, F&T: 11–14.
103. Pirrone, N., Cinnirella, S., Feng, X., Finkelman, R. B., Friedli, H. R., Leaner, J., Mason, R., Mukherjee, A. B., Stracher, G. B., Streets, D. G., & Telmer, K. (2010). Global mercury emissions to the atmosphere from anthropogenic and natural sources. *Atmospheric Chemistry and Physics, 10*, 4719–4752. www.atmos-chem-phys-discuss.net/10/4719/2010/.
104. Polak-Juszczak, L. (2009). Temporal trends in the bioaccumulation of trace metals in herring, sprat, and cod from the southern Baltic Sea in the 1994–2003 period. *Chemosphere, 76*, 1334–1339.
105. Post, D. M. (2002). Using stable isotopes to estimate trophic position: models, methods, and assumptions. *Ecology, 83*(3), 703–718.
106. Qiu, Y. W., Yu, K. F., Zhang, G., & Wang, W. X. (2011). Accumulation and partitioning of seven trace metals in mangroves and sediment cores from three estuarine wetlands of Hainan Island, China. *Journal of Hazardous Materials, 190*, 631–638.

107. Roach, K. A., Jacobsen, N. F., Fiorello, C. V., Stronza, A., & Winemiller, K. O. (2013). Gold mining and mercury bioaccumulation in a floodplain lake and main channel of the Tambopata River, Peru. *Journal of Environmental Protection, 4*, 51–60.

108. Rolfus, K., & Fitzgerald, W. (1995). Linkages between atmospheric mercury deposition and the methylmercury content of marine fish. *Water Air and Soil Pollution, 80*, 291–297.

109. Rutgers Van Der Loeff, M. M., Anderson, L. G., Hall, P. O. J., Iverfeldt, A., Josefson, A. B., Sunby, B., & Westerlund, S. F. G. (1984). The asphyxiation technique: an approach to distinguishing between molecular diffusion and biologically mediated transport at the sediment-water interface. *Limnology and Oceanography, 29*, 675–686.

110. Schmidt, D. (1991). Mercury in Baltic and North Sea waters. *Water, Air, and Soil Pollution, 62*, 43–55.

111. Selin, N. E., Sunderland, E. M., Knightes, C. D., & Mason, R. P. (2010). Sources of mercury exposure for US seafood consumers: implications for policy. *Environmental Health Perspectives, 118*, 137–143.

112. Shi, J. B., Liang, L. N., Yuan, C. G., He, B., & Jiang, G. B. (2005). Methylmercury and total mercury in sediments collected from the East China Sea. *Bulletin of Environmental Contamination and Toxicology, 74*, 980–987.

113. Shi, J.-B., Carman, C. M., Tang, C. W. Y., Zhang, G., Rudolf, S. S., & Dong, X. (2007). Spatial and temporal variations of mercury in sediments from Victoria Harbour, Hong Kong. *Marine Pollution Bulletin, 54*, 464–488.

114. Squadrone, S., Prearo, M., Brizio, P., Gavinelli, S., Pellegrino, M., Scanzio, T., Guarise, S., Benedetto, A., & Abete, M. C. (2013). Heavy metals distribution in muscle, liver, kidney and gill of European catfish (*Silurus glanis*) from Italian Rivers. *Chemosphere, 90*, 358–365.

115. St. Louis, V. L., Rudd, J. W. M., Kelly, C. A., Beaty, K. G., Bloom, N. S., & Flett, R. J. (1994). Importance of wetlands as sources of methyl mercury to boreal forest ecosystems. *Canadian Journal of Fisheries and Aquatic Sciences, 51*, 1065e1076.

116. Stein, E. D., Cohen, Y., & Winer, A. M. (1996). Environmental distribution and transformation of mercury compounds. *Critical Reviews in Environmental Science and Technology, 26*, 1–43.

117. Streets, D. G., Zhang, Q., & Wu, Y. (2009). Projections of global mercury emissions in 2050. *Environmental Science and Technology, 43*, 2983–2988.

118. Strode, S., Jaegle, L., Selin, N., Jacob, D., Park, R., Yantosca, R., Mason, R., & Slemr, F. (2007). Air-sea exchange in the global mercury cycle. *Global Biogeochemal Cycles, 21*, GB1017.

119. Sunderland, E. M., & Mason, R. P. (2007). Human impacts on open ocean mercury concentrations. *Global Biogeochemical Cycles, 21*(4), GB4022.

120. Sunderland, E. M., Gobas, F., Branfireun, B. A., & Heyes, A. (2006). Environmental controls on the speciation and distribution of mercury in coastal sediments. *Marine Chemistry, 102*, 111–123.

121. Sunderland, E. M., Krabbenhoft, D. P., Moreau, J. W., Strode, S. A., & Landing, W. M. (2009). Mercury sources, distribution, and bioavailability in the North Pacific Ocean: insights from data and models. *Global Biogeochemical Cycles, 23*, GB2010.

122. Szefer, P. (2002). Metal pollutants and radionuclides in the Baltic Sea—an overview. *Oceanologia, 44*(2), 129–178.

123. Szefer, P. (2013). Safety assessment of seafood with respect to chemical pollutants in European Seas. *International Journal of Oceanography and Hydrobiology, 42*(1), 110–118.

124. Szumiło-Pilarska, E., Grajewska, A., Falkowska, L., Hajdrycha, J., Meissner, W., Frączek, T., Bełdowska, M., & Bzoma, S. (2016). Species differences in total mercury concentration in gulls from the Gulf of Gdansk (Southern Baltic). *Journal of Trace Elements in Medicine and Biology, 33*, 100–109.

125. Thongraa-Ra, W., & Parkpian, P. (2002). Total mercury concentrations in coastal areas of Thailand: a review. *Science Asia, 28*, 301–312.

126. Trasande, L., Landrigan, P. J., & Schechter, C. (2005). Public health and economic consequences of methyl mercury toxicity to the developing brain. *Environmental Health Perspectives, 113*, 590–596.
127. Tuzen, M. (2009). Toxic and essential trace elemental contents in fish species from the Black Sea, Turkey. *Food and Chemical Toxicology, 47*, 1785–1790.
128. Tyler, G. (1992). *Mercury in soil—distribution, speciation and biological effect.* Copenhagen: Nordic Council of Ministers.
129. Ullrich SM, Tanton TW, Abdrashitova SA (2001) Mercury in the aquatic environment: A review of factors affecting methylation (Review). *Critical Reviews in Environmental Science and Technology 31*(3), 241–293
130. US Environment Protection Agency. (1997). *Mercury Study Report to Congress. EPA-452/R-97.* Washington, DC: Office of Research and Development.
131. Voegborlo, R. B., & Akagi, H. (2007). Determination of mercury in fish by cold vapour atomic absorption spectrometry using an automatic mercury analyser. *Food Chemistry, 100*, 853–858.
132. Walsh, C., Distefano, M., Moore, M., Shewchuk, L., & Vetdine, G. (1988). Molecular basis of bacterial resistance to organomercurial and inorganic mercuric salts. *FASEB Journal, 2*, 124–130.
133. Wang, Q., Kim, D., Dionysiou, D. D., Soriala, G. A., & Timberlake, D. (2004). Sources and remediation for mercury contamination in aquatic systems—a literature review. *Environmental Pollution, 131*, 323–336.
134. Wang, Z. W., Zhang, X. S., Xiao, J. S., Zhijia, C., & Yu, P. Z. (2009a). Mercury fluxes and pools in three subtropical forested catchments, southwest China. *Environmental Pollution, 157*, 801–808.
135. Wang, S. F., Jia, Y. F., Wang, S. Y., Wang, X., Wang, H., Zhao, Z. X., & Liu, B. Z. (2009b). Total mercury and monomethylmercury in water, sediments, and hydrophytes from the rivers, estuary, and bay along the Bohai Sea coast, northeastern China. *Applied Geochemistry, 24*, 1702–1711.
136. Wang, C., Wang, Z., Ci, Z., Zhang, X., & Tang, X. (2016). Spatial-temporal distributions of gaseous element mercury and particulate mercury in the Asian marine boundary layer. *Atmospheric Environment, 126*, 107–116.
137. Weber, J. (1993). Review of possible paths for abiotic methylation of mercury(II) in the aquatic environment. *Chemosphere, 26*(11), 2063–2077.
138. Weber, J. H., Evans, R., Jones, S. H., & Hines, M. E. (1998). Conversion of mercury (II) into mercury (0), monomethylmercury cation, and dimethylmercury in saltmarsh sediment slurries. *Chemosphere, 36*, 1669–1687.
139. Whalin, L., Kim, E.-H., & Mason, R. P. (2007). Factors influencing the oxidation, reduction, methylation and demethylation of mercury species in coastal waters. *Marine Chemistry, 107*, 278–294.
140. Wiener, J., Krabbenhoft, D. P., Heinz, G., & Sheuhammer, A. (2003). Ecotoxicology of mercury. In D. J. Hoffman (Ed.), *Handbook of ecotoxicology* (2nd ed., pp. 407–461). Boca Raton: CRC Press.
141. Wilken, R. D., & Hintelmann, H. (1991). Mercury and methylmercury in sediments and suspended particles from the river Elbe, North Germany. *Water, Air, and Soil Pollution, 56*, 427–437.
142. Winfrey, M. R., & Rudd, J. W. M. (1990). Environmental factors affecting the formation of methylmercury in low pH lakes: a review. *Environmental Toxicology and Chemistry, 9*, 853–869.
143. Wong, A. H. K., Mcqueen, D. J., Williams, D. D., & Demers, E. (1997). Transfer of mercury from benthic invertebrates to fish in lakes with contrasting fish community structures. *Canadian Journal of Fisheries and Aquatic Sciences, 54*, 1320–1330.
144. Zaib, M., Athar, M. M., Saeed, A., & Farooq, U. (2015). Electrochemical determination of inorganic mercury and arsenic—a review. *Biosensors and Bioelectronics, 74*, 895–908.

CHAPTER 9

Nitrate Pollution and Preliminary Source Identification of Surface Water in a Semi-Arid River Basin, Using Isotopic and Hydrochemical Approaches

Ying Xue [1], Jinxi Song [1,2,*], Yan Zhang [1], Feihe Kong [1], Ming Wen [1] and Guotao Zhang [1]

[1]*College of Urban and Environmental Sciences, Northwest University, Xi'an 710127, China*

[2]*State Key Laboratory of Soil Erosion and Dryland Farming on the Loess Plateau, Institute of Soil and Water Conservation, CAS & MWR, Yangling 712100, China*

ABSTRACT

Nitrate contamination in rivers has raised widespread concern in the world, particularly in arid/semi-arid river basins lacking qualified water. Understanding the nitrate pollution levels and sources is critical to control the nitrogen input and promote a more sustainable water management in those basins. Water samples were collected from a typical semi-arid river basin, the Weihe River watershed, China, in October 2014. Hydrochemical assessment and nitrogen isotopic measurement were used to determine the level of nitrogen compounds and identify the sources of nitrate contamination. Approximately 32.4% of the water samples exceeded the World Health Organization (WHO) drinking water standard for NO_3^--N. Nitrate pollution in the main stream of the Weihe River was obviously much more serious than in the tributaries. The $\delta^{15}N$-NO_3^- of water samples ranged from +8.3‰ to +27.0‰. No significant effect of denitrification on the shift in nitrogen isotopic values in surface water was observed by high dissolved oxygen (DO) values and linear relationship diagram between NO_3^--N and $\delta^{15}N$-NO_3^-, except in the Weihe River in Huayin County and Shitou River. Analyses of hydrochemistry and isotopic compositions indicate that domestic sewage and agricultural activities are the main sources of nitrate in the river.

Keywords: nitrate pollution; surface water; nitrogen isotope; hydrochemistry; source

1. INTRODUCTION

Due to the need for more food and energy, the applications of fertilizers, lands and fossil fuels are increasing, which will augment the nitrogen pollution in the environment [1,2]. Pollution of nitrate nitrogen (NO_3^--N) is a major problem in the Earth's surface environments, especially in arid/semi-arid regions [3,4,5]. River systems are vital to terrestrial transformation and transportation of nutrients [4]. Most of the surface water pollution is accompanied by excessive chloride, sulfate, nitrate and other pollutants. Nitrate has been one of the dominant forms of increased N loading since the 1970s [6,7,8]. According to the Global Environment Monitoring System database, concentration of nitrate nitrogen (NO_3^--N) of the most rivers in populated regions is seven times higher than the healthy water quality standards of 10 mg/L suggested by the World Health Organization [9]. Since the 1980s, nitrogen fertilizer consumption in China has increased significantly [9]. High concentration of nitrate can result in many environmental and ecological problems, such as blooms of toxic algae, eutrophication of lakes and reservoirs and extinction of species in the river ecosystem [10,11]. In addition, long-term exposure to high nitrate drinking water may increase human health risks [12,13], which may lead to chronic poisoning linked to methemoglobinemia [14,15,16]. Even the existence of nitrite, another form of nitrogen, can cause cancer [1,17]. Therefore, the nitrogen pollution is a severe environmental problem that should be of high concern.

Nitrogen in surface waters has a variety of sources [18], including atmosphere deposition, dust in rainwater, industrial waste water, domestic sewage, urban garbage, nitrogen chemicals, fertilizers, livestock waste, plant humus, etc. [19]. The traditional method for identifying nitrate pollution sources in water bodies combines investigation of land use type of pollution area with analyses of concentrations of nitrogen compounds in water. However, it is difficult to identify the actual sources of nitrate pollution effectively using the traditional method, since the nitrogen compounds are affected by physical, chemical and biological processes simultaneously [3]. Analysis of nitrogen isotopic composition, i.e., $\delta^{15}N$ in nitrate, can address the problems caused by traditional methods and provide a more direct, effective and accurate method to identify the sources of nitrate in water bodies [20,21]. The nitrogen isotope tracing technology has been widely used to identify the nitrate sources by discriminating values of $\delta^{15}N\text{-}NO_3^-$ in different sources [22,23,24]. Nitrates from different nitrogen sources have different nitrogen isotopic compositions, which can be used to identify the sources of nitrate and trace the nitrogen cycling process. Previous studies have shown that the $\delta^{15}N\text{-}NO_3^-$ sourced from chemical fertilizer is the same as $\delta^{15}N$ in nitrogen in the atmosphere about 0.0‰ to +2.0‰. The $\delta^{15}N\text{-}NO_3^-$ sourced from soil nitrification has the same nitrogen isotopic composition as soil organic nitrogen, which ranges from +2.0‰ to +8.0‰ and is invariant during oxidation and nitrification processes. Organic nitrogen is converted to NH_4^+ by oxidation and then generates NO_3^- with similar nitrogen isotope by nitrification. $\delta^{15}N\text{-}NO_3^-$ in nitrogen from animal waste is generally high, within the range of +8.0‰ to +20.0‰ [25,26,27].

Since nitrate pollution has led to quality-induced water scarcity, it is essential to obtain a more complete understanding of nitrate pollution in semi-arid regions, where both the demands and costs for water and ecosystem restoration are high. The urbanization process in the Silk Road region results in a more fragile physical environment, even renewable water resources cannot meet the growing demands [28]. Water environment at the starting point of the Silk Road is a serious issue, which needs to be protected urgently. The Weihe River has a great effect on the construction of the "Silk Road Economic Belt", which is the biggest tributary of the Yellow River, China. Meanwhile, it is crucial for the development and management of water resources in the Yellow River [29,30]. Natural factors and human activities lead to many environmental problems, e.g., the shortage of water resources, aggravation of water pollution and degradation of vegetation over the past 50 years [29]. Due to an annual sewage discharge yield of more than 9.0×10^8 tons, the Weihe River is one of the largest contributors of sewage discharge flux into the Yellow River, leading to the main water pollution in the Yellow River [30]. A wide range of human activities has changed the original chemical composition of surface waters in the study area. A large amount of domestic sewage in cities along the Weihe River (point source pollution) flows back into irrigations, which exacerbates water pollution, especially the nitrogen pollution, and restricts economic development of the basin [31]. Furthermore, nitrate non-point source pollution is also widespread in the Weihe River basin [32]. However, the potential sources of nitrate in the semi-arid basins are not obvious and needs further study.

Accordingly, in this paper, hydrochemical assessment and stable isotopic measurement ($\delta^{15}N$) are used to (1) identify nitrate pollution level in surface waters; (2) evaluate the spatial variation of nitrate pollution; and (3) trace the main sources and transformations of nitrate. Understanding the nitrate pollution levels and pollution sources is of great importance to control the nitrogen input and could provide more sustainable water management methods for the semi-arid river basin area.

2. MATERIALS AND METHODS

2.1. Study Area

The Weihe River, the largest tributary of the Yellow River, is located in a semi-arid area with a temperate continental monsoon climate. It originates from the Niaoshu Mountain in Gansu province and flows across Shaanxi province with stream length of 818 km and drainage area of 1.35×10^5 km^2 in the Yellow River basin. It runs into the Yellow River at Tongguan County. The mean elevation of the Weihe River is 3485 m above sea level, between 104°00′ E~110°20′ E and 33°50′ N~37°18′ N. Limestone is the dominant rock type in the study area. The southern tributaries originate from the Tsinling Mountains and northern tributaries originate from the Loess Plateau (Figure 1). The Loess Plateau in north of the Weihe River is covered by thick Eolian deposits. The Tsinling Mountains in the south of the Weihe River provide an east–west trending land barrier. The average annual precipitation in the river basin is 558–750 mm (increase from north to south). And approximately 60% of the total annual precipitation occurs from May to September. The average annual temperature is 13.3 °C, while the average annual evaporation varies from 800 mm to 1000 mm (increase from west to east) [29]. The drainage area and annual sediment load of the Weihe River account for 17.9% and 2.5% of the total amount of the Yellow River basin.

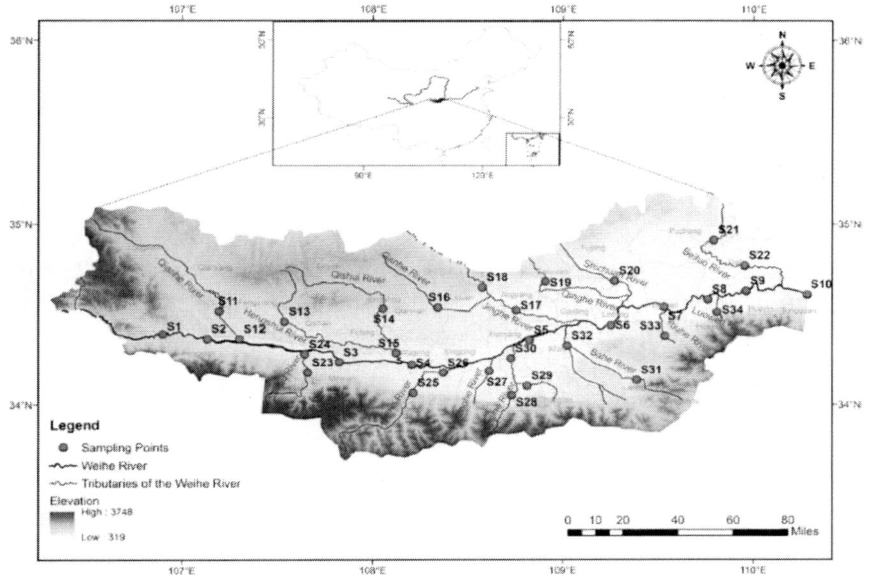

Figure 1. A schematic map showing the location of the study area and water sampling sites.

At present, the Weihe River is facing a predicament of water source shortage and poor water quality [31]. The upstream and lower stream of the Weihe River are divided by the Linjiacun. Water pollution in the region upstream from Linjiacun is less, while there are several middle- and large-sized cities in the lower region where the total agricultural area has reached about 10,000 km^2 [31]. The upstream is not affected by human activities intensively as population in this area is relatively low. Below the Baoji gorge, there are many cities near the main stream including Baoji City, Xianyang City, Xi'an City, and Weinan City, some of which are industrial cities. This portion of the Weihe River is affected by human activities obviously. Furthermore, there are a large number of factories including cotton factories, cement factories and paper mills in the surrounding regions of the Weihe River [33].

2.2. Water Sampling Sites and Measurements

Water samples were collected in the main stream and tributaries of the Weihe River in October 2014. Natural and human factors were considered to deploy the sampling sites including hydrogeological conditions, the principle of water quality control, characterization and

"Water Quality-Technical Regulation on the Design of Sampling Programmes" (HJ 495-2009) [34]. The positions of the sampling sites were identified by GPS (Trimble Juno 3B). The water sites of main stream, northern tributaries and southern tributaries of the Weihe River were designated as S1–S10, S11–S23, S23–S34, which are shown in Figure 1.

In situ water quality parameters including pH, electrical conductivity (EC), dissolved oxygen (DO) and oxidation-reduction potential (ORP) of the water samples were measured with a portable multi-parameter water quality analyzer (HACH HQ40d). Water samples were filtered by 0.45 μm acetate cellulose filter membranes and were stored at 4 °C. Concentrations of SO_4^{2-}, Cl^-, NO_3^-, NO_2^- and NH_4^+ were analyzed by Auto Discrete Analyzers (Clever Chem200, Hamburg, Germany). Concentrations of Ca^{2+}, Mg^{2+}, Na^+ and K^+ were determined using an inductively-coupled plasma optical emission spectrometer (ICP-OES, Optima 5300DV, Shelton, CT, USA). The HCO_3^- and CO_3^{2-} contents were analyzed by acid-base titration. All chemical results were implemented when the charge balance error was within ±0.5.

The water samples with nitrate contents that were higher than the maximum contaminant level (MCL) of 45 mg/L were chosen for nitrogen isotopic analyses. The nitrate was separated by the anion exchange resin method proposed by Silva et al. [35]. The values of $\delta^{15}N\text{-}NO_3^-$ were measured by a Finnigan MAT 253 mass spectrometer with an online Flash Elemental Analyzer (Thermo Fisher Scientific, Bremen, Germany). The values for $\delta^{15}N\text{-}NO_3^-$ are defined as:

$$\delta\ (‰) = \left(R_{\text{sample}}/R_{\text{standard}} - 1\right) \times 1000 \quad (1)$$

where R_{sample} and R_{standard} are the nitrogen isotopic ratios, i.e., $^{15}N/^{14}N$ ratio of water samples and N_2 in the air, respectively. The 1δ analytical precision for $\delta^{15}N\text{-}NO_3^-$ analysis was ±0.3.

3. RESULTS AND DISCUSSION

3.1. Hydrochemistry

The hydrochemistry information of water samples are illustrated in Figure 2. The EC values ranged from 160.1 to 2081 μS/cm, with a mean value of 971.7 μS/cm. The lowest EC value was observed at the site of Fenghe River (S28). All water samples were relatively similar in pH, ranging from 7.2 to 8.6, with a mean value of 7.9, indicating a slight alkaline nature of water samples. The DO ranged from 1.2 to 12.9 mg/L, with a mean value of 9.4 mg/L. The lowest DO content was observed at the site of Shitou River (S24), which might indicate the transformation of nitrogen through denitrification. The values of DO of most samples were higher than 2.0 mg/L, except for the Weihe River in Huayin County (S9) and Shitou River (S24), suggesting that denitrification was not a pivotal transformation process of nitrogen in most area [11]. The ORP ranged from 25.4 to 138.5 mV, with a mean value of 79.9 mV. The Ca^{2+}, Na^{+}, K^{+}, Mg^{2+}, Cl^{-}, SO_4^{2-}, HCO_3^{-} ranged from 21.6 to 159.2 mg/L, 3.7 to 280.8 mg/L, 1.5 to 16.1 mg/L, 3.5 to 71.1 mg/L, 2.2 to 270.3 mg/L, 19.9 to 439.5 mg/L, 48.6 to 413.3 mg/L, respectively. Concentrations of CO_3^{2-} were all below detection limit. The mean values of Ca^{2+}, Na^{+}, K^{+}, Mg^{2+}, Cl^{-}, $SO_4^{2-}$and HCO_3^{-} were 64.1 mg/L, 84.0 mg/L, 6.3 mg/L, 25.9 mg/L, 88.9 mg/L, 152.9 mg/L, and 228.1 mg/L, respectively. Especially, the highest concentrations of Na^{+}, Mg^{2+}, Cl^{-} and SO_4^{2-} were observed at the site of Qinghe River (S20) with low nitrate concentration, indicating the combined effect of geological characteristics and human activities on the water quality [36]. Positive correlations were observed between EC and SO_4^{2-} ($\boldsymbol{r} = 0.648, \boldsymbol{p} < 0.01$), between EC and Cl^{-} ($\boldsymbol{r} = 0.717, \boldsymbol{p} < 0.01$), between Cl^{-} and SO_4^{2-} ($\boldsymbol{r} = 0.916, \boldsymbol{p} < 0.01$). In addition, positive correlations were also detected between Ca^{2+} and NO_3^{-} at the $\boldsymbol{p} < 0.01$ level, between K^{+} and NO_3^{-}, between Cl^{-} and NO_3^{-}, and between EC and NO_3^{-} at $\boldsymbol{p} < 0.05$ level. According to the observed correlation among EC, Cl^{-} and SO_4^{2-}, Cl^{-} and SO_4^{2-} can be regarded as describers of the hydrochemistry of surface waters, which shares a common origin that significantly contributed to the total content of dissolved salts [37].

3.2. Concentrations of NO_2^{-}-N, NO_3^{-}-N and NH_4^{+}-N

Concentrations of NO_2^{-}-N, NO_3^{-}-N, NH_4^{+}-N ranged from BDL (below the detection limit) to 0.015 mg/L, 1.3 to 35.7 mg/L, 0.008 to 8.0 mg/L, respectively (Table 1). The mean values of NO_2^{-}-N, NO_3^{-}-N, NH_4^{+}-N were 0.009 mg/L, 8.6 mg/L, 0.3 mg/L, and the median values were 0.001 mg/L, 7.4 mg/L and 0.05 mg/L, respectively. Concentrations of NH_4^{+}-N were generally low (below the MCL of 1.0 mg/L), except for Luowen River (S34), which was possibly associated with the sewage from paper mills near the river [33]. It also suggests that NH_4^{+}-N in the water sample at the site of Luowen River (S34) could not fully transform into NO_3^{-}-N, due to a large amount of organic nitrogen and ammonium derived from sewage [38]. Concentrations of NO_2^{-}-N in all water samples were below the WHO recommended level of 0.2 mg/L. The concentration of NO_3^{-}-N exceeded the MCL (10 mg/L) accounted for 32.4%, which was much higher than the over standard rate of NH_4^{+}-N (2.9%) and NO_2^{-}-N (0). Consequently, nitrate was a major water pollutant in the semi-arid river basin. It is critical to understand the dynamics of NO_3^--N through further analyses of spatial distribution and sources of nitrate.

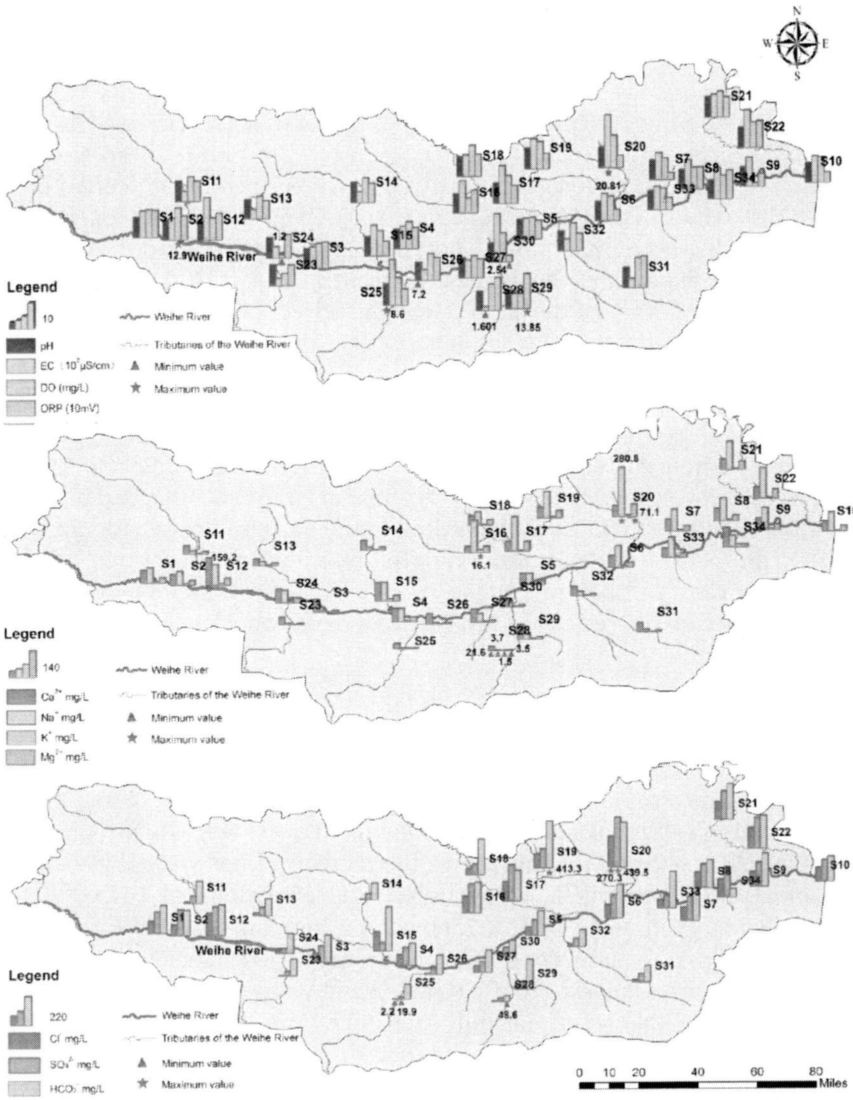

Figure 2. Analyses of chemical parameters in the study area.

Table 1. Statistics of three different forms of nitrogen of all the water samples.

	Mean (mg/L)	Maximum (mg/L)	Minimum (mg/L)	Median (mg/L)	MCL [1] (mg/L)	Exceeded Rate (%)
NO_2^--N	0.009	0.015	BDL [2]	0.01	0.2	0
NO_3^--N	8.6	35.7	1.3	7.4	10	32.4
NH_4^+-N	0.3	8.0	0.008	0.05	1.0	2.9

Note: [1] The maximum contaminant levels (MCL) drinking water standard suggested by the World Health Organization (WHO); [2] BDL stands for below detection limit of 0.0018 mg/L of NO_2^--N.

3.3. Spatial Distribution of Nitrate and Its Controlling Factors

Referring to the WHO drinking water standard and Chinese drinking water standards enacted since 1986, four levels of nitrate concentration (level I of 0–10 mg/L for background level water, level II of 10–45 mg/L for water unpolluted but in a critical condition, level III of 45–90 mg/L for slight polluted water and level IV of the concentration above 90 mg/L for severely polluted water) were generated to evaluate nitrate pollution (Figure 3) [12]. About 8.8%, 58.8%, 29.4%, 3.0% of water samples were within level I, level II, level III and level IV, respectively. According to the spatial distribution of NO_3^-across the semi-arid river basin (Figure 3), nitrate contents in the main stream were obviously greater than those in the tributaries (Figure 3). In addition, nitrate contents of the rivers in mountain areas were lower than in other areas (Figure 3). The mean values of nitrate in the main stream of the Weihe River, northern tributaries and southern tributaries were 10.6 mg/L, 7.9 mg/L and 7.7 mg/L, respectively. Results indicate that the main stream of the river is polluted most seriously, due to the import of waters in tributaries. Approximately 70.0% of the water samples in the main stream were within level III and level IV. Waters in northern tributaries and southern tributaries were less polluted. About 1.7% of water samples were within level III and level IV, both in the northern tributaries and southern tributaries. In addition, southern tributaries near the Tsinling Mountains were slightly less polluted than northern tributaries. Furthermore, concentrations of nitrate in semi-arid regions are influenced by frequent and severe droughts and infrequent but vital floods [39]. The nitrate contents in the river increased gradually from the site of Weihe River in Fengxian County (S2). Around the site of Qianhe River (S12), the main stream flows across the urban area, where the highest content of nitrate and low NH_4^+ concentration existed. The rapid increase of nitrate content here suggests that there is a combination of various pollution sources of nitrate, which is also affected by nitrification processes [38]. The nitrate concentration decreased within the Weihe River in Wugong County (S4), due to the fact that the lower nitrate concentration level in the inflow tributaries (Heihe River, Hengshui River and Qishui River) dilutes the nitrate content in the main stream of the river. Nitrate content decreased slightly from the site of the Weihe River in Xi'an City (S5) to Weinan City (S7). In the sample of the Weihe River in Huaxian County (S8), nitrate content increased again. It was mainly sourced from agricultural activities and industrial wastewater. Pesticide and chemical fertilizer containing nitrate and ammonium nitrogen washed into streams by rainfall [40]. A high spatial variability of nitrate content occurred in some stream waters on the basis of catchment characteristics [41]. Results also indicate that nitrate contamination in the study area has distinct regional differences.

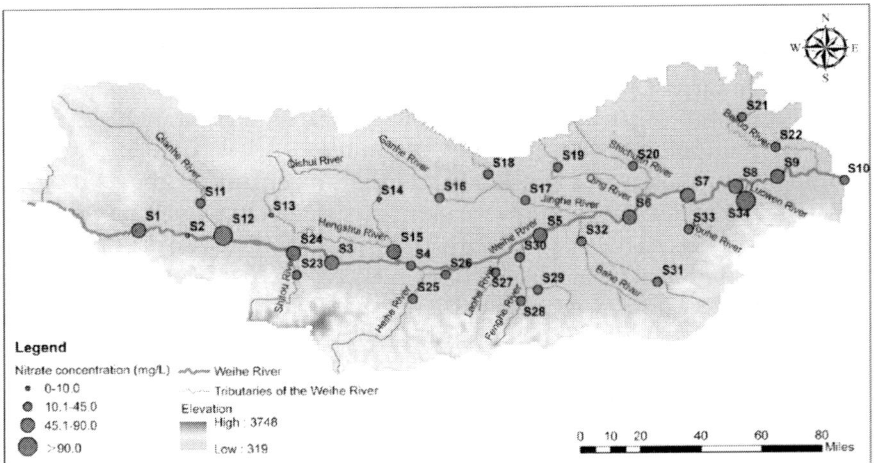

Figure 3. Spatial distribution of nitrate in the rivers.

The above analyses suggest that nitrate pollution in the river basin might be mainly sourced from the use of chemical fertilizer and pesticide in the agricultural areas, domestic sewage and industrial wastewater discharge. Fertilizer is one of the main sources of nitrate pollution in the studied waters, which has a great impact on the water quality in the study area [3,42]. Furthermore, about 5%–10% of applied fertilizer can enter into groundwater [43]. The process of fertilizer application may lead to the cropland being the main contributor to nitrate pollution in groundwater [44,45]. In addition, it is common that polluted water including domestic sewage and industrial wastewater is discharged into the river without any treatment in the study area, which also causes the increase of nitrate content in the study area.

3.4. Sources and Transformations of Nitrate

As shown in Figure 4, the concentration of Cl^- increased with the concentration of Na^+. The slope of the correlation was close to 0.81. Cl^- can be derived from chemical fertilizers, sewage and animal waste. The Cl^- concentration can provide significant information to identify the different input sources [2]. Results suggest that nitrate in the surface water might be also contributed from these sources of Cl^-. Correlation between SO_4^{2-} with Ca^{2+} was not strong. Basically, the concentration of SO_4^{2-} increased with the concentration of Ca^{2+}, which suggested that the input of exogenous substance affected the concentration of SO_4^{2-}. Ca^{2+} might originate from the fertilizer ($Ca(H_2PO_4)_2{\cdot}H_2O$) [46], despite the dissolution of carbonate rocks may also contribute. The SO_4^{2-} in the study area was mainly from agricultural fertilizers. Therefore, one of the main sources of nitrate in the basin was chemical fertilizers.

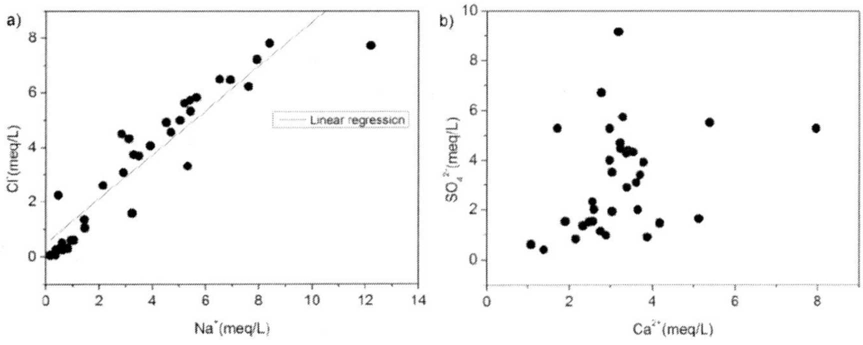

Figure 4. Relationship between Na^+ and Cl^-, SO_4^{2-} and Ca^{2+} concentrations of surface waters.

Generally, there are no apparent correlations between $\delta^{15}N\text{-}NO_3^-$ versus NO_3^- in the basin (Figure 5), which indicates no simple mixing or only one single biological process is responsible for the shifting in the nitrogen isotopic composition of nitrate. In addition, the high DO values and the linear relationship diagram between NO_3^--N and $\delta^{15}N\text{-}NO_3^-$ suggest that denitrification had no significant effect on the shift in nitrogen isotopic values in most surface waters in the basin, except in the water samples at the site of Weihe River in Huayin County (S9) and Shitou River (S24). The $\delta^{15}N\text{-}NO_3^-$ of water samples S9, S24 were high with low concentrations of nitrate (Figure 5) and DO, which means denitrification might have occurred [38].

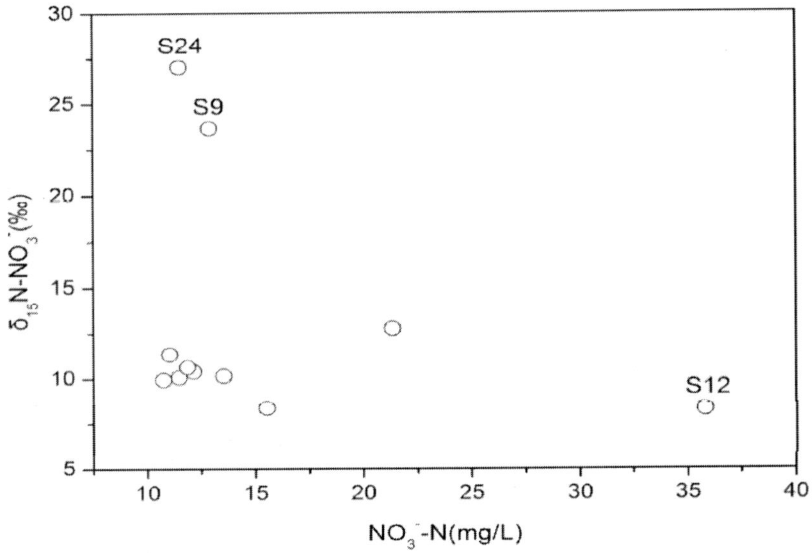

Figure 5. Relationship between NO_3^--N and $\delta^{15}N$-NO_3^- in surface waters.

In the study, 32.4% of water samples with nitrate contents that were higher than the MCL of 45 mg/L were chosen for nitrogen isotopic analyses. The nitrogen isotopic compositions of NO_3^- ranged from +8.3‰ to +27.0‰, which were higher than those results in other rivers, such as in Jinghe River (+3.5‰–+10.3‰) [47], Chanhe River (+1.3‰–+9.2‰) and Laohe River (+2.9‰–+7.4‰) [33]. The mean value of $\delta^{15}N$-NO_3^- was +13.0‰ in the study area. The lowest value of $\delta^{15}N$-$NO_3^-$was +8.3‰ with high concentration of NO_3^- in Qianhe River (S12) (Figure 5), which might be influenced directly by sewage effluent [38]. Domestic sewage usually contains manure, which leads to high value of $\delta^{15}N$-NO_3^-, because nitrogen fractionation has occurred in animal waste and the $\delta^{15}N$-NO_3^- is high in the residual ammonium nitrogen by nitration reaction [48]. About 81.8% of $\delta^{15}N$-NO_3^- values fell between +8.0‰ and +20.0‰, while 18.2% of $\delta^{15}N$-NO_3^- values were above +20.0‰ (Figure 6). Nitrogen isotope values of most samples were within +8‰–+13‰, which suggested that the pollution was mainly sourced from animal waste. The findings were similar to the results reported by Urresti et al. [49,50], which revealed the main origin of the dissolved NO_3^- was related to the use of ammonium fertilizers and manure. Meanwhile, Yue et al. [38] also found similar results that domestic sewage and agricultural activities were the two main sources of nitrate in surface waters. The values of $\delta^{15}N$-NO_3^- in the water samples at the site of Weihe River in Huayin County (S9) and Shitou River (S24) were up to +20‰ and nitrate concentrations were low with DO below 2 mg/L (Figure 5), which mean denitrification occurred [4,38]. Microbial denitrification can usually lead to significant change in nitrogen isotope fractionation, which results in the enrichment of heavy nitrogen isotope, i.e., ^{15}N, in waters [38]. In addition, it is hard to accurately identify the two potential nitrogen sources and the extent of the impact of the biological process, which needs additional studies, due to the $\delta^{15}N\text{-}NO_3^-$ value in atmosphere precipitation being close to that in soil organic nitrogen mineralization and biological processes in the basin is complicated [38,48]. Furthermore, the seasonal phenomena (floods, droughts, high industrial productivity periods, etc.) favor different biogeochemical processes, leading to dynamic variations in the inorganic nitrogen levels. Due to the large riverine input changing with the variation in concentrations and discharge rates, nitrate concentrations were greatly affected by riverine input over all seasons [51], which will be analyzed and assessed in further study.

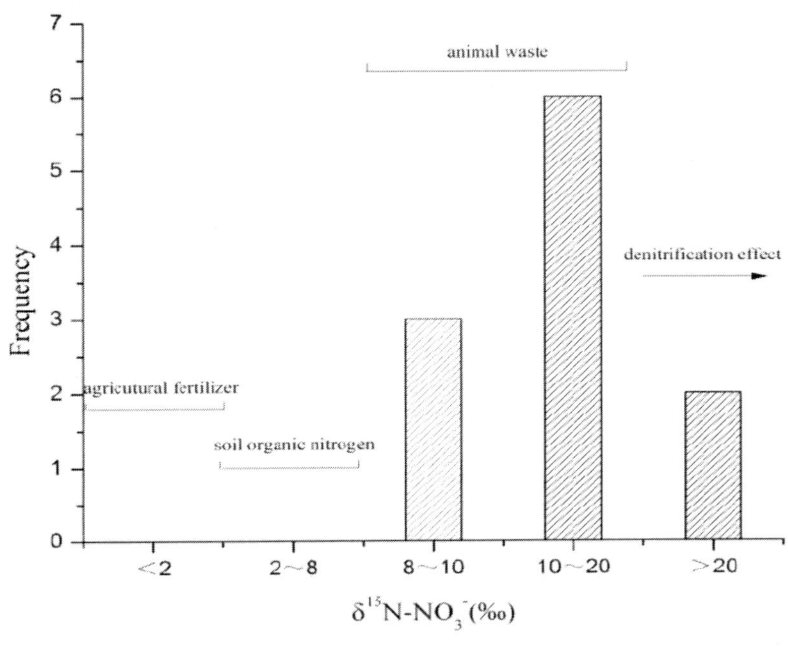

Figure 6. Histograms of $\delta^{15}N\text{-}NO_3^-$ in surface waters of the Weihe River basin.

In view of pollution extent of nitrate caused by intensive agricultural activities and domestic sewage in the semi-arid river basin, attempts should be made as follows: (1) ecological agricultural policy should be carried out to control the non-point source pollution of nitrate, appropriate amount of nitrogen fertilizers should be estimated; (2) the time and position of nitrogen fertilizers application should be strictly controlled, e.g., reducing fertilization frequency and adopting fertigation technology; (3) early winter cover crops should be used to improve the utilization ratio of nitrogen in crop rotation system [3]; (4) direct discharge of domestic sewage should be cut down in some seriously polluted regions, artificial rapid filtration can be used; and (5) chemical and biological processes should be used to reduce nitrate content if necessary.

4. CONCLUSIONS

This study evaluated the sources and variation of nitrate in the Weihe River basin, China. Approximately 32.4% of the water samples did not meet the WHO drinking water standard for nitrate. Results indicate that nitrate contamination of the study area has an apparent spatial pattern. Nitrate contents in the main stream were obviously greater than in the tributaries. In addition, nitrate contents in southern tributaries were slightly lower than in the northern tributaries because southern tributaries near the Tsinling Mountains were less polluted. Nitrate pollution in semi-arid regions was influenced by frequent and severe droughts and infrequent but vital floods. The highest content of nitrate in all surface water samples was observed in Qianhe River, which was related to the acceptance of the domestic sewage. Considering that there are a wide range of agricultural areas along the coast of the river and fertilizers containing nitrate and ammonium nitrogen are washed into streams by rainfall, nitrate is also sourced from agricultural activities. The $\delta^{15}N\text{-}NO_3^-$ of water samples ranged from +8.3‰ to +27.0‰. $\delta^{15}N\text{-}NO_3^-$ in two samples was up to +20‰, indicating that denitrification only occurred in the Weihe River in Huayin County and Shitou River. For the water samples with high nitrate concentrations (above 45 mg/L), domestic sewage is the main source of nitrate. The results of hydrochemistry and isotopic compositions suggest that domestic sewage and agricultural activities are the main sources of nitrate in the semi-arid basin. Additional studies are needed to accurately identify the nitrogen sources and transformations including biological processes. Nitrate pollution in the river basin shows the urgent needs for ecological agricultural policy and reduction of direct sewage discharge in semi-arid river basins.

Acknowledgments: This research was financially supported by the National Natural Science Foundation of China (Grant No. 51379175), the Specialized Research Fund for the Doctoral Program of Higher Education (Grant No. 20136101110001), the Postdoctoral Science Foundation of China (Grant No. 2015M572592), the Scientific Research Plan Projects of Shaanxi Education Department (Grant No. 15JK1762), the Hundred Talents Project of the Chinese Academy of Sciences (Grant No. A315021406), and the Program for Key Science and Technology Innovation Team in Shaanxi Province (Grant No. 2014KCT-27), and the Scientific Research Foundation of Northwest University (Grant No. 14NW01). We are very grateful to three anonymous reviewers and editors for their helpful comments and suggestions, which helped to improve the quality of the manuscript.

REFERENCES

1. Erisman, J.W.; Galloway, J.N.; Seitzinger, S.; Bleeker, A.; Dise, N.B.; Petrescu, A.R.; Leach, A.M.; de Vries, W. Consequences of human modification of the global nitrogen cycle. *Philos. Trans. R. Soc. B Biol. Sci.* **2013**, *368*, 91–97.
2. Lu, L.; Cheng, H.G.; Pu, X.; Liu, X.L.; Cheng, Q.D. Nitrate behaviors and source apportionment in an aquatic system from a watershed with intensive agricultural activities. *Environ. Sci. Process. Impacts* **2015**, *17*, 131–144.
3. Zhang, Y.; Li, F.D.; Zhang, Q.Y.; Li, J.; Liu, Q. Tracing nitrate pollution sources and transformation in surface- and ground-waters using environmental isotopes. *Sci. Total Environ.* **2014**, *490*, 213–222.
4. Li, S.L.; Liu, C.Q.; Li, J.; Liu, X.L.; Chetelat, B.; Wang, B.L.; Wang, F.S. Assessment of the sources of nitrate in the Changjiang River, China using a nitrogen and oxygen isotopic approach. *Environ. Sci. Technol.* **2010**, *44*, 1573–1578.
5. Cook, P.L.M.; Aldridge, K.T.; Lamontagne, S.; Brookes, J.D. Retention of nitrogen, phosphorus and silicon in a large semi-arid riverine lake system. *Biogeochemistry* **2009**, *99*, 49–63.
6. Dise, N.B.; Wright, R.F. Nitrogen leaching from European forests in relation to nitrogen deposition. *For. Ecol. Manag.***1995**, *71*, 153–161.
7. Smith, V.H.; Tilman, G.D.; Nekola, J.C. Eutrophication: Impacts of excess nutrient inputs on freshwater, marine, and terrestrial ecosystems. *Environ. Pollut.* **1999**, *100*, 179–196.
8. Vitousek, P.M.; Aber, J.D.; Howarth, R.W.; Likens, G.E.; Matson, P.A.; Schindler, D.W.; Schlesinger, W.H.; Tilman, D.G. Human alteration of the global nitrogen cycle: Sources and consequences. *Ecol. Appl.* **1997**, *7*, 737–750.
9. He, B.; Kanae, S.; Oki, T.; Hirabayashi, Y.; Yamashiki, Y.; Takara, K. Assessment of global nitrogen pollution in rivers using an integrated biogeochemical modeling framework. *Water Res.* **2011**, *45*, 2573–2586.
10. Li, S.L.; Liu, C.Q.; Li, J.; Xue, Z.C.; Guan, J.; Lang, Y.C.; Ding, H.; Li, L.B. Evaluation of nitrate source in surface water of southwestern China based on stable isotopes. *Environ. Earth. Sci.* **2013**, *68*, 219–228.
11. Gautam, S.; Iqbal, M. Using stable isotopes of nitrogen to study its source and transformation in a heavily farmed watershed. *Environ. Earth. Sci.* **2010**, *60*, 11–20.
12. Li, J.; Li, F.D.; Liu, Q.; Suzuki, Y. Nitrate pollution and its transfer in surface water and groundwater in irrigated areas: A case study of the Piedmont of

South Taihang Mountains, China. *Environ. Sci. Process. Impacts* **2014**, *16*, 2764–2773.

13. Carpenter, S.R.; Caraco, N.F.; Correll, D.L.; Howarth, R.W.; Sharpley, A.N.; Smith, V.H. Nonpoint pollution of surface waters with phosphorus and nitrogen. *Ecol. Appl.* **1998**, *8*, 559–568.
14. Liu, T.; Wang, F.; Michalski, G.; Xia, X.H.; Liu, S.D. Using ^{15}N, ^{17}O, and ^{18}O to determine nitrate sources in the Yellow River, China. *Environ. Sci. Technol.* **2013**, *47*, 13412–13421.
15. Yang, L.; Song, X.; Zhang, Y.; Han, D.; Zhang, B.; Long, D. Characterizing interactions between surface water and groundwater in the Jialu River basin using major ion chemistry and stable isotopes. *Hydrol. Earth Syst. Sci. Discuss.***2012**, *9*, 5955–5981.
16. Itoh, M.; Takemon, Y.; Makabe, A.; Yoshimizu, C.; Kohzu, A.; Ohte, N.; Tumurskh, D.; Tayasu, I.; Yoshida, N.; Nagata, T. Evaluation of wastewater nitrogen transformation in a natural wetland (Ulaanbaatar, Mongolia) using dual-isotope analysis of nitrate. *Sci. Total Environ.* **2011**, *409*, 1530–1538.
17. Huang, G.X.; Fallowfield, H.; Guan, H.D.; Liu, F. Remediation of nitrate-nitrogen contaminated groundwater by a heterotrophic-autotrophic denitrification approach in an aerobic environment. *Water Air Soil Pollut.* **2012**, *223*, 4029–4038.
18. Fogg, G.E.; Rolston, D.E.; Decker, D.L.; Louie, D.T.; Grismer, M.E. Spatial variation in nitrogen isotope values beneath nitrate contamination sources. *Gound Water* **1998**, *36*, 418–426.
19. Stevenson, F.J. Origin and distribution of nitrogen in soil. *Agron. Monogr.* **1982**, *22*, 1–42.
20. Hübner, H. Isotope effects of nitrogen in the soil and biosphere. In *Handbook of Environmental Isotope Geochemistry*; Fritz, P., Fontes, J.C., Eds.; Elsevier: New York, NY, USA, 1986; pp. 361–425.
21. Mayer, B.; Boyer, E.W.; Goodale, C.; Jaworski, N.A.; Van Breemen, N.; Howarth, R.W.; Seitzinger, S.; Billen, G.; Lajtha, K.; Nadelhoffer, K.; et al. Sources of nitrate in rivers draining sixteen watersheds in the northeastern U.S.: Isotopic constraints. *Biogeochemistry* **2002**, *57*, 171–197.
22. Macko, S. Pollution studies using stable isotopes. In *Stable Isotopes in Ecology and Environmental Science*; Lajtha, K., Michener, R.M., Eds.; Blackwell Sci.: Malden, MA, USA, 1994; pp. 45–62.
23. Böhlke, J.K.; Denver, J.M. Combined use of groundwater dating, chemical, and isotopic analyses to resolve the history and fate of nitrate contamination in two agricultural watersheds, Atlantic coastal plain, Maryland. *Water Resour. Res.* **1995**, *31*, 2319–2339.
24. BryantMason, A.; Xu, Y.J.; Altabet, M. Isotopic signature of nitrate in river waters of the lower Mississippi and its distributary, the Atchafalaya. *Hydrol. Process.* **2013**, *27*, 2840–2850.
25. Heaton, T. Isotopic studies of nitrogen pollution in the hydrosphere and atmosphere: A review. *Chem. Geol. Isot. Geosci. Sect.* **1986**, *59*, 87–102.
26. Kendall, C. Tracing nitrogen sources and cycling in catchments. *Isot. Tracers Catchment Hydrol.* **1998**, *1*, 519–576.
27. Liu, C.Q.; Li, S.L.; Lang, Y.C.; Xiao, H.Y. Using $\delta^{15}N$- and $\delta^{18}O$-values to identify nitrate sources in karst ground water, Guiyang, Southwest China. *Environ. Sci. Technol.* **2006**, *40*, 6928–6933.
28. Fang, C.L.; Xie, Y. Sustainable urban development in water-constrained Northwest China: A case study along the mid-section of Silk-Road—He-Xi corridor. *J. Arid Environ.* **2010**, *74*, 140–148.
29. Zuo, D.P.; Xu, Z.X.; Wu, W.; Zhao, J.; Zhao, F.F. Identification of streamflow response to climate change and human activities in the Wei River Basin, China. *Water Resour. Manag.* **2014**, *28*, 833–851.

30. Song, J.X.; Xu, Z.X.; Liu, C.M.; Li, H.E. Ecological and environmental instream flow requirements for the Wei River—The largest tributary of the Yellow River. *Hydrol. Process.* **2007**, *21*, 1066–1073.
31. Li, J.K.; Li, H.E.; Shen, B.; Li, Y.J. Effect of non-point source pollution on water quality of the Weihe River. *Int. J. Sediment Res.* **2011**, *26*, 50–61.
32. Liu, Q.; Li, F.D.; Guo, J.Q.; Wu, X.D.; Song, S.; Zhao, G.S. Source and variation of nitrogen along middle and down stream of Weihe River and its main tributaries. *J. Water Resour. Water Eng.* **2011**, *22*, 128–132.
33. Xing, M.; Liu, W.G.; Hu, J. Using nitrate isotope to trace the nitrogen pollution in Chanhe and Laohe River. *Environ. Sci.* **2010**, *10*, 2305–2310.
34. Water Quality-Technical Regulation on the Design of Sampling Programmes. HJ 495-2009.
35. Silva, S.; Kendall, C.; Wilkison, D.; Ziegler, A.; Chang, C.; Avanzino, R. A new method for collection of nitrate from fresh water and the analysis of nitrogen and oxygen isotope ratios. *J. Hydrol.* **2000**, *228*, 22–36.
36. Wang, S.Q.; Tang, C.Y.; Song, X.F.; Yuan, R.Q.; Wang, Q.X.; Zhang, Y.H. Using major ions and delta $^{15}N\text{-}NO_3^-$ to identify nitrate sources and fate in an alluvial aquifer of the baiyangdian lake watershed, north China plain. *Environ. Sci. Process. Impacts* **2013**, *15*, 1430–1443.
37. Di Lorenzo, T.; Brilli, M.; Del Tosto, D.; Galassi, D.M.P.; Petitta, M. Nitrate source and fate at the catchment scale of the Vibrata River and aquifer (central Italy): An analysis by integrating component approaches and nitrogen isotopes. *Environ. Earth. Sci.* **2012**, *67*, 2383–2398.
38. Yue, F.J.; Liu, X.L.; Li, J.; Zhu, Z.Z.; Wang, Z.L. Using nitrogen isotopic approach to identify nitrate sources in waters of Tianjin, China. *Bull. Environ. Contam. Toxicol.* **2010**, *85*, 562–567.
39. Jolly, I.D.; McEwan, K.L.; Holland, K.L. A review of groundwater-surface water interactions in arid/semi-arid wetlands and the consequences of salinity for wetland ecology. *Ecohydrology* **2008**, *1*, 43–58.
40. Yang, J.M.; Liu, J.H.; Yu, X.B.; Yang, W.B.; Zhou, M.P. Variation of nitrogen resource in groundwater and river water in Changchun Region. *J. Soil Water Conserv.* **2010**, *24*, 255–257.
41. Tabayashi, Y.; Koba, K. Heterogeneous atmospheric nitrogen deposition effects upon the nitrate concentration of stream waters in a forested mountain area. *Water Air Soil Pollut.* **2011**, *216*, 105–115.
42. Koszelnik, P.; Gruca-Rokosz, R. Determination of nitrate isotopic signature in waters of different sources by analysing the nitrogen and oxygen isotopic ratio. *Environ. Sci. Process. Impacts* **2013**, *15*, 751–759.
43. Gu, B.J.; Ge, Y.; Chang, S.X.; Luo, W.D.; Chang, J. Nitrate in groundwater of China: Sources and driving forces. *Glob. Environ. Chang.* **2013**, *23*, 1112–1121.
44. Ju, X.T.; Xing, G.X.; Chen, X.P.; Zhang, S.L.; Zhang, L.J.; Liu, X.J.; Cui, Z.L.; Yin, B.; Christie, P.; Zhu, Z.L.; et al. Reducing environmental risk by improving N management in intensive Chinese agricultural systems. *Proc. Natl. Acad. Sci. USA* **2009**, *106*, 3041–3046.
45. Bakhsh, A.; Kanwar, R.S.; Baker, J.L. N-application methods and precipitation pattern effects on subsurface drainage nitrate losses and crop yields. *Water Air Soil Pollut.* **2010**, *212*, 65–76.
46. Jin, Z.F.; Pan, Z.Y.; Jin, M.T.; Li, F.L.; Wan, Y.; Gu, B. Determination of nitrate contamination sources using isotopic and chemical indicators in an agricultural region in China. *Agric. Ecosyst. Environ.* **2012**, *155*, 78–86.
47. Liu, W.G.; Xing, M. Isotopic indicators of carbon and nitrogen cycles in river catchments during soil erosion in the arid Loess Plateau of China. *Chem. Geol.* **2012**, *296–297*, 66–72.
48. Zhang, D.; Yang, W.; Zhao, J.L. Tracing nitrate sources of the Yellow River and its tributaries with nitrogen isotope. *J. Ecol. Rural Environ.* **2012**, *28*, 622–627.

49. Urresti-Estala, B.; Vadillo-Pérez, I.; Jiménez-Gavilán, P.; Soler, A.; Sánchez-Garcia, D.; Carrasco-Cantos, F. Application of stable isotopes ($\delta^{34}S-SO_4$, $\delta^{18}O-SO_4$, $\delta^{15}N-NO_3$, $\delta^{18}O-NO_3$) to determine natural background and contamination sources in the Guadalhorce River Basin (southern Spain). *Sci. Total Environ.* **2015**, *506–507*, 46–57.

50. Urresti-Estala, B.; Jiménez-Gavilán, P.; Vadillo-Pérez, I.; Carrasco-Cantos, F. Assessment of hydrochemical trends in the highly anthropised Guadalhorce River Basin (southern Spain) in terms of compliance with the European groundwater directive for 2015. *Environ. Sci. Pollut. Res. Int.* **2016**.

51. Yu, H.Y.; Yu, Z.M.; Song, X.X.; Cao, X.H.; Yuan, Y.Q.; Lu, G.Y. Seasonal variations in the nitrogen isotopic composition of dissolved nitrate in the Changjiang River estuary, China. *Estuar. Coast. Shelf Sci.* **2015**, *155*, 148–155.

CHAPTER 10

Assessing Historical Mining and Smelting Effects on Heavy Metal Pollution of River Systems over Span of Two Decades

Magdalena Strzebońska Elżbieta Jarosz-Krzemińska Ewa Adamiec

M. Strzebońska E.Jarosz-Krzemińska E. Adamiec AGH University of Science and Technology, 30 Mickiewicza Ave., 30-059 Kraków, Poland

ABSTRACT

Research was conducted on the most polluted river system in Poland, impacted by active and historical mining. Bottom sediment, suspended particulate matter and river water were collected in 2014 from Przemsza river and its tributaries. Sampling points remained the same as those chosen in a 1995 study. This allowed the comparison of heavy metal accumulation in bottom sediment over a span of almost two decades. It was concluded that Przemsza river water and its tributaries are heavily contaminated with the following (in μg/dm^3): Pb (0.99–145.7), Zn (48–5020), and Cd 0.12–12.72). Concentrations of metals in bottom sediment exceeded the background values by a factor of several hundred (100 times for Zn, 150 times for Pb, and 240 times for Cd). The arithmetic mean for metal concentration in fractions <63 μm sampled in 2014 has remained comparable to the level found in 1995 (in mg/kg): Zn 16,918 and 13,505,

Pb 4177 and 4758, and Cd 92 and 134. It was determined that 20–50% more metals have accumulated in suspended matter, rather than in bottom sediment (in mg/kg): 20,498 Zn, Pb 5170, and 164 Cd. This exceeds the limits of the most polluted LAWA Class IV classification. Since the concentrations of Zn, Pb, and Cd increase drastically after the outlet of the Przemsza into the Vistula, it was concluded that river Przemsza is the cause of significant degradation of Vistula's bottom sediment and suspended matter. A two-decade legacy of extremely high contamination of the Przemsza river sediments has persisted despite decreasing mining and smelting activity in the vicinity.

Keywords: Historical mining River bottom sediment Suspended particulate matter Correlation coefficient

1 INTRODUCTION

Multiple sources of mostly anthropogenic origin are causing pollution of aquatic environments with heavy metals. The most recognizable are current and historical mining and smelting industries, industrial effluents, leaks from dumping site leakage as well as common use of fertilizers and pesticides, or other surface runoffs. Heavy metals from atmospheric emissions also enter rivers. As indicated by Church and Scudlark (1998), they are responsible for the introduction of approximately 7% Cd, 4% Cu, 3% Zn, and 2% both Co and Ni into aquatic environments. Since heavy metals are persistent and not degradable under normal conditions, they pose a serious threat to aquatic environment even for hundreds of years after contamination. When introduced into rivers, they could be transported as compounds dissolved in water, be adsorbed onto suspended particulate matter, or be deposited in river bottom sediment. According to Martin and Meybeck 1979, almost 90% of contaminants are incorporated in the sediment-associated form rather than aqueous; this is mostly due to the sorptive nature of fine-grained suspended particles (Miller et al. 2007; Taylor and Hudson-Edwards 2008). Most riverbed sediments have fractions below 63 μm and are composed mainly of quartz as well as minor amounts of feldspar and carbonates. Finer fractions of sediments mostly compose of clay minerals, carbonates, or Mn and Fe hydroxides and oxides. The bottom sediment heavy metals are not permanently fixed but can be periodically remobilized, desorpted, or redistributed along the river course. Sediments act both as a sink and secondary source of metal contamination in surface water systems (Singh et al. 2005). The process of remobilization of heavy metals from both suspended particulate matter (SPM) and bottom sediment is potentially more dangerous than the accumulation itself. Mobility of contaminates is governed by changes of various physico-chemical parameters in which pH, redox potential, salinity, or the presence of complexation substances are of predominant concern. For instance, lowering pH by one unit can cause an increase of metal solubility by a factor of 10 (Förstner et al. 1990), which for example results in the dissolution of carbonates and hydroxides. Moreover, changing the redox condition from reducing to oxidizing also poses a great threat to the river environment, since it promotes transformation of stable metal sulfides into unstable metal sulfates. The use of river bottom sediment as environmental monitoring tool is often difficult as has been indicated by many authors (Regnier and Wollast 1993; Larsen et al. 1988; Eisma et al. 1989). Suspended particulate matter on the other hand was reported to be much better suited for monitoring purposes (Ji et al. 2016; Onderka and Pekárová 2008; Lartiges et al. 2001; Adamiec and Helios-Rybicka 2002). The role of suspended particulate matter in the uptake, release, and transport of heavy metals is obvious. Therefore, as stated by Turner and Millward 2002, SPM is a crucial link in heavy metal cycle between the water column, bed sediment, and consequently food chain. Assessing the quality of SPM provides the information about load of contaminants released from river bottom sediment but also from other more recent point sources of pollution. Contaminants are more likely immobilized in bottom sediment, and they are regarded as a potential source of contamination, in suspended matter; however, they are considered a real threat (Barbusiński et al. 2012). This component of the river system is also sensitive to the changes of physico-chemical parameters; for instance, lowering the pH (as a result of, e.g., active mining) promotes much easier transport of heavy metals from suspended matter into river water causing metals to become more available to living aquatic organisms and to further travel up to the end of the food chain.

River bottom sediment is commonly considered as a legacy of the past, especially where historical metal mining activity being the contamination source is concerned. Most studies regarding contamination of aquatic environment with heavy metals sourced from active or historical mining activities focus on examination of only bottom sediment (Jabłońska-Czapla et al. 2016; Dadová et al. 2016; Hogarh et al. 2016). However, in order to assess other than historical sources of river contamination and to differentiate between historical and present anthropogenic pollution sources, the investigation of suspended particulate matter should also be considered.

The research involved examining the bottom sediment, suspended particulate matter, and river water, collected from the Przemsza river and its tributaries in 2014. This is the most severely contaminated river in Poland. Its degradation can mostly be attributed to current and historical mining activities. The sampling points remained the same as these were chosen during sampling campaign in 1995 and carried out by authors (Helios-Rybicka et al. 1998). This allowed for a comparison of heavy metal accumulation in bottom sediment over a span of almost two decades. Moreover, an examination of suspended particulate matter was carried out in order to differentiate between recent sources of anthropogenic pollution and historical ones.

2 MATERIALS AND METHODS

2.1 Study Area

Przemsza river is one of the most polluted rivers in Poland and consequently in Europe. It comprises of Biała and Czarna Przemsza and its total length equals 87.6 km. It flows through highly industrialized and anthropogenically transformed area of the Upper Silesia, and it enters the Vistula river near Oświęcim (Auschwitz) (Fig. 1).

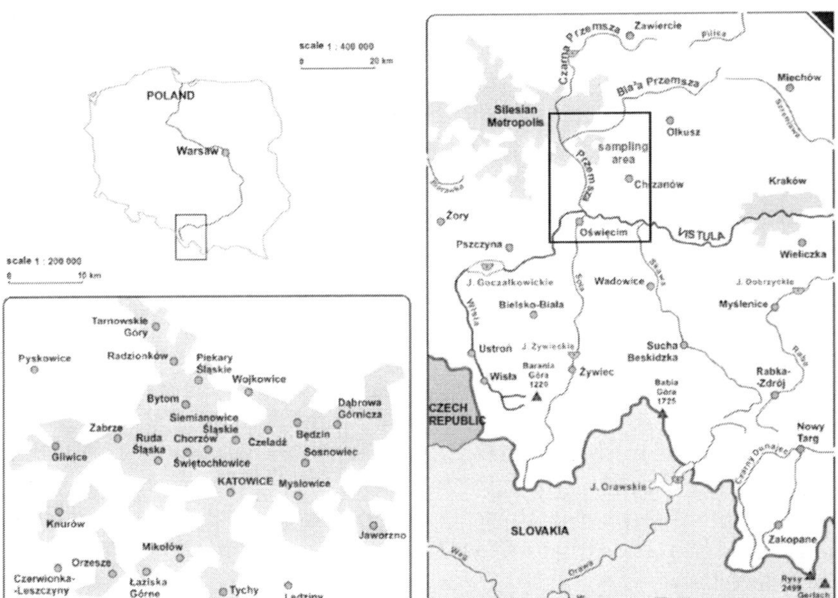

Figure. 1 Research area (Olczyk 2016; source: www.tyflomapy.pl)

Multitude of factors of both natural and anthropogenic origin influences this river's pollution with heavy metals. The most important ones are geological formations, lead-zinc and hard coal deposits, mining, industry, urbanization, and air pollution. Pb-Zn ores are deposited at the depth of 50 to 400 m and are accompanied by silver and cadmium deposits. This region is one of the biggest areas of zinc and lead deposits of so-called MVT (Mississippi Valley type), with parameters similar to deposits from the Mississippi River valley. The presence of these deposits led to the influx of population in the area of Upper Silesia and the rise of the largest and most densely populated agglomerations in Poland. These zinc-lead-silver ores are one of the oldest in Poland. They were mined in the area of Bytom from the Middle Ages, but now, they are mostly of historical significance.

Despite the fact that in this region, there is still approximately 80 million tons of zinc and lead ores, exploitation of only 12.5 million tons of these geological resources is profitable (data from 2010). The annual extraction rate of the last active mine—Pomorzany—is around 2.5 million tons of ore, and soon, the resources will be completely exhausted. Geochemical background of Upper Silesia region is naturally elevated since prehistorical times, and due to weathering, abnormally high concentrations of Zn, Pb, and Cd were recorded. However, according to the research of Klimek (1993), concentration of these metals spiked up in overbank sediments in the last 100–150 years. It was found that the amount of Zn, Cd, and Pb in the fraction <63 μm of overbank river sediments of the Przemsza river and its tributaries were one to two orders of magnitude higher when compared to the concentration of these metals determined in the layers of sediments before industrialization (anthropogenic pollution).

2.2 Materials and Methods

Samples of river water and bottom sediments were collected from 13 locations along Przemsza and Vistula rivers and were stored in the polyethylene containers. Bottom sediments were sampled using probe from the depth of 20 cm. Physical parameters such as pH and conductivity were determined in situ in all samples. In the laboratory, samples of river water were filtered through <0.45 μm, acidified with nitric acid, and analyzed for heavy metal content using inductively coupled plasma-optical emission spectroscopy (ELAN 6100; Perkin Elmer). Samples of suspended matter, which were obtained as fraction <0.45 μm, were dried in 105 °C for 2 h and then digested with conc. HNO_3 according to EPA 3051. Concentrations of Zn, Pb, and Cd were determined using ICP-MS, according to EN-ISO 17294-1 (2006). Bottom sediment samples were homogenized, mixed, and sieved through <63 μm; dried in 105 °C for 2 h; and furthermore, digested with conc. HNO_3 according to EPA 3051.

Concentrations of heavy metals were determined using atomic absorption spectroscopy (AAS) (F-AAS Thermo Scientific IC 3500).

3 RESULTS AND DISCUSSION

3.1 River Water

Waters of Przemsza river and its tributaries are heavily contaminated with lead (0.99–145.7 μg/dm³), zinc (48–5020 μg/dm³), and cadmium (0.12–12.72 μg/dm³) (Table 1).

Table 1 Concentration of Zn, Pb, and Cd in water, suspended matter, and bottom sediment of the Przemsza river and its tributaries (2014 campaign)

River	Water (μg/dm³)			Suspended particulate matter (mg/kg)			Bottom sediment (mg/kg)		
	Zn	Pb	Cd	Zn	Pb	Cd	Zn	Pb	Cd
Biała	5020	8.57	12.72	39,097	21,848	340	21,822	8872	137
Biała Przemsza	48	6.90	0.27	2431	1206	37	765	670	13
Biała Przemsza	4076	11.92	10.26	34,666	18,170	269	28,158	7137	148
Sztoła	1570	145.70	3.37	26,211	9056	221	26,318	7033	167
Kozi Bród	249	23.65	0.68	8396	3064	71	2730	549	17
Biała Przemsza	1945	8.19	0.79	27,012	10,704	232	26,845	9451	154
Kanał Matylda	184	3.68	0.12	10,999	4112	79	7907	2976	64
Przemsza	698	1.10	0.32	19,881	4843	152	16,897	2133	93
Przemsza	821	1.34	0.27	20,587	5497	177	16,254	3051	74
Przemsza	667	0.99	0.19	20,409	4142	140	12,589	2620	67

The pH of the Przemsza river and its tributaries varies between 7.01 and 8.19 and does not exceed the threshold value corresponding to the first water quality class. Electrical conductivity ranges between 346 and 2395 μS/cm, but an average (median) values respond to the second purity class (Table 2).

Zinc, lead, and cadmium content in river waters varies significantly and depends on the geological bedrock and the point sources of contamination. According to Kabata-Pendias and Pendias (1999), natural concentration of zinc in river water should be approximately 10 μg/dm³. Research revealed, however, that the average content of this element in Biała Przemsza and Przemsza waters was as high as 1406 μg/dm³. That is ten times more than

Table 2 Statistical parameters of pH, conductivity, and concentration of Zn, Pb, and Cd in river water (2014 campaign)

Parameter	pH	Conductivity (μS/cm)	Zn (μg/dm^3)	Pb (μg/dm^3)	Cd (μg/dm^3)
Biała Przemsza and Przemsza river					
Maximum	8.06	2395	4076	11.92	10.26
Minimum	7.50	346	48	0.99	0.19
Arithmetic mean	7.76	1404	1376	5.07	2.02
Median	7.73	1406	760	4.12	0.29
Geometric mean	7.75	1127	726	3.16	0.57
Przemsza river tributaries					
Maximum	8.19	1756	5020	145.70	12.72
Minimum	7.01	550	184	3.68	0.12
Arithmetic mean	7.55	931	1756	45.40	4.22
Median	7.49	708	910	16.11	2.03
Geometric mean	7.53	829	776	18.15	1.37
Vistula before Przemsza tributary					
	7.04	2690	153	9.13	0.54
Vistula after Przemsza tributary					
	7.09	3600	450	1.77	0.72

the geochemical background established for the Upper Silesia rivers, which equals 86 μg/ dm^3(Pasieczna et al. 2010; Lis and Pasieczna 1995). Since lead compounds are poorly soluble at the pH around 7, the natural content of this element in river water should be as low as 0.2 μg/dm^3(Kabata-Pendias and Pendias 1999). In the Biała Przemsza and Przemsza river, however, an average concentration of Pb was found to be 4.12 μg/dm^3, and it was lower than that previously reported in 2005 by Pasieczna et al. (2010) (median—34.3 μg/dm^3). Natural background concentration of Cd in surface water should be as low as 0.02 μg/dm^3 (Kabata-Pendias and Pendias 1999), but in both Biała Przemsza and Przemsza river, concentration of Cd was elevated and equaled to an average of 0.29 μg/dm^3. Significant amount of Cd is introduced into those two rivers by its tributaries (median of 2.03 μg/dm^3), especially by Biała river, which is the serious source of Cd contamination (12.7 μg/dm^3).

These research findings are consistent with the chemical monitoring report provided in 2014 by the Regional Inspectorate of Environmental Protection in Katowice. It was reported that concentrations of priority substances and the so-called other impurities in the water exceeded the quality standards with respect to cadmium and lead (Biała river, Biała Przemsza from Ryczówka to Kozi Bród and from Kozi Bród to estuary); hexachlorocyclohexane and the sum of aldrin, dieldrin, endrin, and isodrin (Przemsza from Biała Przemsza to the estuary) as well as PAHs (Biała Przemsza, from Kozi Bród to the estuary and Przemsza, from Biała Przemsza to the estuary).

3.2 Suspended Particulate Matter

Suspended particulate matter of the Przemsza river and its tributaries is severely contaminated with all of the investigated heavy metals (Tables 1 and 3). Concentrations of metals vary within sampling points (2431–39,097 mg/kg of Zn, 1206–21,898 mg/kg of Pb, and 37–340 mg/kg of Cd) reaching the highest levels in the suspended matter from the Biała tributary.

Table 3 Statistical parameters of Zn, Pb, and Cd concentration in suspended matter of the Przemsza, its tributaries, and Vistula river (2014 campaign). Classification of suspended matter according to LAWA

Parameter	Zn (mg/kg)	Pb (mg/kg)	Cd (mg/kg)
Biała Przemsza and Przemsza River			
Minimum	2431	1206	37
Maximum	34, 666	18, 170	269
Arithmetic mean	20, 831	7427	168
Median	20, 498	5170	164
Geometric mean	16, 338	5438	143
Przemsza River tributaries			
Minimum	8396	3064	71
Maximum	39, 097	21, 848	340
Arithmetic mean	21, 176	9520	178
Median	18, 605	6584	150
Geometric mean	17, 539	7066	143
Vistula before Przemsza tributary			
	2858	478	17

Vistula after Przemsza tributary			
	8458	2387	61
Classification LAWA 1998			
I	≤100	≤25	≤0,3
I-II	≤200	≤50	≤0,6
II	≤400	≤100	≤1,2
II-III	≤800	≤200	≤2,4
III	≤1600	≤400	≤4,8
III-IV	≤3200	≤800	≤9,6
IV	>3200	>800	>9,6

Despite the fact that suspended matter plays a very important role in the transport of heavy metals in the river system, Polish regulations do not specify legal requirements for its quality. For the purpose of this study, comparison with threshold concentrations for individual metals, set by German guidelines—LAWA, was employed (Irmer 1997; LAWA 1998) (Table 3).

The results revealed that SPM collected from almost all sampling locations along the Przemsza rivers and its tributaries is exceeding the threshold values for the most polluted Class IV according to LAWA classification with respect to Zn, Pb, and Cd concentration. Only samples from the beginning coarse of the Biała Przemsza river contain an amount of Zn corresponding to the class III–IV, and the sample from the Vistula river before the outlet to Przemsza could be assigned to III purity class with respect to Pb contamination.

It was concluded that the Przemsza's river suspended matter is excessively more contaminated with heavy metals than Oder river, the second biggest and cross-border river in Poland. Concentrations of heavy metals reported by Adamiec (2004) in 2000 in the Oder river's suspended matter were found to be 1618 mg/kg Zn, 64.5 mg/kg Pb, and 9.24 mg/kg Cd.

3.3 River Bottom Sediment

Bottom sediment of the Przemsza river and its tributaries is severely contaminated with heavy metals exceeding hundreds of times the background values established by Turekian & Wedepohl 1961 (Table 4). Maximum metal concentrations in the fraction <63 μm sampled in 2014 and in 1995 were as follows respectively (in mg/kg): Zn 28,158 and 29,580, Pb 9451 and 19,510, and Cd 154 and 269. The arithmetic mean for concentration of Zn, Pb and Cd in the fraction <63 μm of all investigated rivers has remained comparable in both collection periods and equals respectively to the following (in mg/kg): Zn 16,918 and 13,505, Pb 4177 and 4758, and Cd 92 and 134. Very high load of metals is introduced to the Przemsza river by its tributaries. In 2014, the concentrations of heavy metals in the bottom sediment of the most polluted rivers Biała Przemsza (Fig. 2) and Sztola were as high as follows (in mg/kg): Zn 28,158 and 26,318, Cd 154 and 167, and Pb 9451 and 7033.

The highest contamination of bottom sediment with Pb was reported in Biała river tributary (8872 mg/kg). The geometric averages for concentration of Zn, Pb, and Cd are comparable for the Przemsza river and its tributaries, and they are found to be respectively as follows (in mg/kg): Zn 11,225 and 10,552, Pb 30,120 and 3177, and Cd 72 and 71. Concentrations of metals in all investigated bottom sediment samples are exceeding geochemical background concentrations established by Turekian and Wedepohl (1961) by a hundredfold or more: 100 times for Zn, 150 times for Pb, and for 240 times with respect to Cd.

Table 4 Statistical parameters of Zn, Pb, and Cd concentration in bottom sediment of the Przemsza, its tributaries, and Vistula river. Classification of bottom sediment according to LAWA

Parameter		<63 μm fr. %wt.		Zn		Pb mg/kg		Cd	
year		1995	2014	1995	2014	1995	2014	1995	2014
Przemsza River									
Minimum		0.20	4.83	139	765	28	670	6	13
Maximum		79.17	45.27	29, 580	28, 158	19, 510	9451	269	154
Arithmetic mean		33.87	16.70	13, 505	16, 918	4758	4177	134	92
Median		25.21	13.13	11, 960	16, 575	1918	2835	152	83
Geometric mean		14.77	12.34	7815	11, 225	1663	3028	82	72
Przemsza River tributaries									
Minimum		0.74	11.49	5640	2730	451	549	20	17
Maximum		36.03	53.55	33, 860	26, 318	8555	8872	140	167
Arithmetic mean		13.82	26.22	17, 650	14, 694	3536	4857	80	96
Median		9.25	19.92	15, 550	14, 865	2569	5004	81	101
Geometric mean		6.91	22.09	14, 584	10, 552	2217	3177	65	71
Classification LAWA 1998				Vistula before Przemsza tributary					
I	I-II	67.57	43.82	2680	507	128	311	14	4
II	II-III			Vistula after Przemsza tributary					
III	III-IV	62.19	19.10	2980	3881	194	1009	23	23
IV		Geochemical background Turekian & Wedepohl 1961		95		20		0.3	

Figure. 2 Biała Przemsza river after Biała river tributary

Research revealed that bottom sediment of the Przemsza river and its tributaries has been classified as severely contaminated with heavy metals according to LAWA IV purity class. Only Biała Przemsza in its initial run, before the estuary of the Biała river, and river Kozi Bród are less contaminated with zinc and lead. Nonetheless, outlet of the Przemsza into Vistula river has caused the deterioration of its bottom sediment to IV quality class with respect to all metals.

3.4 Correlation Calculated for All Components of the River System

Comparison of the concentrations of individual metals in the bottom sediment, suspended matter, and river water along the Biała Przemsza and Przemsza rivers (Fig. 3) revealed that there is a strong correlation between all components of the river system. The correlation coefficients calculated for each pair of variables showed strong, very strong, or almost full correlation for most couples, and the average correlation for only five pairs of variables (Table 5). This proves the existence of the same sources of heavy metal contamination for the examined rivers. Elements coexist with each other, and their content in the sediment, water, and the suspension is distributed proportionally along the entire course of Biała Przemsza and Przemsza rivers.

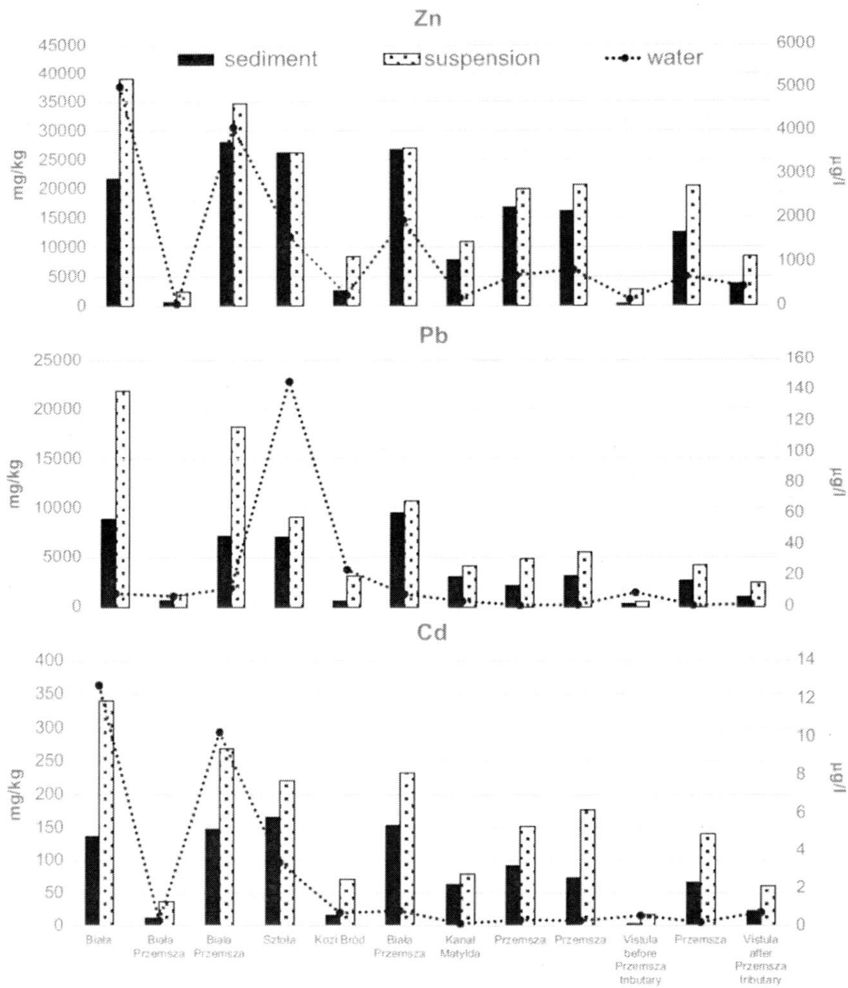

Figure. 3 Comparison of Zn, Pb, and Cd concentrations in the bottom sediment, suspended matter, and river water along the Biała Przemsza and Przemsza rivers

Table 5 Correlation coefficients for river water, suspended particulate matter, and bottom sediment of the Przemsza and Biała Przemsza rivers

	Sed. Zn	Sed. Pb	Sed. Cd	Susp. Zn	Susp. Pb	Susp. Cd	Water Zn	Water Pb	Water Cd
Sed. Zn	–	0.885	0.988	0.963	0.879	0.987	0.831	0.426	0.665
Sed. Pb	0.885	–	0.923	0.784	0.812	0.854	0.763	0.631	0.475
Sed. Cd	0.988	0.923	–	0.927	0.870	0.955	0.821	0.488	0.557
Susp. Zn	0.963	0.784	0.927	–	0.884	0.982	0.853	0.352	0.655
Susp. Pb	0.857	0.812	0.870	0.884	–	0.904	0.995	0.540	0.885
Susp. Cd	0.987	0.854	0.955	0.982	0.904	–	0.864	0.427	0.641
Water Zn	0.831	0.763	0.821	0.853	0.995	0.864	–	0.884	0.982
Water Pb	0.426	0.631	0.488	0.352	0.540	0.427	0.884	–	0.904
Water Cd	0.655	0.475	0.557	0.655	0.855	0.641	0.982	0.904	–

4 CONCLUSIONS

The 2014 study revealed that Przemsza river is the cause of a significant increase in metal contamination in both river bottom sediment and suspended particulate matter of Poland's main river, the Vistula. Determined concentrations of Zn, Pb and Cd increased drastically downstream from the outlet of the Przemsza to the Vistula river. All of the Przemsza's river components, water, suspended particulate matter, and bottom sediment, are severely contaminated with heavy metals. Extremely high concentrations of metals in the Przemsza river bottom sediments persisted for over two decades on the comparable level or even higher even despite of decrease in activity of mining and metal casting industries in the vicinity. The arithmetic mean of metal concentration in size fraction <63 μm in 1995 and 2014 (in mg/kg) was as follows respectively: Zn 16,918 and 13,505, Pb 4177 and 4758, and Cd 92 and 134. Results confirm other authors' reports (Graham Bird 2016; Audry et al. 2004) that elevated concentrations of heavy metals in river sediments continue to persist in river catchments despite mining having ceased for over 10 to a 100 years, sometimes even exceeding guideline concentrations hundreds of times. Przemsza's bottom sediment is enriched with heavy metals exceeding geochemical background values established by Turekian and Wedepohl (1961): 100 times in terms of Zn, 150 times for Pb, and for 240 times with respect to Cd. However, it turns out that even more heavy metals are accumulated in suspended particulate matter, up to 20–50% more than in the river bottom sediment. This is in line with the findings of other authors' claims that heavy metals are transported in the river system mostly bound with the suspended sediment. Such high concentrations of heavy metals in the suspended particulate matter are not only the resultant of their release from bottom sediments but must also be attributed to other constant point sources of pollution. Regardless of whether we take into account the arithmetic average, median, or geometric mean with regards to metal concentration in both suspended matter and bottom sediment, these parameters classify Przemsza river and its tributaries into the worst IV purity class according to the LAWA classification.

ACKNOWLEDGEMENTS

This study was carried out under research activities of the Faculty of Geology, Geophysics and Environmental Protection at the AGH University of Science and Technology No. 11.11.140.199.

REFERENCES

1. Adamiec, E., & Helios-Rybicka, E. (2002). Distribution of pollutants in the Odra River system part V. Assessment of total and mobile heavy metals content in the suspended matter and sediments of the Odra river system and recommendations for river chemical monitoring. *Polish Journal of Environmental Studies, 11*(6), 675–688.
2. Adamiec E. (2004). Rola zawiesiny w zanieczyszczeniu metalami śladowymi rzeki Odry. AGH Uczelniane Wydawnictwa Naukowo-Dydaktyczne, 1–88. (in Polish). Kraków.
3. Audry, S., Schäfer, J., Blanc, G., & Jouanneau, J. M. (2004). Fifty-year sedimentary record of heavy metal pollution (Cd, Zn, Cu, Pb) in the Lot River reservoirs (France). *Environmental Pollution, 132*(3), 413–426.
4. Barbusiński, K., Nocoń, W., Nocoń, K., & Kernert, J. (2012). The role of suspended solids in the transport of heavy metals in surface water, exemplified by the Kłodnica River (Upper Silesia). *Ochrona Środowiska, 34*(2), 33–38 (in Polish).
5. Bird, G. (2016). The influence of the scale of mining activity and mine site remediation on the contamination legacy of historical metal mining activity. *Environmental Science and Pollution Research.* doi:10.1007/s11356-7400-z.
6. Church, T.M., & Scudlark, J.R. (1998). Trace metals in estuaries: a Delaware Bay synthesis.[in] Allen H.E., Garrison A.W., Luter IIIG.W., *Metals in Surface Waters*, 1–21.
7. Dadová, J., Andráš, P., Kupka, J., Krnáč, J., Andráš, J. P., Hroncová, E., & Midula, P. (2016). Mercury contamination from historical mining territory at Malachov Hg-deposits (Central Slovakia). *Environmental Science and Pollution Research, 23*(3), 2914–2927.
8. Eisma, D., Berger, G. N., Chen W.-Y., & Shen, J. (1989). Pb-210 as a tracer for sediment transport and deposition in the Dutch-German Waddensea. In: Proc. KNGMG Symp. Coastal Lowlands, Geology and Geotechnology, Den Haag, 1987. Dordrecht: Kluwer Academic, pp. 237–253.
9. Förstner, U., Ahlf, W., Calmano, W., & Kersten, M. (1990). Sediment criteria development—contributions from environmental geochemistry to water quality management. In D. Heling, P. Rothe, U. Förstner, & P. Stoffers (Eds.), *Sediments and environmental geochemistry. Selected aspects and case histories* (pp. 311–338). New York: Springer.

10. Helios-Rybicka, E., Strzebońska, M. & Budek, L. (1998). Contaminated alluvial as the sources of heavy metals (example Przemsza River, Upper Silesia, Poland). Enviweath96 Environmental Aspects of Weathering. Proc. of an international conference held on the occasion of the start of IGCP Project # 405 – Anthropogenic Impact on Weathering Processes" (1996–2000). Brno, Czech Republik, pp 75–82, 1996.
11. Hogarh, J. N., Adu-Gyamfi, E., Nukpezah, D., Akoto, O., & Adu-Kumi, S. (2016). Contamination from mercury and other heavy metals in a mining district in Ghana: discerning recent trends from sediment core analysis. *Environ Syst Res, 5*, 15. doi:10.1186/s40068-016-0067-0.
12. Irmer, U. (1997). Bedeutung Von Hintergrundwerten für Qualitätsanforderung an Oberflächenwassern. IKSE – Workshop: Bewertung der Ergebnisse aus der Elbeschadstofforschung, Geesthacht, 36–40.
13. Jabłońska-Czapla, M., Nocoń, K., Szopa, S., & Łyko, A. (2016). Impact of the Pb and Zn ore mining industry on the pollution of the Biała Przemsza River, Poland. *Environmental Monitoring and Assessment, 188*, 262. doi:10.1007/s10661-016-5233-3.
14. Ji, H., Ding, H., Tang, L., Li, C., Gao, Y., & Briki, M. (2016). Chemical composition and transportation characteristic of trace metals in suspended particulate matter collected upstream of metropolitan drinking water source, Beijing. *Journal of Geochemical Exploration, 169*, 123–136.
15. Kabata-Pendias, A., Pendias, H. (1999). Biochemistry of trace elements. Państwowe Wydawnictwo Naukowe (in Polish). Warsaw, Poland.
16. Klimek, K. 1993. Antropogenne aluwia Przemszy i Wisły Śląskiej. Georama 1. (in Polish).
17. Larsen, B., Madsen, P. P., & Jensen, A. (1988). Sensitivity of sediments in pollution monitoring. In: report of the 75th Statutory Meeting of the Working Group of Marine Sediments in Relation to Pollution. Copenhagen: International Council for the Exploration of the Sea, pp. 61–78.
18. Lartiges, B. S., Deneux-Musrin, S., Villemin, G., Mustin, C., Barres, O., Chamerois, M., Gerard, B., & Babut, M. (2001). Composition, structure and size distribution of suspended particulates from the Rhine River. *Water Research, 35*(3), 808–816.
19. LAWA - LÄNDERARBEITSGEMEINSCHAFT WASSER (1998). Beurteilung der Wasserbeschaffenheit von Fließgewässern in der Bundesrepublik Deutschland – Chemische Gewässergüteklassifikation – Band 2, Berlin, 10,26.
20. Lis, J., & Pasieczna, A. (1995). Geochemical atlas of Upper Silesia. Wydawnictwo Kartograficzne Polskiej Agencji Ekologicznej S.A., (in Polish). Warsaw: Poland.
21. Martin, J. M., & Meybeck, M. (1979). Elemental mass balance of material carried by major world rivers. *Marine Chemistry, 7*, 173–206.
22. Miller, J. R., Lechler, P. J., Mackin, M. G., Germanoski, D., & Villarroel, L. F. (2007). Evaluation of particle dispersal from mining and milling operations using lead isotopic fingerprinting techniques, Rio Pilcomayo Basin, Bolivia. *The Science of the Total Environment, 384*, 355–373.
23. Olczyk, M. (2016). http://tyflomapy.pl/Aglomeracja_ Gornoslaska_i_ zrodla_ Wisly.html. Opracowanie tyflokartograficzne (in Polish). Accessed 13.05.2016.
24. Onderka, M., & Pekárová, P. (2008). Retrieval of suspended particulate matter concentrations in the Danube River from Landsat ETM data. *Science of the Total Environment, 397*(1–3), 238–243.
25. Pasieczna, A., Dusza-Dobek, A., & Markowski, W. (2010). Effect of coal mining and steel industry on Biała Przemsza and Bobrek water quality. *Górnictwo i Geologia, 5*(4), 181–190.
26. Regnier, P., & Wollast, R. (1993). Distribution of trace metals in suspended matter of the Scheldt estuary. *Marine Chemistry, 43*, 3–19.
27. Singh, K. P., Malik, A., Sinha, S., Singh, V. K., & Murthy, R. C. (2005). Estimation of source of heavy metal contamination in sediments of Gomti River (India) using principal component analysis. *Water Air Soil Pollution, 166*(1–4), 321–341.

28. Taylor, M. P., & Hudson-Edwards, K. A. (2008). The dispersal and storage of sediment-associated metals in an and river system: the Leichhardt River, Mount Isa, Queensland, Australia. *Environmental Pollution, 152*, 193–204.
29. Turekian, K. K., & Wedepohl, K. H. (1961). Distribution of the elements in some major units of the Earth's crust. *Geological Society of America Bulletin, 72*, 175–192.
30. Turner, A., & Millward, G. E. (2002). Suspended particles: their role in estuarine biogeochemical cycles. *Estuarine, Coastal and Shelf Science, 55*(6), 857–883.

CHAPTER 11

Influence of The Interaction of Calcium Carbonate Particles with Surfactants on The Degree of Water Pollution in Small Rivers

Petru Spataru[1], Francisco Fernandez[2], Joseph W. Sista[3], Tudor Spataru[3], Oxana Spinu[1] and Igor Povar[1]

[1]Institute of Chemistry of the Academy of Sciences of Moldova, 3 Academiei str., Chisinau, MD 2028, Republic of Moldova.

[2]Natural Sciences Department, Hostos Community College, 500 Grand Concourse, Bronx, NY 10451, USA.

[3]Chemistry Department, Columbia University, 3000 Broadway, New York, NY 10027, USA

ABSTRACT

Introduction

The influence of the interaction of calcium carbonate ($CaCO_3$) and surface-active substances (SAS; surfactants) with different chain lengths and cationic and anionic hydrophilic centers has been analyzed.

Results

Laboratory simulations indicate reduced negative influences on cationic SAS nitrification/self-purification processes in the presence of anionic species. This suggests the role of complex ionic formation [anionic SAS * cationic SAS] as a cause of this effect. UV-Vis spectra of lauryl sulfate (LS) and of cetyltrimethylammonium (CTMA), as well as of their mixtures in ratios of 2:1 and 1:1, treated by fine particles of $CaCO_3$, display decreased amounts of SAS in analyzed solutions and their presence on the surface of $CaCO_3$ nanoparticles. UV-Vis spectra reveal the decomposition of the complex [anionic SAS (SAS-An) * cationic SAS (SAS-Ct)] in solutions when $CaCO_3$ is added. CTMA can be bonded by LS through hydrophobic chains, on the surface of $CaCO_3$ particles. Therefore, $CaCO_3$modifies the nature of LS and CTMA interactions. This leads to an increased degree of toxicity of cationic SAS in aquatic environment. The amounts of CTMA in aqueous solutions are diminished in the presence of ammonium ion NH_4^+ (2 mg/L). In the presence of two orders higher concentration of ammonium ion, this effect strongly increases, making the association obvious. The structure of cationic SAS does not influence this effect. The obtained results have been confirmed both by timed natural aquatic sample analysis and laboratory simulations using water from Moldovan small rivers (Isnovat, Raut, and Bic).

Conclusions

UV-Vis spectra and laboratory simulations demonstrate the change due to the addition of calcium carbonate. Simulations and laboratory tests of water samples from Isnovat, Bic, and Raut Rivers, establish the cationic SAS negative influence on treatment and self-purification processes.

Keywords: Anionic and cationic surface-active substances Calcium carbonate Small rivers Cetyltrimethylammonium Lauryl sulfate Tetrabuty lammonium Toxicity Water pollution

INTRODUCTION

Toxic surfactants from pesticides that have polluted natural sources of water have become a serious problem in recent decades (Lechuga et al. 2016; Thomas et al. 2009). It is known that the presence of surfactants in water can also lead to the solubilization of other oil-soluble pollutants such as DDT and trichlorobenzene (Scott and Jones 2000).The reported toxicities of surfactants vary widely over several orders of magnitude. Consequently, the generalization and extrapolation of such types of data are difficult to complete. At the same time, anionic and nonionic surfactant as well as surface-active substance (SAS) builders are relatively non-toxic when compared to various cationic mono-alkyl and di-alkyl quaternary ammonium salts (Scott and Jones 2000). The influence of pesticides and metal ions on surfactant toxicity is hard to predict, although oil-surfactant mixtures are often found to be more toxic than what one would expect. Increases in water hardness and temperature typically lead to higher surfactant toxicity. The fate of all organic pollutants in waste water is determined by several processes including sorption to suspended solids. The adsorption of the cationic and anionic surfactants on carbonates has been investigated by many authors (Ma et al. 2013; Ying 2005). The presence of suspended solid particles and naturally occurring dissolved substances decreases the bioavailability of cationic surfactants but not that of anionic and nonionic ones (Lechuga et al. 2016; Thomas et al. 2009). In the presence of carbonate and silicate suspended particles, the SAS solutions manifest a specific behavior (Cui et al. 2010; Cui et al. 2012; Rosen and Kunjappu 2012; Spataru et al. 2015).

The Nistru River, downstream of Soroca town (Section Varancau), is polluted with pesticides via runoff by storm drains and sewage pipes. The presence of cationic detergents (Sandu et al. 2007; Sandu et al. 2008) can explain the difference between the samples from the Nistru River in Varancau village and the Cunicea village in terms of calcium carbonate impact (Fig. 1). In the sample without the carbonate substrate, nitrification is significant, obtaining high concentrations (12–14 mg/L) of nitrites. Formation of considerable quantities of nitrites shows that from day 11 to day 30 the rate of oxidation of ammonia and the $NO_3^- \rightarrow NO_2^-$ reduction exceeds the nitrite ion oxidation $NO_2^- \rightarrow NO_3^-$ (Philip et al. 2002; Spataru P 2011). In the sample with calcium carbonate ($CaCO_3$), the ammonium ion oxidation is stopped. The nitrite ion concentration is 0.4 mg/L while the nitrate ion concentration remains small in the analyzed sample within the studied period of time. In the same figure, the dynamics of the concentrations of NH_4^+ and NO_2^- ions for the river in the Cunicea village, whose waters do not contain synthetic surfactants, and the oxidation of ammonia is similar in both samples (with and without $CaCO_3$).

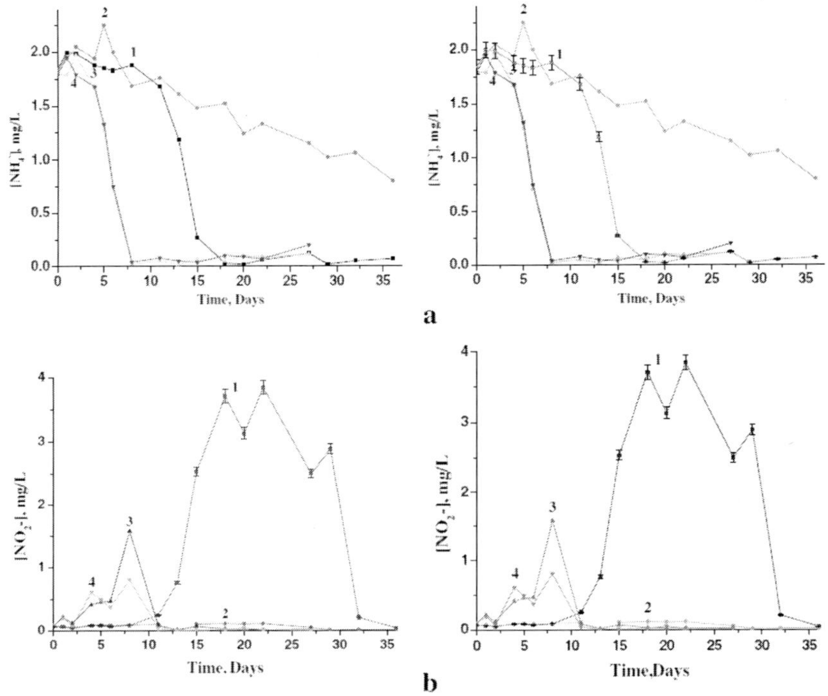

Figure. 1 Dynamics of ammonium (**a**) and nitrite (**b**) ion concentrations in the Nistru River (Spataru et al. 2015). *1*—Nistru River, Varancau village. *2*—Nistru River, Varancau village + $CaCO_3$. *3*—Cunicea village, small river. *4*—Cunicea village, small river + $CaCO_3$

For the samples taken from downstream Nistru and Prut Rivers, where massive amounts of pesticides from sewage treatment plants reach their waters, this phenomenon is revealed. Wastewaters accumulated in these treatment plants are mostly of domestic origin. In this case, the large amounts of pollutants in wastewaters constitute synthetic detergents. In urban centers, where machine laundering is increasingly practiced, the detergent solubility is not affected by the water hardness. This paper presents a continuation of our research of the water pollution in the Nistru and Prut Rivers with a series of surface-active substances (surfactants, SAS) for small river sections with a different urban influence. The need to investigate the impact of SAS on self-purification and, in particular, on the oxidation of reduced forms of nitrogen in rivers is related to their different influence on the abovementioned processes in river waters. The aim of this paper is to highlight the changes caused by the similar anthropic pollution for several most affected small rivers by this type of pollution adjacent to urban centers in the Republic of Moldova. At the same time, the influence of carbonate suspended particles on the redox processes of water-soluble nitrogen species has been investigated. Using laboratory simulations, the cationic SAS transfer from the soluble form on the suspended $CaCO_3$ particles in the presence of an anionic surfactant has been studied.

METHODS

A number of water samples were taken from the Isnovat River during October 2015. The tests of natural water were performed according to ISO methods published in specialized literature (ISO 7150–1 2001; ISO 8466–1 1990; SR ISO 7890–3 2000; SM SR EN 26777 2006; Coquery et al. 2005). Laboratory trials were initiated in glass vessels and respected the minimum recommended water-sample model volume (3 L). The same volume and conditions for the whole series of samples from laboratory simulations is essential (Sandu et al. 2014). A solution of $(NH_4)_2SO_4$ or NH_4Cl was added to each sample to obtain the ammonium ion concentration between 1.7 and 2.5 mg/L. In the samples, 2 g of chemically pure fine powder of $CaCO_3$ was added.

The purity of all the used substances corresponded to ISO (ISO 7150–1 2001; ISO 8466–1 1990; SR ISO 7890–3 2000; SM SR EN 26777 2006) requirements. The surface-active substances (lauryl sulfate (LS), cetyltrimethylammonium (CTMA), tetrabutylammonium (TBA)) with at least a purity of 98.5% were used. Water samples from the model were kept in natural lighting and away from direct sunlight. Laboratory simulations were performed under the static conditions. Stirring was performed after each series of tests. The tests were completed at the same time of day. The testing of samples was performed by the UV spectroscopy using Perkin Elmer Lambda 25 and LOMO SFD-46 at the Institute of Chemistry of the Academy of Sciences of Moldova. The contents of ammonium and nitrites in natural water were taken into account in all laboratory simulations.

RESULTS AND DISCUSSION

Spring samples from the Nistru River and the rivers of the Cunicea village were investigated by means of a model described in Spataru et al. 2015, using water samples both in the presence and absence of $CaCO_3$. Braking of the nitrogen oxidation process is known for cationic surfactants and amines (Spataru P 2011). This process has also been studied in the case of the mixture of cationic and anionic detergents. In the presence of anionic surfactants, the toxicity of cationic surfactants decreases (Sandu et al. 2007;

Sandu et al. 2008; Spataru P 2011). The changes described in Fig. 2a show that in the presence of $CaCO_3$ the braking of the ammonium ion oxidation takes place. The standard deviations of the ammonium and nitrite ion test methods are in the ranges of 0.036–0.045 and 0.004–0.027 mg/L, respectively. In the figures showing the dynamics of these indices, the error ranges are only indicated at one of the curves, avoiding the picture overload.

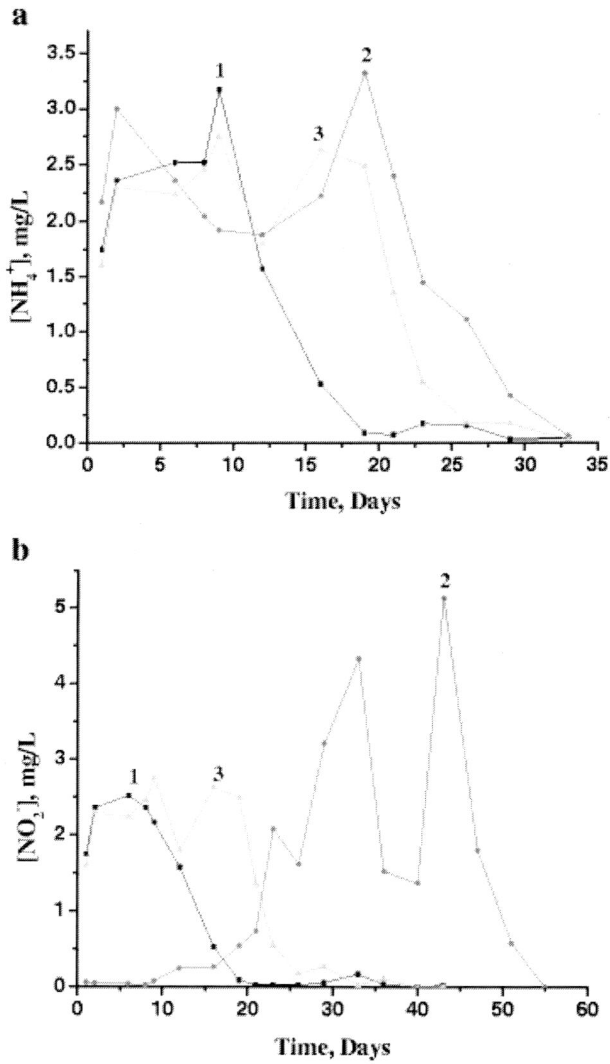

Figure. 2 Dynamics of ammonium (**a**) and nitrite (**b**) ion concentrations in the Nistru River. *1*—Reference sample. *2*—SAS-Ct. *3*—SAS-An + SAS-Ct

The study of dynamic processes in stable soluble forms of nitrogen by laboratory simulations demonstrates diminishing braking effect of nitrification processes in the Nistru River caused, perhaps, by decreasing toxicity in the presence of anionic SAS-Ct (Fig. 2). The cause of this effect may be explained by the fact of the complex [anionic SAS (SAS-An) * cationic SAS (SAS-Ct)] formation. On the contrary, the toxicity of cationic surfactants amplifies with increasing water hardness (Lechuga et al. 2016; Thomas et al. 2009). The phenomenon of the nitrification braking in the presence of $CaCO_3$ is characteristic for the water in sections downstream of wastewater treatment plants (WWTP) of cities. In the presence of mixtures of cationic and anionic surfactants, the preferential binding of anionic surfactants to calcium carbonate particles has been found (Cui et al. 2010; Cui et al. 2012). Consequently, it can be assumed that the decomposition of the complex of anionic and cationic surfactants contributes to the braking of nitrogen oxidation. Soluble surface-active cationic substances, which have bactericidal effect, remain. Enzyme concentration, which accelerates nitrogen oxidation, decreases. Thus, the $CaCO_3$ influence can be understood. Additionally, it is expected to suppose the formation of associates of $CaCO_3$ with anionic organic species, NH_4^+ and NH_2OH, which could also lead to braking of the $NH_4^+ \rightarrow NO_2^-$ process (Sandu et al. 2007; Sandu et al. 2008). The decomposition of the complex by treatment with fine powder of $CaCO_3$ was tested by the UV spectroscopy method.

The assessment of these changes was performed using the ability of the complex [SAS-Ct * SAS-An] to have similar but slightly different UV-Vis spectra to that of SAS-Ct. Thus, the complex [CTMA * LS] has a slightly higher spectral line configuration than the 210–250-nm range for CTMA. Using this difference, the change in concentration of this complex can be determined.

Rosen 2004 indicated elsewhere on complex formation between an ionic surfactant and an oppositely charged surfactant of approximately equal hydrophobic chain length with a composition 1:1 (e.g., $C_{10}H_{21}N(CH_3)_3^+$ * $C_{10}H_{21}SO_4^-$ or $C_{12}H_{25}N(CH_3)_3OH^+$ * $C_{12}H_{25}SO_3^-$). Two values of the ratio [CTMA]/[LS] were used in our experiments, 1:1 and 1:2, e.g., corresponding to that of complex stoichiometric composition and to excess of SAS-An. We studied spectra for solutions of CTMA (10^{-5} mol/L) alone and together with LS in the presence of $CaCO_3$particles, demonstrating considerable changes among the various aqueous solutions. The spectra movement of complex [CTMA * LS] solution in the absence of $CaCO_3$ particles and in presence of substrate when compared to that of CTMA solution alone demonstrates almost complete breakdown of the combination [CTMA * LS] caused by the presence of $CaCO_3$ nanoparticles.

Thus reducing spectrum solution [CTMA * LS] values in the presence of $CaCO_3$ and their proximity to the spectrum of only CTMA solution proves breakdown of the combination [CTMA * LS] caused by the presence of $CaCO_3$nanoparticles. It is necessary to note that in the solution remains SAS-Ct. The change of the combination may be by SAS orientation binding on the particulate carbonate surface, wherein the LS formed a first layer (Cui et al. 2010).

The UV-Vis spectra (Fig. 3) of the same quaternary ammonium (in combination with LS taken in different proportions) solution without and with $CaCO_3$ particles in their presence proves the impact of lauryl sulfate on the registered changes.

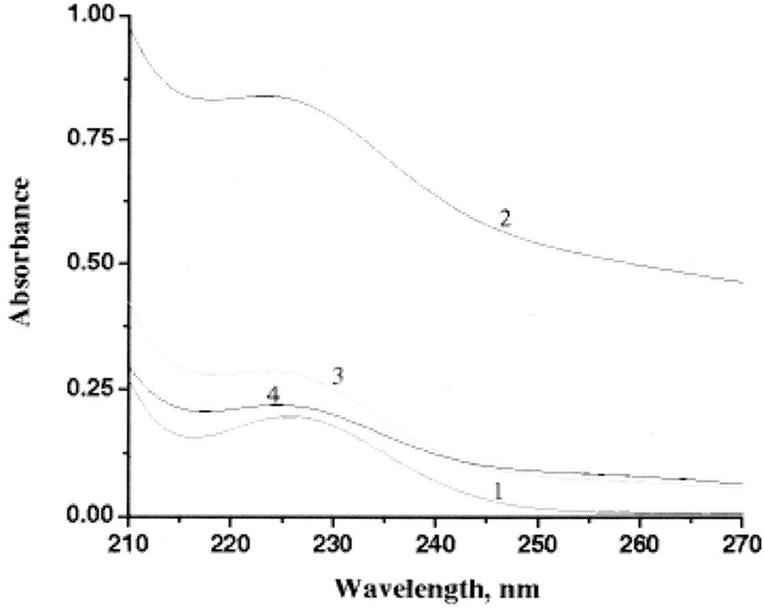

Figure. 3 UV-Vis spectra of CTMA solutions in the ratio 1:1 and 1:2 with LS, and the addition of $CaCO_3$. *1*—CTMA. *2*—CTMA + LS (1:1). *3*—CTMA + LS (1:1) + $CaCO_3$. *4*—CTMA + LS (1:2) + $CaCO_3$

UV-Vis spectra of lauryl, tetrabutylammonium solution with and without carbonate nanoparticles show a different behavior, which is not evident in any of the LS impact or $CaCO_3$ nanoparticles. Thus, the formation of the SAS-Ct * SAS-An and changes occurring under the impact of calcium carbonate particles is due to geometry, imposed by the mostly hydrophobic chain length of quaternary ammonium species.

In the simulation laboratory study, a surface change of $CaCO_3$ particles was highlighted differently on UV-Vis spectra of aqueous solution of CTMA and TBA, with and without NH_4^+. Both in solutions containing long chain of cetyltrimethylammonium and in those containing tetrabutylammonium with the same dimensions, upon adding ammonium ion (10^2 mg/L), the change of SAS concentration is evident.

UV-Vis spectra of aqueous solutions of only cationic SAS do not show the change in concentration upon addition of fine calcium carbonate particles. Instead, the presence of ammonium ion spectra demonstrates decreasing concentrations of these SAS. Very small effects are observed at 2 mg/L of NH_4^+; however, if the concentration of ammonium ions is increased by two orders of magnitude, the effect is marked. Thus, the surface of the calcium carbonate in the absence of ammonium ion selects for SAS-An only; however, upon adding an aqueous solution of NH_4^+, the surface of the $CaCO_3$ particles may adsorb both the anionic and the cationic SAS. This effect does not depend on the structure of cationic SAS. One can understand that change occurs at the interface of carbonate particles/solution.

Studying the dynamics of concentrations of ammonium and nitrite ions into the Isnovat River after a massive rain gives the opportunity to conduct complex research. Laboratory simulations include CTMA and LS samples separately as well as added together, and the same samples with suspended particles of $CaCO_3$ (Figs. 4 and 5). SAS sample (Fig. 4a) oxidation of ammonium ion dynamics highlights the CTMA and LS impact braking. This is without a contribution to reduce the harmful effects mentioned above. This effect can be caused by various substances on the cationogen sample's active surface until the adding the cationic surfactant provided in laboratory simulations. Thus, LS can be linked to other more stable units with substances reached by washing in water catchment areas. The SAS effect is evident only due to the dynamics of nitrite ions and organic matter, which stimulates the oxidation of nitrite ion comparing with ammonium ion oxidation. Due to the effect of the same SAS, and that of carbonate particles, the dynamics of concentrations of ions NH_4^+ and NO_2^- has a different runway (Fig. 5). In the sample with LS and $CaCO_3$, the retardation of oxidation as compared to the reference sample without calcium carbonate particles is evident. In this case we could speak about a braking effect of suspended $CaCO_3$ particles. This effect is characteristic only for natural waters polluted with organic cationogen substances.

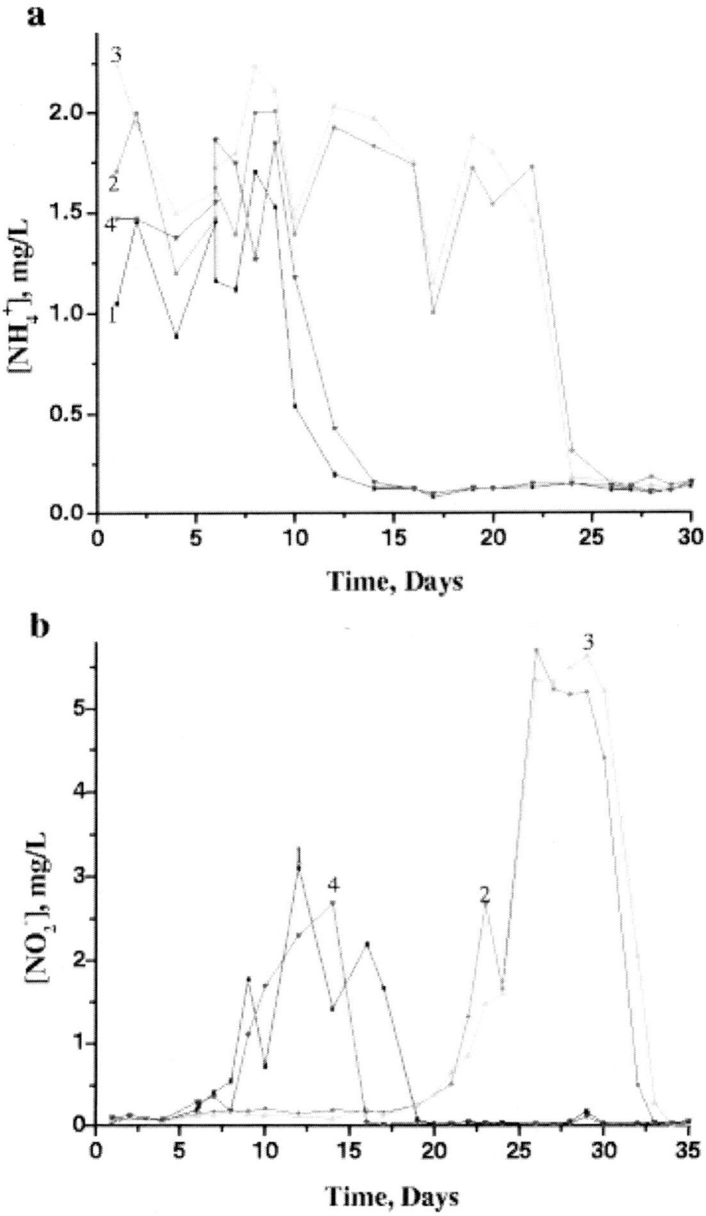

Figure. 4 Dynamics of ammonium (**a**) and nitrite (**b**) ion concentrations in the Isnovat River (October 2015) in the presence of CTMA, LS special, and their combination. *1*—Reference sample. *2*—10^{-5} mol/L CTMAB + LS. *3*—10^{-5} mol/L CTMAB. *4*—10^{-5} mol/L LS

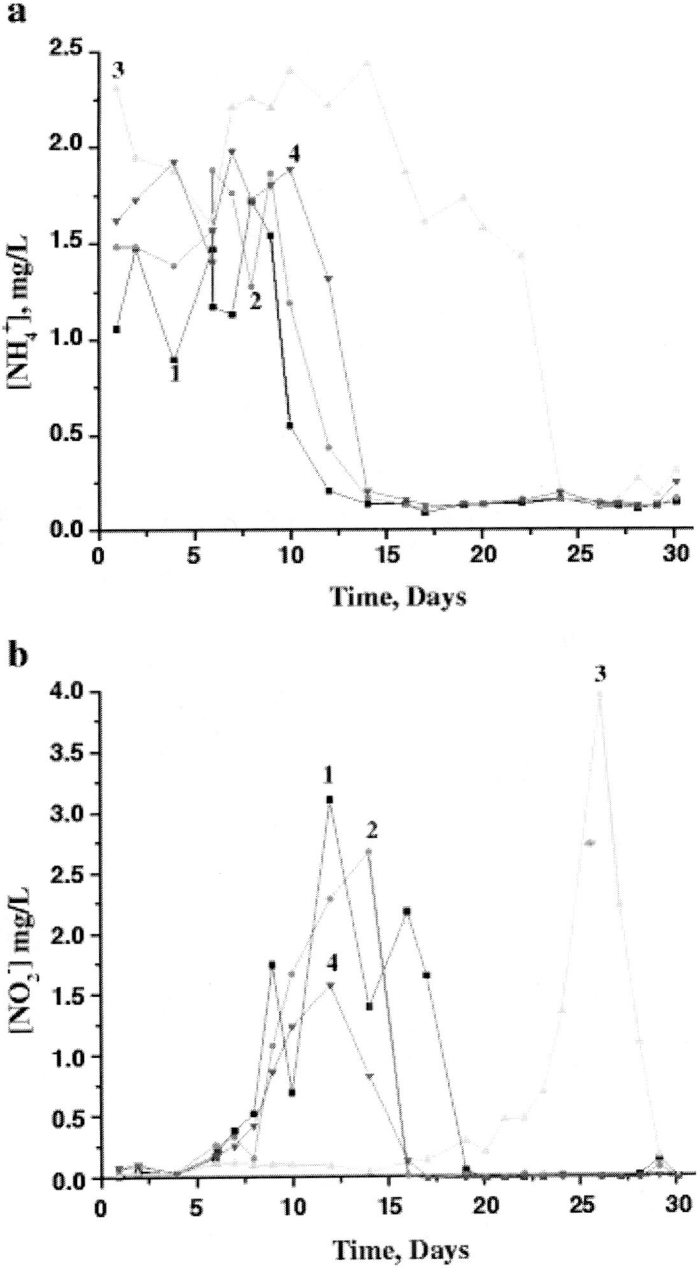

Figure. 5 Dynamics of ammonium (**a**) and nitrite (**b**) ion concentrations in the Isnovat River in the presence of CTMA, LS, and $CaCO_3$ nanoparticles. *1*—Reference sample. *2*—10^{-5} mol/L LS. *3*—10^{-5} mol/L CTMAB + $CaCO_3$. *4*—10^{-5} mol/L LS + $CaCO_3$

It should be noted that carbonaceous particles in suspension reduces formation time and oxidation of nitrite ions. Thus, the peak area of the dynamic NO_2^- is 2–10 times less than in samples with $CaCO_3$ particles as compared to samples with the same composition except that $CaCO_3$ particles are missing (Figs. 4b and 5b). Using published research of the persistence of relatively high nitrite ion concentrations (Philip et al. 2002), one could admit two causes that influence NO_2^- dynamics: competition of organic matter in the redox reactions for potential oxidants and substances which are toxic to bacteria generating nitrite-oxidative enzymes.

In the polluted aqueous environment of the Isnovat River, both versions are possible (Thomas et al. 2009; Spataru P 2011). They assume washing farmland producing secondary pollution by pesticides as well as the SAS from wastewater drain into the river.

Bic and Raut Rivers are more affected by an urban phenomenon than Isnovat. Data from the State Hydrometeorological Service (SHS) is available for mixing river water with water from a wastewater treatment plant (WWTP). In 2013, the Bic River physico-chemical tests showed that maximum allowable concentration (MAC) for the ammonium content exceeded from 40.4 to 78.1 times (in 10 of 12 samples); BOD_5 MAC (in 5 samples) exceeded from 50.3 to 79.67 times, and dissolved oxygen (in 11 samples) was below the fish subsistence level.

In 2014 exceedances of MAC for ammonium were from 40.2 to 84.6 times (in 11 from 12 samples) and MAC for biochemical oxygen demand after 5 days (CBO_5) was exceeded (in 9 samples) from 50.3 to 80.0 times, the dissolved oxygen in 10 samples of water of this river was below the subsistence fish. Braking self-purification/nitrification processes occur over several tens of kilometers to the confluence with the Nistru River. Table 1 shows some of the quality indicators that prove the deplorable state of the Bic River water. The Bic River situation starts from unpurified water shedding from WWTP of Chisinau into it and their impact on the Nistru River. Mixing of the Bic water with the Nistru one devolves on a more larger portion of the river than 200 m. However, the large impact of these pollutants and pesticides is marked.

Table 1 Indicators of water quality of the Nistru and Bic Rivers at their confluence

Indic./sample	Nistru River, upstream to the confluence of the Bic River	Bic River, upstream to the confluence of the Nistru River	Nistru River, 200 m downstream of the confluence with the Bic River
pH	8.18	7.62	7.71
$[NH_4^+]$, mg/L	0.257	51.0	42.3
$[NO_2^-]$, mg/L	0.156	0.253	0.242
$[NO_3^-]$, mg/L	5.6	0	0
Oxygen, mgO/L	8.71	0	0
CBO_5, mgO/L	0.67	97.8	93.4
CCO_{Cr}, mgO/L	17.0	150	144

Bic River water downstream of the treatment plant contains large amounts of organic matter (chemical oxygen demand (COD) = 433–480 mg/L). After shedding of water passed through the technology WWTP of Chisinau in Bic (Table 1) and derived from channeling the city's predominantly domestic wastewater (from residential apartments), one could assume that the population of bacteria specific to this medium has developed to the estuary inhabited of the Bic River in the Nistru River. The large concentration of organic substances (most of which being easily degradable) will cause further reduction of dissolved in water nitrogen to its reduced form, ammonia. Ratio CBO_5/COD confirms clearly that about 65% of organic substances in water are easily degradable biochemically in the section of the Bic River confluence with the Nistru River (Table 1). The cationic organic and organic halogen compounds (with impact braking processes of

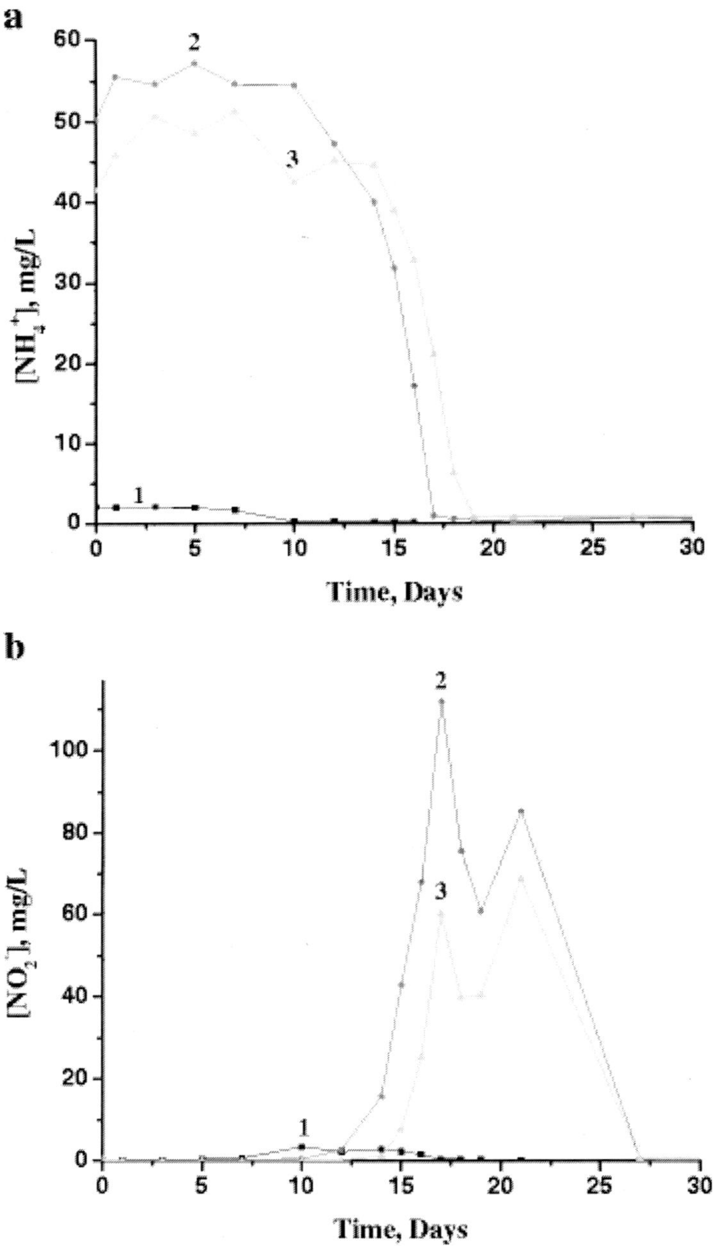

Figure. 6 Dynamics of ammonium (**a**) and nitrate (**b**) ion concentrations in the Bic and Nistru Rivers in natural samples. *1*—Nistru River upstream Bic River. *2*—Bic River watering place in Nistru River. *3*—Nistru River 200 m downstream of Bic River

nitrification/self cleaning) have relatively low organic matter river water. Therefore, the related associations with other species of the organic component may cause a toxic effect more pronounced only after a period of time in which a large consumption of the naturally organic matter occurs. Thus, it is logical that the braking effect is stronger in nitrite dynamic changes which occur after a certain time. That is why in samples of polluted water from the Bic and Nistru Rivers (diluted with the Nistru River water) relatively high concentrations persist longer than for nitrite water samples from the river. In ammonium ion dynamics, this difference does not exist (Fig. 6).

Operation below the required standards of WWTP Balti has the same cause. From our data, the most obvious and serious pollution occurs upstream and downstream of the Balti city, causing great concern for high ammonia content in the Raut River water downstream of Balti. There are multiple cases of upper limit of average concentrations of 7.03 mgN/L (18.02 MAC) and the maximum value detected being 22.3 mgN/L (57.18 MAC). These maximum concentrations are recorded during the spring and summer, when juvenile fish development is very sensitive even to low concentrations of ammonia. This impact in WWTP activity is caused by changing the composition of organic matter and especially in the increasingly significant SAS-Ct.

CONCLUSIONS

Dynamics of ammonium and nitrite ion concentrations in the Isnovat River highlights braking nitrification/self cleaning of the water samples containing a cationic surfactant as CTMA. The influence of $CaCO_3$ on the toxicity and SAS interaction with different long chain (hydrophobic) groups of organic radicals and electrostatic groups is obvious. UV-Vis spectra exhibit the interaction described above, which leads to a reduction of STA in the solution. Experiences have proved the changes in the SAS interaction with $CaCO_3$ particles, depending on the proportions of these substances in water. The higher the concentration ratio of SAS-An in aqueous solution (LS) compared to that in the cationic, the separation of the latter occurs completely. So, the transfer and the reorientation of CTMA to carbonate particles takes place depending on the amount of LS. So, the decomposition occurs in combining CTMA * LS and depositing them on nanoparticles, wherein the first bonding layer is formed with LS, and the rest are oriented based on the hydrophobic chain and group electrostatic interaction. This effect of shift results in a change of environmental toxicity. Both UV-Vis spectra and laboratory simulations establish the change due to the addition of calcium carbonate. Simulations and laboratory tests of water samples from Isnovat, Bic, and Raut Rivers demonstrate the SAS-Ct has a significant influence on treatment and self-purification processes.

ABBREVIATIONS

$CaCO_3$: Calcium carbonate

CBO_5 : Biochemical oxygen demand after 5 days

COD: Chemical oxygen demand

CTMA: Cetyltrimethylammonium

LS: Lauryl sulfate

MAC: Maximum allowable concentration

SAS: Surface-active substances (surfactants)

SAS-An: Anionic SAS

SAS-Ct: Cationic SAS

TBA: Tetrabutylammonium

WWTP: Wastewater treatment plant

ACKNOWLEDGEMENTS

The study is produced as part of the Diaspora Professional Return Program, which is a part of the Diaspora Engagement Hub, implemented by the Diaspora Relations Bureau of the State Chancellery of the Republic of Moldova in partnership with the International Organization for Migration, Mission to Moldova, in the framework of the "Consolidating Moldova's Migration and Development Institutional Framework" project, funded by the Swiss Agency for Development and Cooperation. Also, this research was supported, in part, under National Science Foundation Grants CNS-0958379 and CNS- 0855217 and the City University of New York High Performance Computing Center at the College of Staten Island and by the National Science Foundation through TeraGrid resources provided by the TeraGrid Science Gateways program under grants CHE090082 and CHE0000036.

REFERENCES

1. Coquery M, Morin A, Becue A, Lepot B (2005) Priority substances of the European Water Framework Directive: analytical challenges in monitoring water quality. TrAC Trends Anal Chem 24(2):117–127. doi:10.1016/j.trac.2004.11.004
2. Cui Z, Cui Y, Cui C, Chen Z, Binks B (2010) Aqueous foams stabilized by in situ surface activation of $CaCO_3$ nanoparticles via adsorption of anionic surfactant. Langmuir 26:12567–12574. doi:10.1021/la1016559
3. Cui Z, Li W, Qi J, Wang H (2012) Individual and mixed adsorption of alkylcarboxylbetaines and fatty amide ethoxylates at Daqing sandstone/water interface. Colloids Surf A Physicochem Eng Asp 414:180–189. doi:10.1016/j.colsurfa.2012.08.013
4. ISO 7150–1 (2001) Water quality—determination of ammonium—spectrometric method
5. ISO 8466–1 (1990) Water quality—calibration and evaluation of analytical methods and estimation of performance characteristics—part 1: statistical evaluation of the linear calibration function
6. Lechuga M, Fernández-Serrano M, Jurado E, Núñez-Olea J, Ríos F (2016) Acute toxicity of anionic and non-ionic surfactants to aquatic organisms. Ecotox Environ Safe 125:1–8. doi:10.1016/j.ecoenv.2015.11.027
7. Ma K, Leyu C, Dong Y, Wang T, Da C, Hirasaki GJ, Biswal SL (2013) Adsorption of cationic and anionic surfactants on natural and synthetic carbonate materials. J Coll Interface Sci 408:164–172
8. Philip S, Laanbroek HJ, Verstraete W (2002) Origin, causes and effects of increased nitrite concentrations in aquatic environments. Rev Environ Sci Biotechnol 1:115–141. doi:10.1023/A:1020892826575
9. Rosen M (2004) Surfactants and interfacial phenomena, 3rd edn. John Wiley & Sons, Inc., New Jersey
10. Rosen M, Kunjappu J (2012) Adsorption of surface-active agents at interfaces: the electrical double layer, in surfactants and interfacial phenomena, 4th edn. John Wiley & Sons, Inc., New Jersey
11. Sandu M, Spataru P, Negru M, Mosanu E, Dragutan D, Goreacioc T (2007) The dynamic of nitrification process in the presence of cationic surfactants. Paper presented at the International symposium "The Environment and Industry", V. 1, Bucharest, In Romanian
12. Sandu M, Spataru P, Mosanu E, Tarata A (2008) Nitrification capacity water tributaries of the Prut River and the factors that influence. Paper presented at the XXX-th Romanian Chemistry Conference, Romanian Academy, Calimanesti-Caciulata, In Romanian
13. Sandu M, Lupascu T, Tarita A, Goreacioc T, Turcan S, Mosanu E (2014) Methods for nitrate determination in water in the presence of nitrite. Chem J Moldova 9:8–13
14. Scott MJ, Jones MN (2000) The biodegradation of surfactants in the environment. Biochim Biophys Acta 1508:235–251
15. SM SR EN 26777 (2006) Water quality—determination of the content of nitrites. The method of the spectrometry of molecular absorption
16. Spataru P (2011) Transformations of organic substances in surface waters of Republic of Moldova. Dissertation, State University of Moldova., in Romanian
17. Spataru P, Povar I, Mosanu E, Trancalan A (2015) Study of stable nitrogen forms in natural surface waters in the presence of mineral substrates. Chem J Moldova 10:26–32
18. SR ISO 7890–3 (2000) Water quality—the determination of the content of nitrates. The part 3: the spectrometric method with sulfosalicylic acid
19. Thomas P, Velthoven K, Geurts M, van Wijk D (2009) Bioavailability and detoxification of cationics: II. Relationship between toxicity and CEC of cationic

surfactants on Caenorhabditis elegans (Nematoda) in artificial and natural substrates. Chemosphere 75:310–318. doi:10.1016/j.chemosphere.2009.02.003

20. Ying GG (2005) Fate, behavior and effects of surfactants and their degradation products in the environment. Environ Int 32:417–431

CHAPTER 12

Monitoring Recreational Waters: How to Integrate Environmental Determinants

Patricia Turgeon[1,2]

[1]Laboratory for Foodborne Zoonoses, Public Health Agency of Canada, Saint-Hyacinthe, Canada; [2]Groupe de Recherche en Épidémiologie des Zoonoses et Santé Publique, Université de Montréal, Saint-Hyacinthe, Canada.

ABSTRACT

Recreational waters are associated with a higher risk of disease for people engaged in activities that bring them into contact with these waters. The primary cause of contamination of recreational waters is fecal microorganisms, which may originate from various sources and involve several modulating factors, making it a complex public health and environmental issue. Monitoring recreational water quality should include two key components: Microbial water testing and monitoring environmental determinants associated with higher risks of contamination. Conducting both activities provides the foundation for a comprehensive assessment according to risk and the actual level of fecal pollution and thus could promote good management actions to ensure safe water quality. Nevertheless, monitoring of environmental determinants is rarely fully integrated in monitoring programs and is also harder to achieve, especially when water pollution is mainly associated with nonpoint sources. In order to achieve identification and monitoring of environmental determinants associated with fecal contamination of recreational waters, some specific steps should be followed and some questions must be answered. The objective of this review article is to present current knowledge on this topic and to suggest and discuss recommendations. Potential sources of contamination and factors able to modulate them should be identified and measured after the geographical area influencing fecal contamination of recreational water has been delineated. Statistical models have been developed to identify the relative importance of these environmental characteristics on fecal pollution of recreational waters but they do not allow for a full comprehension of the exact processes leading to this pollution, thus other methods should also be used to better understand these processes.

Keywords: Recreational waters; fecal pollution; environmental determinants; monitorin

1. INTRODUCTION

In the last few years, health risks associated with activeties in recreational waters were a growing preoccupation for public health communities all around the world, but more specifically for industrialized countries. This interest can be explained in part by the growing attendance on public beaches, driven by climatic and demographic changes. In the next decades, global changes, including climate changes, will cause important perturbations to aquatic ecosystems. Among these, degradation of microbiological quality of surface waters is expected [1]. Climate changes will bring an increase in the frequency and intensity of rain events which will be followed by an increasing amount of water reaching watercourses by runoff [2]. This increased runoff will carry a larger amount of microorganisms. Moreover, the growing frequency of rain events could overload the sewer systems and the wastewater treatment plants, which were not designed to work with and treat this amount of water. These overloads could prevent the appropriate performance of treatment plants and thus lead to the dumping of wastewater directly into watercourses.

Human populations can be exposed to impaired surface water through drinking water but also through recreational waters. Different types of pollution can affect recreational waters, such as fecal contamination, cyanobacteria, and chemical pollutants. Fecal contamination can result in a large amount of microorganisms that lead to various types of illnesses including gastrointestinal illnesses (GI) and respiratory, skin, and ear infections [3]. GI illnesses are the most frequent diseases associated with activities in recreational waters but they are usually mild and self-limiting, which can lead to some difficulties in their monitoring. Nonetheless, epidemiological studies have shown positive associations between sporadic cases of GI illness and activities in recreational waters and between fecal contamination of these waters and GI outbreaks [4-8].

Surveillance of the microbial quality of recreational waters is mainly done by measuring levels of fecal microbial indicators such as Escherichia coli or fecal coliforms. This measurement can provide an indication of water quality relatively quickly, usually in 24 - 48 hours following the sampling. Different microbiological technologies like QPCR have been developed to palliate this time limit, providing results in only a few hours [9,10]. Moreover, predictive models have been elaborated from various meteorological parameters such as rainfall and wind speed in an attempt to predict the level of fecal contamination according to these conditions [11,12]. In general, these methods aim to answer the question: Are recreational waters safe for people today or in the next few days? Although these measures are crucial for quick decision making on the likelihood of a microbial hazard, they do not allow a complete evaluation of the risk of fecal contamination.

To account for a broader assessment of risk, the World Health Organization (WHO) recommends the evaluation and monitoring of sources of fecal contamination and environmental characteristics (environmental determinants) influencing this contamination in addition to water testing [13]. Unlike rainfall events and temperature, these characteristics are relatively stable in time and could explain as much as 40% of the fecal contamination variation for a beach [14]. Knowing and monitoring these determinants could help ascertain which beaches have a higher risk of fecal contamination. Integrating a sound assessment of both risk and determinants into the monitoring of fecal pollution using microbial indicators would provide the base for a more comprehensive and accurate evaluation and thus would promote good management actions to ensure safe water quality.

Monitoring of the sources of fecal contamination can be performed in different ways. The WHO recommends the annual census of every potential source of contamination of recreational waters by field inspection activities [13]. This precise and relevant method for point source pollution can be very difficult to apply to water bodies associated mostly with non-point source (diffuse) pollution, like most fresh and inland water bodies. Moreover, annual field campaigns can be time consuming and also very demanding in terms of financial and human resources, resulting in a big challenge when trying to apply them across large territories. Therefore, there is a need to develop new efficient methods able to characterize the proximal environment of beaches and thus contribute to the identification and monitoring of environmental characteristics associated with a higher risk of fecal contamination.

In order to achieve this identification and monitoring for recreational waters, some specific steps should be followed and some questions must be answered. The objective of this article is to conduct a review of knowledge on this topic and to suggest and discuss four steps that should be followed:

- Identify potential sources of fecal contamination of recreational waters.
- Identify factors able to modulate this pollution.
- Delineate geographical area influencing the fecal pollution of water bodies.
- Identify and apply methods for quantitatively measuring these sources and factors and assessing their association with fecal pollution of recreational waters in order to establish best monitoring practices.

2. IDENTIFY POTENTIAL SOURCES OF FECAL CONTAMINATION OF RECREATIONAL WATERS

Sources of fecal contamination are various and can include urban, wildlife, and agricultural activities. Water contaminated by both animal and human sources of fecal microorganisms can represent a health risk for people engaged in recreational activities. Although it is assumed that, in general, animal sources of fecal contamination represent a smaller risk to human health than human sources [13], animals can carry zoonotic pathogens such as *Salmonella, Campylobacter*, and *E. coli* O157:H7, which may lead to gastrointestinal illness and sometimes severe sequelae [15,16]. Studies have indicated that risk of gastrointestinal illness associated with exposure to recreational waters impacted by animal feces may not be different from waters impacted by human sources. Nonetheless, human health risk from exposure to recreational waters impacted by non-human sources is still not well understood; some studies did not find statistically significant associations between illness risk and fecal contamination of water by animal sources [17,18].

2.1. Urban Activities

Urban fecal pollution can occur through different means. During rain events, waters can runoff by two major ways. They can reach watercourses via pluvial sewer systems, carrying many pollutants such as organic components and fecal pathogens coming from domestic wastes, domestic fauna, and wildlife [19]. Waters can also be directed to combined sewer systems that collect domestic industrial and pluvial waters [20]. However, during heavy rainfalls and rapid snow melts, water discharge coming into these systems can overwhelm wastewater treatment plant capacities and overload. Water containing pluvial waters and wastewater from domestic and industrial use can be discharged into watercourses by overload drains called combined sewer overflows (CSOs) [21]. CSOs may represent a health risk for people in contact with these waters since the principal source of bacteria it contains is from human waste, which can contain large quantitiesof fecal pathogens [22]. E. coli concentration associated with CSOs can reach more than 10^6/100ml, which it is much higher than recreational water guidelines [23]. Wastewater treated by wastewater treatment plants can also contain large amounts of fecal bacteria if these waters are not disinfected before being discharged into watercourses, placing people swimming in waters in close proximity to these discharges at higher risk for health problems [24].

Fecal pollution coming from urban activities and populations can also occur directly on the beaches. Bathers can be a source of fecal microorganisms via fecal accident, mainly where the proportion of children among bathers is high and if there are babies and toddlers in diapers [25,26]. Some outbreaks of *Shigella sonnei* (human pathogen) associated with activities in recreational freshwaters have been reported. In these outbreaks, the bacteria source mostly implicated were bathers who had been attending the beaches [27-29]. Bathers can also carry fecal microorganisms from the sand to the water and also stir up bottom sediment, making contact between bathers and the microorganisms trapped in sediment more frequent [30].

Domestic fauna such as dogs and cats can also be a source of fecal contamination. These animals can carry pathogens like species of *Campylobacter, Salmonella, Giardia,* and *Cryptosporidium* [31,32]. These microorganisms could reach watercourses mainly by runoff following rain events.

2.2. Wildlife

Wild animals can contribute to fecal pollution of recreational waters. Waterfowl have specifically been studied in the last decade [33,34]. These birds can be carriers of many fecal pathogens including species of *Campylobacter, Salmonella,* and *Cryptosporidium* and studies suggest they can be a source of contamination of recreational waters, particularly in rural regions [35-38]. Birds can pollute water by direct deposit of fecal material but also by contaminating the beach sand. Wild mammals can also carry fecal pathogens communicable to humans, although prevalence is usually low [39-41]. These animals can bring fecal microorganisms into watercourses by soil leaching and runoff during rain events and by direct deposit of fecal material.

2.3. Agricultural Activities

Agricultural activities can contribute to fecal pollution of waters and eventually recreational waters in many ways. Fecal pollution can come from animals and manure piles on farms and animal production sites. Livestock can carry and excrete in their feces zoonotic agents which may lead to various health problems from self-limiting GI disturbances to severe diseases that are potentially deadly. Zoonotic pathogens most often associated with farm animals are *Salmonella Enterica, Campylobacter jejuni* and coli, *E. coli* O157:H7, Giardia spp. and Cryptosporidium spp. [42,43]. These pathogens can reach watercourses mainly following rain events. Water contamination by stored manure on farms can result from leaking storage, overflows, or manure piles. A significant GI outbreak caused by *E. coli* O157:H7 and Campylobacter occurred in Walkerton, Canada in 2000 following a contamination of a municipal water well. Much of the evidence suggests that the origin of the contamination was runoff coming from a manure pile close to a cattle farm [44]. Manure spread on crop lands can also contain a large amount of fecal microorganisms [45]. As an example, fecal coliform concentrations measured in runoff waters following cattle manure spreading can reach 1.9×10^4 to 1.1×10^6/100ml, depending on the initial manure concentration [46]. Therefore, runoff waters coming from agricultural fields often exceed the drinking and microbiological guidelines for recreational waters [47]. Even if they can survive for many weeks on the soil after manure spreading, risk of water pollution by fecal microorganisms is highest soon after the spreading. On the other hand, some procedures can reduce the concentration of microorganisms in manure before spreading, such as aerobic or anaerobic digestion and composting [48,49].

In addition to manure spreading, animals at pasture can be responsible for water contamination through microorganisms present in their feces [50]. For instance, some agents can persist in soil for many months after exiting an animal's gastrointestinal tract [51,52]. The impact of animals at pasture can vary according to animal density. A high density can lead to a large quantity of urine and feces on a relatively small surface, increasing the probability of nutrients and microorganisms reaching surface waters through runoff and also getting through the soil to reach the ground waters [53]. High animal density can also accelerate soil erosion which can increase runoff during rain events. The impact of animals at pasture could also be associated with their distribution across the watershed and if they have access to a watercourse [54].

3. IDENTIFY FACTORS ABLE TO MODULATE FECAL POLLUTION OF RECREATIONAL WATERS

Some factors can modulate fecal contamination of waters by influencing the passage of fecal microorganisms to surface waters, including vegetation, climatic conditions, soil type, and topography.

3.1. Vegetation

Riparian zones can reduce the concentration of fecal microbes coming from agricultural lands and entering surface waters by more than 90% by preventing soil erosion and contributing to absorption of runoff water [55-57]. However, no scientific consensus exists on the minimal required width for an efficient reduction of microorganism transfer, although it is known that recommended width is affected by ground slope and vegetation type [58]. Wetlands can capture many elements brought by runoff such as sediments, microorganisms, and nutrients, reducing water pollution [59,60]. According to a study carried out in an agricultural watershed, wetlands could capture as much as 68% of *E. coli* coming from runoff waters [60]. Forest areas can also be associated with better water quality given their capacity to reduce runoff, sediments, and nutrients in surface waters. They could also have a positive influence on surface water quality by promoting water infiltration to the water tables [61].

3.2. Climatic Conditions

Rain events and their intensity are an important factor affecting the transport of microorganisms at the soil level. Runoff can carry bacteria a long distance downstream and thus contribute to water pollution [62-64]. As an example, risk of fecal contamination of watercourses following manure spreading during dry periods would be significantly lower than when manure is spread within a few days before rain events [64]. If there is no rain during days following manure spreading, bacteria movement to watercourses will diminish and under some environmental conditions—UV rays and desiccation—bacteria count will also diminish [65]. Moreover, during rainfall some bacteria such as *E. coli* would be quickly carried away by runoff waters and would have less chance to interact with the soil matrix [66]. Furthermore, studies have shown positives associations between rain events and waterborne disease outbreaks [67-69].

Exposure to solar rays could also be an important element in the bacterial inactivation process in water and soil [62]. Bacteria count in water decreases more rapidly during sunny days than during cloudy days, even at various depths [70]. In general, microorganism survival is prolonged in cold temperatures [71].

3.3. Soil Type and Topography

The two major modes of microorganism movement with water in soil are the infiltration to ground waters and runoff to surface waters. Relative proportions of these two modes depend on various elements including the type of soil [72,73]. Soil composition such as clay and soils with high levels of organic matter can influence the movement and survival of microorganisms. High clay concentrations can influence this survival by offering a protection against environmental stresses such as UV and desiccation. For example, *E. coli* counts in soil could decrease more rapidly in sandy soil than in clay soil [74]. Humidity level can also influence the microorganism movement in soil. The higher the soil humidity, the more contaminated the runoff will be following a rain event because the high humidity facilitates microorganism transport to watercourses [63].

Land slope also plays a role in microorganism movement. The more steep the slope, the more runoff will occur and at faster speeds, which promotes soil erosion and particle movement, which includes microorganisms [62,73].

4. DELINEATE GEOGRAPHICAL AREA INFLUENCING THE FECAL POLLUTION OF WATER BODIES

There are two ways by which surface waters and eventually recreational waters can be fecally polluted: they can be contaminated by point sources such as wastewater or storm water discharges; or they can be contaminated by non-point or diffuse sources. These sources can include some land uses as mentioned above—agricultural and urban lands. Diffuse pollution is harder to estimate and manage since it can be influenced by various characteristics of the drainage area such as topography and soil type and also by interactions between these characteristics and land uses.

Therefore when it is the time to identify and quantitatively evaluate the environmental characteristics or determinants that influence water quality, questions about the area influencing it the most remain not fully unanswered [75,76]. Those answers will inevitably vary according to the type of environment and type of water and will be different if we study surface waters in small vs. large watersheds, lakes vs. rivers, fresh vs. marine waters, or chemical vs. microbiological pollution. Despite the fact that a single answer is impossible to reach, since each water body and watershed has its own and unique combination of characteristics, few studies have tried to find definitive answers applicable to particular situations in microbiological pollution [14,76,77].

The study of Sliva *et al.* [76] compared two approaches to studying the impact of land use on chemical and microbiological water quality in three watersheds. Water sampling stations were located on rivers and microbiological quality was evaluated according to the fecal coliform level. The first approach was to measure landscape characteristics in an entire catchment area and to determine if there was a correlation with water quality measurements in the corresponding catchment. The second approach was to measure those same characteristics but only for a 100 m buffer around each sampling station. Results showed that the correlation between fecal coliforms and landscape predictors was different depending on the season of sampling. During the summer the correlation was better with the 100 m buffer zone data but during the fall, when the overland runoff is increased, the correlation was better with data on the entire catchment area.

Results of the study of Crowther *et al.* [77] also showed a difference in the relationship between land use and river water quality depending on if the sampling was done under baseor high-flow conditions. During baseflow conditions, *E. coli* levels seemed to be influenced more by land use within 2 km of the sampling station than the land within the whole catchment. Under highflow conditions, although the best correlation between E. coli levels and landscape data was seen at the whole catchment level, land use within 5 km was still able to explain more than 70% of the contamination variation. The study of Turgeon et al. [14] focused on recreational freshwaters associated with a lake and sampling was performed during the bathing season (mid-June to end of August). According to the results shown in Crowther et al., two buffer zones were delineated within the catchment area of each lake, one 2 km in size and another 5 km from the sampling point (beach) (Figure 1).Various land use and geo-hydrological characteristics were measured for each zone and the relationship between these characteristics and the fecal coliform levels was assessed. According to their results, the level of fecal coliform was influenced by agricultural activities and the additional 3 km of the 5 km buffer zone did not bring any additional information on the risk of contamination, suggesting that the most influence on water quality is from within the first 2 km.

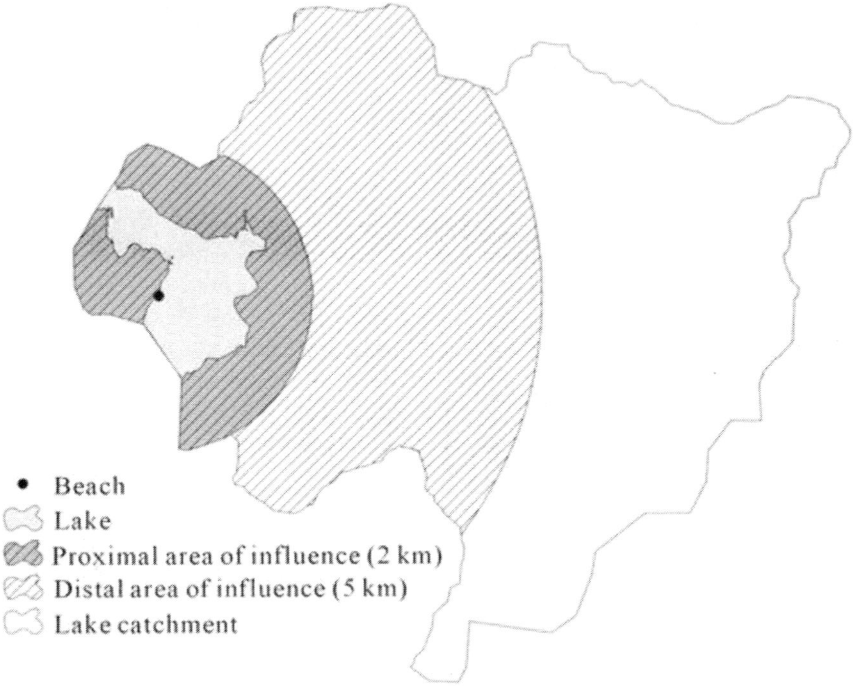

Figure 1. Two buffer zones used to delineate the geographical area influencing the fecal pollution of water bodies in the article of Turgeon et al. (With kind permission from Springer Science +Business Media: Water Quality, Exposure and Health, Fecal Contamination of Recreational Freshwaters: the Effect of Time-Independent Factors, Vol. 3 (2), 2011, 109-118, P. Turgeon et al., Figure 1).

Although results of these studies differ according to their objectives and methods, one trend seems to emerge. During a dry season, microbiological surface water quality would be more influenced by land use characteristics located nearest the sampling station versus by the whole catchment, reflecting a possible die-off or sedimentation process along watercourses during this period.

5. QUANTITATIVELY MEASURING THESE SOURCES AND FACTORS AND THEIR ASSOCIATION WITH FECAL POLLUTION OF RECREATIONAL FRESHWATER

After having identified potential sources of fecal contamination and their modulating factors and delineating the area where they should be monitored more closely, the next step should be to quantitatively measure them in the areas of interest. Various sources of data can be used for this such as census data, regional or national thematic maps, and field survey data [14,78-80]. National census data from human populations can be useful for evaluating density or some characteristics such as age or socioeconomic distribution [81]. Census data from animal populations can also provide information on population density and species distribution or contribute to the evaluation of some management practices such as manure spreading [82]. The advantages of this data source are that data are usually standardized at the national level, available freely and for a long period of time. However, census data are collected most of the time by administrative regions or districts or centralized by farms or businesses.

These methods can bring spatial uncertainty which can be very relevant in the context of a hydrological process where the location of the source of contamination in regard to runoff direction and topography can play an important role [83,84]. Moreover, census data are usually constrained in regards to timeline. As an example, in Canada, population census is done every four years and agricultural census every five years, which can prevent availability of up-to-date data [81,82].

Data on landscape, land use, or on some environmental characteristics such as climatic conditions or topography can be acquired from regional or national thematic maps [80,85]. One advantage of using these maps is that one map is usually able to provide data on wide territories for more than one characteristic. Most of the time, data can be relatively easy to extract and thus can be ready to use with little manipulation. Nonetheless, like census processes, thematic maps are usually not performed every year and the timeline can be restrictive in a monitoring context. Furthermore, spatial resolution of these maps may limit their use or contribute to some measurement errors when data are needed for small areas [80]. Field surveys can provide accurate, precise, and up-to-date data on small areas, but one downside of this method is the high cost in terms of time and human and financial resources.

Lastly, land usage and landscape characteristics can also be provided by satellite imagery and more specifically Earth observation imagery [86,87].

These products can provide data on land use and land cover known to influence surface and recreational water quality such as agricultural lands, impervious (built) surfaces, wetlands, and forest areas [88-91]. This source of data has many advantages over more conventional sources. First, satellites can have a very large coverage and thus provide information on wide territories [92,93]. Satellite data from a particular sensor are always collected the same way and usually for a long period of time, allowing for a large amount of repeatable and consistent data which can be very relevant for a monitoring program [94]. Data from Earth observation can be highly precise depending on the sensor; knowing the precise data on characteristics associated with fecal contamination can be very relevant in the context of a hydrological matter. However, some factors can make their use difficult for recreational water monitoring, such as image cost and the expertise and technical resources needed for the gathering, processing, and analysing of those data [95-97]. As all sources of data present advantages and limitations, a combination of more than one source is probably the best way to tackle this step depending on the environment and the recreational waters involved.

Once the environmental characteristics associated with potential sources of fecal contamination and their modulator factors have been measured, it can be pertinent to try to identify which of these determinants have the most influence on recreational waters in the territory of interest. Most of the time, statistical methods that were used to achieve this step were multivariate regression modelling and Pearson's correlation coefficient [14,77-79,98]. Knowledge resulting from this step could provide the answer to the question: Which beaches have a higher risk of fecal contamination? Identification of those beaches combined with water sampling would provide the basis for an overall evaluation according to the risk and the actual level of fecal pollution, allowing for a better targeted monitoring of recreational waters.

6. CONCLUSION AND FUTURE DIRECTIONS

Based on a literature review, this paper aimed to suggest and discuss steps that should be followed for the integration of environmental determinants in the monitoring of recreational waters. Initially, potential sources of contamination and factors able to modulate them should be identified. Various sources and factors can be involved and depend on the region of interest. Measuring these sources and factors for monitoring purposes can only be done after delineating the geographical area influencing fecal contamination of recreational waters. No scientific consensus exists on this matter and few studies have participated in the debate. Nevertheless, in dry periods, which largely correspond to the bathing season, the proximal environment seems to play a more important role in the pollution process than the whole catchment does. Studies using mathematical and hydrological modelling would certainly help to bring new knowledge on this topic. Various sources of data can be used to provide information on sources of contamination and modulating factors and could be used to measure them on the geographical area of interest. Since none of them is sufficient in itself to provide all the information needed to fully characterize the proximal environment of recreational waters, a combination of more than one would be the best way to measure environmental characteristics. Some studies have used statistical models to identify the relative importance of environmental characteristics on fecal pollution of recreational waters. Although these models provide relevant information on environmental characteristics associated with a higher risk of fecal contamination and thus can serve as a good working basis for the integration of the monitoring of the environmental determinants of recreational water quality, they do not allow for a full comprehension of the exact processes leading to this pollution. Source attribution methods should be used to better understand these processes and to provide information on a better estimation of the precise contribution of each source of contamination.

Monitoring environmental determinants of water quality should be integrated with monitoring programs of recreational waters. In addition to water testing, it would provide an overall evaluation according to the risk of contamination and the actual level of fecal pollution, promoting in turn good water management actions to ensure safe water quality [13]. Moreover, knowing and monitoring these determinants could promote preventive actions to diminish their impact on water quality. In addition to contributing to better protecting the public, this preventive approach would align with the multi-barrier approach promoted in the domain of water quality [99, 100].

7. ACKNOWLEDGEMENTS

Patricia Turgeon, The author wishes to thank Dr. Pascal Michel of the Public Health Agency of Canada for his helpful comments on this paper.

REFERENCES

1. J. B. Rose, P. R. Epstein, E. K. Lipp, B. H. Sherman, S. M. Bernard and J. A. Patz, "Climate Variability and Change in the United States: Potential Impacts on Waterand

Foodborne Diseases Caused by Microbiologic Agents [Review]," Environmental Health Perspectives, Vol. 109, Suppl. 2, 2001, pp. 211-221.

2. Intergovernmental Panel on Climate Change, "Climate Change 2007: Synthesis Report," Valencia, Spain, 2007. http://www.ipcc.ch/pdf/assessment-report/ar4/syr/ar4_syr.pdf
3. Pruss, "Review of Epidemiological Studies on Health Effects from Exposure to Recreational Water," International Journal of Epidemiology, Vol. 27, No. 1, 1998, pp. 1-9. doi:10.1093/ije/27.1.1.
4. K. A. Feldman, J. C. Mohle-Boetani, J. Ward, K. Furst, S. L. Abbott, D. V. Ferrero, A. Olsen and S. B. Werner, "A Cluster of Escherichia coli O157: Nonmotile Infections Associated with Recreational Exposure to Lake Water," Public Health Reports, Vol. 117, No. 4, 2002, pp. 380- 385.
5. M. G. Bruce, M. B. Curtis, M. M. Payne, R. K. Gautom, E. C. Thompson, A. L. Bennett and J. M. Kobayashi, "Lake-Associated Outbreak of Escherichia coli O157:H7 in Clark County, Washington, August 1999," Archives of Pediatrics & Adolescent Medicine, Vol. 157, No. 10, 2003, pp. 1016-1021. doi:10.1001/archpedi.157.10.1016
6. Wiedenmann, P. Kruger, K. Dietz, J. M. Lopez-Pila, R. Szewzyk and K. Botzenhart, "A Randomized Controlled Trial Assessing Infectious Disease Risks from Bathing in Fresh Recreational Waters in Relation to the Concentration of Escherichia coli, Intestinal Enterococci, Clostridium perfringens and Somatic coliphages," Environmental Health Perspectives, Vol. 114, No. 2, 2006, pp. 228-236. doi:10.1289/ehp.8115
7. Sartorius, Y. Andersson, I. Velicko, B. De Jong, M. Löfdahl, K.-O. Hedlund, G. Allestam, C. Wangsell, O. Bergstedt, P. Horal, P. Ulleryd and A. Soderstrom, "Outbreak of Norovirus in Västra Götaland Associated with Recreational Activities at Two Lakes during August 2004," Scandinavian Journal of Infectious Diseases, Vol. 39, No. 4, 2007, pp. 323-331. doi:10.1080/00365540601053006
8. J. M. Fleisher, L. E. Fleming, H. M. Solo-Gabriele, J. K. Kish, C. D. Sinigalliano, L. Plano, S. M. Elmir, J. D. Wang, K. Withum, T. Shibata, M. L. Gidley, A. Abdelzaher, G. Q. He, C. Ortega, X. F. Zhu, M. Wright, J. Hollenbeck and L. C. Backer, "The BEACHES Study: Health Effects and Exposures from Non-Point Source Microbial Contaminants in Subtropical Recreational Marine Waters," International Journal of Epidemiology, Vol. 39, No. 5, 2010, pp. 1291-1298. doi:10.1093/ije/dyq084
9. J. W. Santo Domingo, S. C. Siefring and R. A. Haugland, "Real-Time PCR Method to Detect Enterococcus faecalis in Water," Biotechnology Letters, Vol. 25, No. 3, 2003, pp. 261-265. doi:10.1023/A:1022303118122
10. T. J. Wade, R. L. Calderon, E. Sams, M. Beach, K. P. Brenner, A. H. Williams and A. P. Dufour, "Rapidly Measured Indicators of Recreational Water Quality Are Predictive of Swimming-Associated Gastrointestional Illness," Environmental Health Perspectives, Vol. 114, No. 1, 2006, pp. 24-28. doi:10.1289/ehp.8273
11. G. A. Olyphant and R. L. Whitman, "Elements of a Predictive Model for Determining Beach Closures on a Real Time Basis: The Case of 63rd Street Beach Chicago," Environmental Monitoring & Assessment, Vol. 98, No. 1-3, 2004, pp. 175-190. doi:10.1023/B:EMAS.0000038185.79137.b9
12. D. S. Francy, R. A. Darner and E. E. Bertke, "Models for Predicting Recreational Water Quality at Lake Erie Beaches," US Geological Survey. Scientific Investigations Report 2006-5192, 2006.
13. World Health Organization, "Guidelines for Safe Recreationnal Water Environments. Coastal and Fresh Waters," Geneva, Switzerland, 2003, http://www.who.int/water_sanitation_health/bathing/srwe1/en/
14. P. Turgeon, P. Michel, P. Levallois, M. Archambault and A. Ravel, "Fecal Contamination of Recreational Freshwaters: The Effect of Time-Independent Agroenvironmental Factors," Water Quality, Exposure and Health, Vol. 3, No. 2, 2011, pp. 109-118. doi:10.1007/s12403-011-0048-5

15. M. L. Hutchison, L. D. Walters, S. M. Avery, B. A. Synge and A. Moore, "Levels of Zoonotic Agents in British Livestock Manures," Letters in Applied Microbiology, Vol. 39, No. 2, 2004, pp. 207-214. doi:10.1111/j.1472-765X.2004.01564.x
16. E. M. Moriarty, L. W. Sinton, M. L. Mackenzie, N. Karki and D. R. Wood, "A Survey of Enteric Bacteria and Protozoans in Fresh Bovine Faeces on New Zealand Dairy Farms," Journal of Applied Microbiology, Vol. 105, No. 6, 2008, pp. 2015-2025. doi:10.1111/j.1365-2672.2008.03939.x
17. R. Calderon, M. E. and A. Dufour, "Health Effects of Swimmers and Non-Point Sources of Contaminated Water," International Journal of Environmental Health Research, Vol. 1, 1991, pp. 21-31.
18. J. M. Colford Jr., T. J. Wade, K. C. Schiff, C. C. Wright and J. F. Griffith, "Water Quality Indicators and the Risk of Illness at Beaches with Non-Point Sources of Fecal Contamination," Epidemiology, Vol. 18, 2007, pp. 27-35. doi:10.1097 /01.ede.0000249425.32990.b9
19. V. P. Olivieri, K. Kawata and S. H. Lim, "Microbiological Impacts of Storm Sewer Overflows: Some Aspects of the Implication of Microbiological Indicators for Receiving Waters," In: J. B. Ellis, Ed., Urban Discharges and Receiving Water Quality Impacts, Pergamon Press, Oxford, 1989, pp. 47-54. [Citation Time(s):1]
20. Environment Canada, "Wastewater Management," Environment Canada, 2009. http://www.ec.gc.ca/eu-ww/default.asp?lang=En&n=0FB32EFD-1
21. United States Environmental Protection Agency, "Report to Congress. Impacts and Control of CSOs and SSOs," Washington DC, 2004. http://cfpub.epa.gov/npdes/cso/cpolicy_report2004.cfm[Citation Time(s):1]
22. J. A. Castro-Hermida, I. Garcia-Presedo, A. Almeida, M. Gonzalez-Warleta, J. M. C. Da Costa and M. Mezo, "Contribution of Treated Wastewater to the Contamination of Recreational River Areas with Cryptosporidium spp. and Giardia duodenalis," Water Research, Vol. 42, No. 13, 2008, pp. 3528-3538. doi:10.1016/j.watres.2008.05.001
23. J. Marsalek and Q. Rochfort, "Urban Wet-Weather Flows: Sources of Fecal Contamination Impacting on Recreational Waters and Threatening Drinking-Water Sources," Journal of Toxicology and Environmental Health, Vol. 67, No. 20-22, 2004, pp. 1765-7717. doi:10.1080/15287390490492430
24. P. Payment, R. Plante and P. Cejka, "Removal of Indicator Bacteria, Human Enteric Viruses, Giardia Cysts, and Cryptosporidium Oocysts at a Large Wastewater Primary Treatment Facility," Canadian Journal of Microbiology, Vol. 47, No. 3, 2001, pp. 188-193.
25. C. P. Gerba, "Assessment of Enteric Pathogen Shedding by Bathers during Recreational Activity and its Impact on Water Quality," Quantitative Microbiology, Vol. 2, 2000, pp. 55-68.
26. D. Sunderland, T. K. Graczyk, L. Tamang and P. N. Breysse, "Impact of Bathers on Levels of Cryptosporidium parvum Oocysts and Giardia lamblia Cysts in Recreational Beach Waters," Water Research, Vol. 41, No. 15, 2007, pp. 3483-3489. doi:10.1016/j.watres.2007.05.009
27. J. Blostein, "Shigellosis from Swimming in a Park Pond in Michigan," Public Health Reports, Vol. 106, No. 3, 1991, pp. 317-322.
28. W. E. Keene, J. M. McAnulty, F. C. Hoesly, L. P. Williams, K. Hedberg, G. L. Oxman, T. J. Barrett, M. A. Pfaller and D. W. Fleming, "A Swimming-Associated Outbreak of Hemorrhagic Colitis Caused by Escherichia coli O157:H7 and Shigella Sonnei," New England Journal of Medicine, Vol. 331, No. 9, 1994, pp. 579-584. doi:10.1056/NEJM199409013310904
29. M. Iwamoto, G. Hlady, M. Jeter, C. Burnett, C. Drenzek, S. Lance, J. Benson, D. Page and P. Blake, "Shigellosis among Swimmers in a Freshwater Lake," Southern Medical Journal, Vol. 98, No. 8, 2005, pp. 774-778. doi:10.1097/ 01.smj.00001 72764.14147.e5
30. T. K. Graczyk, D. Sunderland, G. N. Awantang, Y. Mashinski, F. E. Lucy, Z. Graczyk, L. Chomicz and P. N. Breysse, "Relationships among Bather Density, Levels of Human Waterborne Pathogens, and Fecal Coliform Counts in Marine Recreational Beach

Water," Journal of Parasitology Research, Vol. 106, No. 5, 2010, pp. 1103-1108. doi:10.1007/s00436-010-1769-2.

31. S. Hill, J. M. Cheney, G. F. Taton-Allen, J. S. Reif, C. Bruns and M. R. Lappin, "Prevalence of Enteric Zoonotic Organisms in Cats," Journal of the American Vetrinary Medical Association, Vol. 216, No. 5, 2000, pp. 687-692. doi: 10.2460/javma.2000.216.687

32. T. Hackett and M. R. Lappin, "Prevalence of Enteric Pathogens in Dogs of North-Central Colorado," Journal of the American Animal Hospital Association, Vol. 39, 2003, pp. 52-56.

33. L. R. Fogarty, S. K. Haack, M. J. Wolcott and R. L. Whitman, "Abundance and Characteristics of the Recreational Water Quality Indicator Bacteria Escherichia coli and Enterococci in Gull Faeces," Journal of Applied Microbiology, Vol. 94, No. 5, 2003, pp. 865-878. doi:10.1046/j.1365-2672.2003.01910.x

34. S. K. Haack, L. R. Fogarty and C. Wright, "Escherichia coli and Enterococci at Beaches in the Grand Traverse Bay, Lake Michigan: Sources, Characteristics, and Environmental Pathways," Environmental Science & Technology, Vol. 37, No. 15, 2003, pp. 3275-3282. doi:10.1021/es021062n

35. Z. Hubalek, "An Annotated Checklist of Pathogenic Microorganisms Associated with Migratory Birds [Review]," Journal of Wildlife Diseases, Vol. 40, No. 4, 2004, pp. 639-659.

36. K. J. Meyer, C. M. Appletoft, A. K. Schwemm, K. Uzoigwe and E. J. Brown, "Determining the Source of Fecal Contamination in Recreational Waters," Journal of Environmental Health, Vol. 68, No. 1, 2005, pp. 25-30.

37. T. A. Edge and S. Hill, "Multiple Lines of Evidence to Identify the Sources of Fecal Pollution at a Freshwater Beach in Hamilton Harbour, Lake Ontario," Water Research, Vol. 41, No. 16, 2007, pp. 3585-3594. doi:10.1016/j.watres.2007.05.012

38. T. K. Graczyk, A. C. Majewska and K. J. Schwab, "The Role of Birds in Dissemination of Human Waterborne Enteropathogens," Trends in Parasitology, Vol. 24, No. 2, 2008, pp. 55-59. doi:10.1016/j.pt.2007.10.007

39. V. R. Simpson, "Wild Animals as Reservoirs of Infectious Diseases in the UK," The Veterinary Journal, Vol. 163, No. 2, 2002, pp. 128-146. doi:10.1053/tvjl.2001.0662

40. K. Handeland, L. L. Nesse, A. Lillehaug, T. Voikoren, B. Djonne and B. Bergsjo, "Natural and Experimental Salmonella typhimurium Infections in Foxes (Vulpes vulpes)," Veterinary Microbiology, Vol. 132, No. 1-2, 2008, pp. 129-134. doi:10.1016/j.vetmic.2008.05.002

41. C. Jardine, R. J. Reid-Smith, N. Janecko, M. Allan and S. A. McEwen, "Salmonella in Racoons (Proycon lotor) in Southern Ontario, Canada," Journal of Wildlife Diseases, Vol. 47, No. 2, 2011, pp. 344-351.

42. R. Majdoub, C. Côté, M. Labadi, K. Guay and M. Généreux, "Impact de l'Utilisation des Engrais de Ferme sur la Qualité Microbiologique de l'eau Souterraine," Instituts de Recherche et de Développement en AgroEnvironnement, Québec, 2003.

43. P. Chevalier, P. Levallois and P. Michel, "Infections Entériques d'Origine Hydrique Potentiellement Associées à la Production Animale: Revue de la Littérature," Vecteur Environnement, Vol. 37, No. 2, 2004, pp. 90-106.

44. Government of Canada, "Waterborne Outbreak of Gastroenteritis Associated with a Contaminated Municipal Water Supply, Walkerton, Ontario, May-June 2000," Canada Communicable Disease Report, Vol. 26, No. 20, 2000, pp. 170-173.

45. J. A. Thurston-Enriquez, J. E. Gilley and B. Eghball, "Microbial Quality of Runoff Following Land Application of Cattle Manure and Swine Slurry," Journal of Water & Health, Vol. 3, No. 2, 2005, pp. 157-171.

46. M. C. Ramos, J. N. Quinton and S. F. Tyrrel, "Effects of Cattle Manure on Erosion Rates and Runoff Water Pollution by Faecal Coliforms," Journal of Environmental Management, Vol. 78, No. 1, 2006, pp. 97-101. doi:10.1016/j.jenvman.2005.04.010

47. Health Canada, "Guidelines for Canadian Recreationnal Water Quality," Ottawa, 1992. http://www.hc-sc.gc.ca/ewh-semt/pubs/water-eau/guide_water-1992-guide_eau/index-eng.php

48. X. P. Jiang, J. Morgan and M. P. Doyle, "Fate of Escherichia coli O157:H7 during Composting of Bovine Manure in a Laboratory-Scale Bioreactor," Journal of Food Protection, Vol. 66, No. 1, 2003, pp. 25-30.
49. L. M. Avery, K. Killham and D. L. Jones, "Survival of E. coli O157:H7 in Organic Wastes Destined for Land Application," Journal of Applied Microbiology, Vol. 98, No. 4, 2005, pp. 814-22. doi:10.1111/j.1365-2672.2004.02524.x
50. J. A. Vinten, J. T. Douglas, D. R. Lewis, M. N. Aitken and D. R. Fenlon, "Relative Risk of Surface Water Pollution by E. coli Derived from Faeces of Grazing Animals Compared to Slurry Application," Soil Use & Management, Vol. 20, No. 1, 2004, pp. 13-22. doi:10.1079/SUM2004214
51. S. M. Avery, A. Moore and M. L. Hutchison, "Fate of Escherichia coli Originating from Livestock Faeces Deposited Directly onto Pasture," Letters in Applied Microbiology, Vol. 38, No. 5, 2004, pp. 355-359. doi:10.1111/j.1472-765X.2004.01501.x
52. L. W. Sinton, R. R. Braithwaite, C. H. Hall and M. L. Mackenzie, "Survival of Indicator and Pathogenic Bacteria in Bovine Feces on Pasture," Applied & Environmental Microbiology, Vol. 73, No. 24, 2007, pp. 7917- 7925. doi:10.1128/AEM.01620-07
53. R. K. Hubbard, G. L. Newton and G. M. Hill, "Water Quality and the Grazing Animals," Journal of animal science, Vol. 82, 2004, pp. E255-E263.
54. P. Rodgers, C. Soulsby, C. Hunter and J. Petry, "Spatial and Temporal Bacterial Quality of a Lowland Agricultural Stream in Northeast Scotland," The Science of the total Environment, Vol. 314-316, 2003, pp. 289-302. doi:10.1016/S0048-9697(03)00061-5
55. J. A. Entry, R. K. Hubbard, J. E. Thies and J. J. Fuhrmann, "The Influence of Vegetation in Riparian Filterstrips on Coliform Bacteria: I. Movement and Survival in Water," Journal of Environmental Quality, Vol. 29, No. 4, 2000, pp. 12061214. doi:10.2134/jeq2000.00472425002900040026x
56. R. M. Roodsari, D. R. Shelton, A. Shirmohammadi, Y. A. Pachepsky, A. M. Sadeghi and J. L. Starr, "Fecal Coliform Transport as Affected by Surface Condition," Transactions of the ASAE, Vol. 48, No. 3, 2005, pp. 1055-1061.
57. T. J. Sullivan, J. A. Moore, D. R. Thomas, E. Mallery, K. U. Snyder, M. Wustenberg, J. Wustenberg, S. D. Mackey and D. L. Moore, "Efficacy of Vegetated Buffers in Preventing Transport of Fecal Coliform Bacteria from Pasturelands," Environmental Management, Vol. 40, No. 6, 2007, pp. 958-965. doi:10.1007/s00267-007-9012-3
58. É. Gagnon and G. Gangbazo, "Efficacité des Bandes Riveraines: Analyse de la Documentation Scientifique et Perspectives," Ministère du Développment Durable, de l'Environnement et des Parcs du Québec, Québec, 2007.
59. C. Kao and M. Wu, "Control of Non-Point Source Pollution by a Natural Wetlands," Water Science and Technology, Vol. 43, No. 5, 2001, pp. 169-174.
60. K. Knox, A. R. Dahlgren, K. W. Tate and E. R. Atwill, "Efficacy of Natural Wetlands to Retain Nutrient, Sediment and Microbial Polluants," Journal of Environmental Quality, Vol. 37, 2008, pp. 1837-1846. doi:10.2134/jeq2007.0067
61. M. Matteo, T. Randhir and D. Bloniarz, "WatershedScale Impacts of Forest Buffers on Water Quality and Runoff in Urbanizing Environment," Journal of Water Resources Planning and Management, Vol. 132, No. 3, 2006, pp. 144-152. doi:10.1061/(ASCE)0733-9496(2006)132:3(144)
62. J. Abu-Ashour and H. Lee, "Transport of Bacteria on Sloping Soil Surfaces by Runoff," Environmental Toxicology, Vol. 15, No. 2, 2000, pp. 149-153. doi:10.1002/(SICI)1522-7278(2000)15:2<149::AID-TOX11>3.0.CO;2-O
63. C. Ferguson, A. M. D. Husman, N. Altavilla, D. Deere and N. Ashbolt, "Fate and Transport of Surface Water Pathogens in Watersheds [Review]," Critical Reviews in Environmental Science & Technology, Vol. 33, No. 3, 2003, pp. 299-361. doi:10.1080/10643380390814497
64. D. Ogden, D. R. Fenlon, A. J. A. Vinten and D. Lewis, "The fate of Escherichia coli O157 in Soil and Its Potential to Contaminate Drinking Water," International Journal of Food Microbiology, Vol. 66, No. 1-2, 2001, pp. 111-117. doi:10.1016/S0168-1605(00)00508-0

65. D. Trevisan, J. Y. Vansteelant and J. M. Dorioz, "Survival and Leaching of Fecal Bacteria after Slurry Spreading on Mountain Hay Meadows: Consequences for the Management of Water Contamination Risk," Water Research, Vol. 36, No. 1, 2002, pp. 275-283. doi:10.1016/S0043-1354(01)00184-1
66. R. W. Muirhead, R. P. Collins and P. J. Bremer, "Interaction of Escherichia coli and Soil Particles in Runoff," Applied and Environmental Microbiology, Vol. 72, No. 5, 2006, pp. 3406-3411. doi:10.1128/AEM.72.5.3406-3411.2006
67. F. C. Curriero, J. A. Patz, J. B. Rose and S. Lele, "The Association between Extreme Precipitation and Waterborne Disease Outbreaks in the United States, 1948- 1994," American Journal of Public Health, Vol. 91, No. 8, 2001, pp. 1194-1199. doi:10.2105/AJPH.91.8.1194
68. D. F. Charron, M. K. Thomas, D. Waltner-Toews, J. J. Aramini, T. Edge, R. A. Kent, A. R. Maarouf and J. Wilson, "Vulnerability of Waterborne Diseases to Climate Change in Canada: A Review," Journal of Toxicology & Environmental Health Part A, Vol. 67, No. 20-22, 2004, pp. 1667-1677. doi:10.1080/15287390490492313
69. J. A. Patz, S. J. Vavrus, C. K. Uejio and S. L. McLellan, "Climate Change and Waterborne Disease Risk in the Great Lakes Region of the US," American Journal of Preventive Medicine, Vol. 35, No. 5, 2008, pp. 451-458. doi:10.1016/j.amepre.2008.08.026
70. R. L. Whitman, M. B. Nevers, G. C. Korinek and M. N. Byappanahalli, "Solar and Temporal Effects on Escherichia coli Concentration at a Lake Michigan Swimming Beach," Applied and Environmental Microbiology, Vol. 70, No. 7, 2004, pp. 4276-4285. doi:10.1128/AEM.70.7.4276-4285.2004
71. S. R. Crane and J. A. Moore, "Modeling Enteric Bacterial Die-Off: A Review," Water Air soil Pollution, Vol. 27, 1986, p. 411. doi:10.1007/BF00649422
72. J. Abu-Ashour, D. M. Joy, H. Lee, H. R. Whiteley and S. Zelin, "Transport of Microorganisms through Soil," Water, Air, & Soil Pollution, Vol. 75, 1994, pp. 141-158. doi:10.1007/BF01100406
73. R. C. Jamieson, R. J. Gordon, K. E. Sharples, G. W. Stratton and A. Madani, "Movement and Persistence of Fecal Bacteria in Agricultural Soils and Subsurface Drainage Water: A Review," Canadian Biosystems Engineering, Vol. 44, No. 1, 2002, pp. 1-9.
74. M. M. Lau, S. C. Ingham and A. R. Arment, "Survival of Faecal Indicator Bacteria in Bovine Manure Incorporated into Soil," Letters in Applied Microbiology, Vol. 33, 2001, p. 131.
75. L. Johnson and S. Gage, "Landscape Approaches to the Analysis of Aquatic Ecosystems," Freshwater Biology, Vol. 37, No. 1, 1997, pp. 113-132. doi:10.1046/j.1365-2427.1997.00156.x
76. L. Sliva and D. D. Williams, "BUFFER zone versus Whole Catchment Approaches to Studying Land Use Impact on River Water Quality," Water Research, Vol. 35, No. 14, 2001, pp. 3462-3472. doi:10.1016/S0043-1354(01)00062-8 [Citation Time(s):2]
77. J. Crowther, M. D. Wyer, M. Bradford, D. Kay, C. A. Francis and W. G. Knisel, "Modelling Faecal Indicator Concentrations in Large Rural Catchments Using Land Use and Topographic Data," Journal of Applied Microbiology, Vol. 94, No. 6, 2003, pp. 962-973. doi:10.1046/j.1365-2672.2003.01877.x
78. J. Crowther, D. Kay and M. D. Wyer, "Faecal-Indicator Concentrations in Waters Draining Lowland Pastoral Catchments in the UK: Relationships with Land Use and Farming Practices," Water Research, Vol. 36, No. 7, 2002, pp. 1725-1734. doi:10.1016/S0043-1354(01)00394-3
79. D. Kay, M. Wyer, J. Crowther, C. Stapleton, M. Bradford, A. McDonald, J. Greaves, C. Francis and J. Watkins, "Predicting Faecal Indicator Fluxes Using Digital Land Use Data in the UK's Sentinel Water Framework Directive Catchment: The Ribble Study," Water Research, Vol. 39, No. 16, 2005, pp. 3967-3981. doi:10.1016/j.watres.2005.07.006
80. D. Kay, S. Anthony, J. Crowther, B. J. Chambers, F. A. Nicholson, D. Chadwick, C. M. Stapleton and M. D. Wyer, "Microbial Water Pollution: A Screening Tool for Initial Catchment-Scale Assessment and Source Apportionment," Science of the Total

Environment, Vol. 408, No. 23, 2010, pp. 5646-5656. doi:10.1016/j.scitotenv.2009.07.033
81. Statistics Canada, "2006 Census," Statistics Canada, 2007. http://www12.statcan.gc.ca/census-recensement/index-eng.cfm
82. Statistics Canada, "2006 Agriculture Census," Statistics Canada, 2008. http://www.statcan.gc.ca/ca-ra2006/index-eng.htm
83. P. D'Arcy and R. Carignan, "Influence of Catchment Topography on Water Chemistry in Southeastern Quebec Shield Lakes," Canadian Journal of Fisheries & Aquatic Sciences, Vol. 54, 1997, pp. 2215-2227.
84. P. Turgeon, P. Michel, P. Levallois, P. Chevalier, D. Daignault, B. Crago, R. Irwin, S. A. McEwen, N. F. Neumann and M. Louie, "Agroenvironmental Determinants Associated with the Presence of AntimicrobialResistant Escherichia coli in Beach Waters in Quebec, Canada," Zoonoses and Public Health, Vol. 58, No. 6, 2011, pp. 432-439.
85. D. W. McKenney, M. F. Hutchinson, J. L. Kesteven and L. A. Venier, "Canada's Plant Hardiness Zones Revisited Using Modern Climate Interpolation Techniques," Canadian Journal of Plant Sciences, Vol. 81, 2001, pp. 129- 143.
86. [87] J. D. Boone, K. C. McGwire, E. W. Otteson, R. S. DeBaca, E. A. Kuhn, P. Villard, P. F. Brussard and S. C. St Jeor, "Remote Sensing and Geographic Information Systems: Charting Sin Nombre Virus Infections in Deer Mice," Emerging Infectious Diseases, Vol. 6, No. 3, 2000, pp. 248-258.
87. Leblond, A. Sandoz, G. Lefebvre, H. Zeller and D. J. Bicout, "Remote Sensing Based Identification of Environmental Risk Factors Associated with West Nile Disease in Horses in Camargue, France," Preventive Veterinary Medicine, Vol. 79, No. 1, 2007, pp. 20-31.
88. L. R. Beck, B. M. Lobitz and B. L. Wood, "Remote Sensing and Human Health: New Sensors and New Opportunities," Emerging Infectious Diseases, Vol. 6, No. 3, 2000, pp. 217-227.
89. S. Kalluri, P. Gilruth, D. Rogers and M. Szczur, "Surveillance of Arthropod Vector-Borne Infectious Diseases Using Remote Sensing Techniques: A Review Art. No. e116 [Review]," PLoS Pathogens, Vol. 3, No. 10, 2007, pp. 1361-1371.
90. J. B. Campbell, "Introduction to Remote Sensing," The Guilford Press, New York, 2007.
91. Y. Zhang, B. Guindon, K. Sun and L. Sun, "Remote Sensing for Improving Understanding on Canadian Urbanization," Canada Centre for Remote Sensing, Natural Resources Canada, 2010. http://www.ccrs.nrcan.gc.ca/optical/curlus_e.php
92. J. P. Messina and K. A. Crews-Meyer, "A Historical Perspective on the Development of Remotely Sensed Data as Applied to Medical Geography," In: D. P. Albert, W. M. Geslier and B. Levrgood, Eds., Spatial Analysis, GIS, and Remote Sensing Applications in the Health Sciences, Ann Arbor Press, Chelsea, 2000, pp. 129-146.
93. United Nations Office for Outer Space Affairs, "Space Solutions for the World's Problems," Vienna, 2005. http://www.oosa.unvienna.org/pdf/ publications/IAM2005E.pdf
94. S. J. Goetz, S. D. Prince and J. Small, "Advances in Satellite Remote Sensing of Environmental Variables for Epidemiological Applications," In: S. I. Hay, S. E. Randolph and D. J. Rogers, Eds., Remote Sensing and Geographical Information Systems in Epidemiology, Elsevier Sciences, Oxford, 2002, pp. 289-309.
95. V. R. M. Correia, M. S. Carvalho, P. C. Sabroza and C. H. Vacsoncelos, "Remote Sensing as a Tool to Survey Endemic Diseases in Brazil," Cad. Saûde Pûblica, Vol. 20, No. 4, 2004, pp. 891-904.
96. B. C. Rundquist, C. J. Henrie and E. J. Grewe, "Internet Acess to Remotely Sensed Data: Satellite Imaging Made Commonplace," Journal of Map & geography Libraries, Vol. 2, No. 2, 2006, pp. 21-30.
97. S. N. V. Kalluri and T. J. Schmugge, "Application of Remote Sensing in Agriculture and Soil Science," In: K. R. Krishna, Ed., Soil Fertility and Crop Production, Chapter 18, Science Publisher, Enfield, 2002, 465 p.

98. D. Hampson, J. Crowther, I. Bateman, D. Kay, P. Posen, C. Stapleton, M. Wyer, C. Fezzi, P. Jones and J. Tzanopoulos, "Predicting Microbial Pollution Concentrations in UK Rivers in Response to Land Use Change," Water Research, Vol. 44, No. 16, 2010, pp. 4748-4759.

99. Canadian Federal-Provincial-Territorial Committee on Environmental and Occupational Health and the Canadian Council of Ministers of the Environment, "From Source to Tap: The Multi-Barrier Approach to Safe Drinking Water," 2002. http://www.hc-sc.gc.ca/ewh-semt/alt_formats/hecs-sesc/pdf/water-eau/tap-source-robinet/tap-source-robinet-eng.pdf

100. Ministry for the Environment, New Zealand, "Draft Users' Guide: National Environmental Standard for Sources of Human Drinking Water," 2012http://www.mfe.govt.nz/publications/rma/nes-draft-sources-human-drinking-water/html/index.html

CHAPTER 13

Biomarkers in Nile Tilapia *Oreochromis niloticus niloticus* (Linnaeus, 1758) to Assess the Impacts of River Nile Pollution: Bioaccumulation, Biochemical and Tissues Biomarkers

Alaa G. M. Osman[1,2]

[1]*Department of Zoology, Faculty of Science, Al-Azhar University (Assiut Branch), Assiut, Egypt;* [2]*Department of Ecophysiology and Aquaculture, Leibniz-Institute of Freshwater Ecology and Inland Fisheries, Berlin, Germany.*

ABSTRACT

The use of biomarkers has become an important tool for modern environmental assessment as they can help to predict pollutants involved in the monitoring program. Here I present data on bioaccumulation, biochemical and tissues biomarkers in Nile tilapia as early warning indicators of river Nile pollution. Nile tilapia sampled from downstream sites accumulated higher levels of all the detected heavy metals than those collected from upstream sites. Heavy metal residues in the tissues of Nile tilapia exhibited different patterns of accumulation and distribution among the selected tissues. Remarkable alterations in the activities of glucose-6-phosphate dehydrogenase (G6PDH) and lactate dehydrogenase (LDH) in the tissues of Nile tilapia were detected. These alterations were followed, in the present study, by the occurrence of histological lesions in liver and gill tissues of fish collected from the same sites. Alterations in bioaccumulation patterns, in enzyme activities and in histology go in parallel with the elevation in the levels of water chemical parameters detected in the downstream sites as a result of pollution stress in these areas. These results provide evidence that bioaccumulation, biochemical and tissues biomarkers can be sensitive indicators of exposure to mixed pollutants in surface waters.

1. INTRODUCTION

The river Nile can be considered as a model polluted aquatic ecosystem for assessment of biomarker responses in fish because it receives untreated wastes from various drain outlets [1,2]. The use of biomarkers has become an important tool for modern environmental assessment as they can help to predict pollutants involved in the monitoring program. Currently, the use of biomarkers for monitoring environmental quality has gained considerable interest in the assessment of river condition in many places around the world [3,4]. There are many different biomarkers that occur at many different levels of organization from sub-cellular to whole-organisms. The biomarker may be the chemical itself (bioaccumulation). It may also be molecular biomarkers, biochemical biomarkers and tissues biomarkers.

Metal bioaccumulation is a major route, through which increased levels of the pollutants are transferred across the food chain and finally assimilated by human consumers resulting in health risks [5]. Therefore, it is important to always determine the bioaccumulation capacity for heavy metals by fishes in order to assess potential risk to human health and take appropriate action to protect public health and the environment [5,6]. Contaminant residues in fish tissues may ultimately reach concentrations hundreds or thousands of times above those measured in the water [1]. For this reason, monitoring fish tissue contamination serves an important function as an early warning biomarker of aquatic pollution [7].

Monitoring changes in biochemical parameters may provide early warning indicators of general as well as specific toxicological responses. Biochemical biomarkers of contamination are important indices used in fish toxicity tests and for field monitoring of aquatic pollution [8]. In toxicological studies of acute exposure, changes in concentrations and activities of some enzymes may reflect cell damage in specific organs [9]. Glucose-6-phosphate dehydrogenase (G6PDH) has long been recognized as an antioxidant enzyme [10] and also as a biomarker of pollution-induced carcinogenesis in fish. The cytoplasmic enzyme lactate dehydrogenase (LDH) is widely used as marker of organ or tissue lesions in toxicology and in clinical chemistry [11,12]. G6PDH and LDH are key factors in the metabolism with high sensitivity to pollutants [13].

Tissue biomarkers are closely related to other biomarkers of stress since many pollutants have to undergo metabolic activation in order to be able to provoke tissues change in the exposed organism. The advantage of histopathology as an excellent biomarker lies in its intermediate location between molecular and individual levels [14-16]. Most biomarkers are narrow in their expression whereas pathology is broad in its evaluation [17].

Fish are widely used to evaluate the health of aquatic systems and their biological responses serve as biomarkers of environmental contaminants. Besides its ecological value, Nile tilapia, Oreochromis niloticus niloticus, also economically important as a food source by low income families along the river Nile. Nile tilapia is one of the most common freshwater fish used in toxicological studies [18-20], because it presents a number of characteristics that may make it an appropriate model that can be used as indicator species in biomonitoring programs [18,21].

Although biomarker studies in relation to water bodies in many countries have been extensively documented, scientific reports of pollutant induced biomarker responses in the river Nile are meager [1,2,22]. In the river Nile, contaminants are present as very complex mixtures and there is no single biomarker that can give a complete diagnosis of environmental degradation. To overcome this difficulty, a battery of biomarkers comprising molecular, bioaccumulation, biochemical and tissues biomarkers in Nile tilapia were used to assess stressful environmental conditions on the river Nile as well as point out which biomarkers to use during future monitoring programs of this and similar aquatic systems. In earlier publication, the author presented the results of molecular biomarker measurements in Nile tilapia from the river Nile [2,23]. Therefore as a consequence, in this study I present an investigation of bioaccumulation, biochemical and tissue biomarkers in Nile tilapia collected from the whole course of the Egyptian river Nile.

2. MATERIALS AND METHODS

2.1. Sampling Sites

Six sites were selected along the whole course of the Egyptian river Nile from its start at Aswan to its estuaries at Rosetta and Damietta (Figure 1). Water and tissues samples from the selected sites were sampled bimonthly

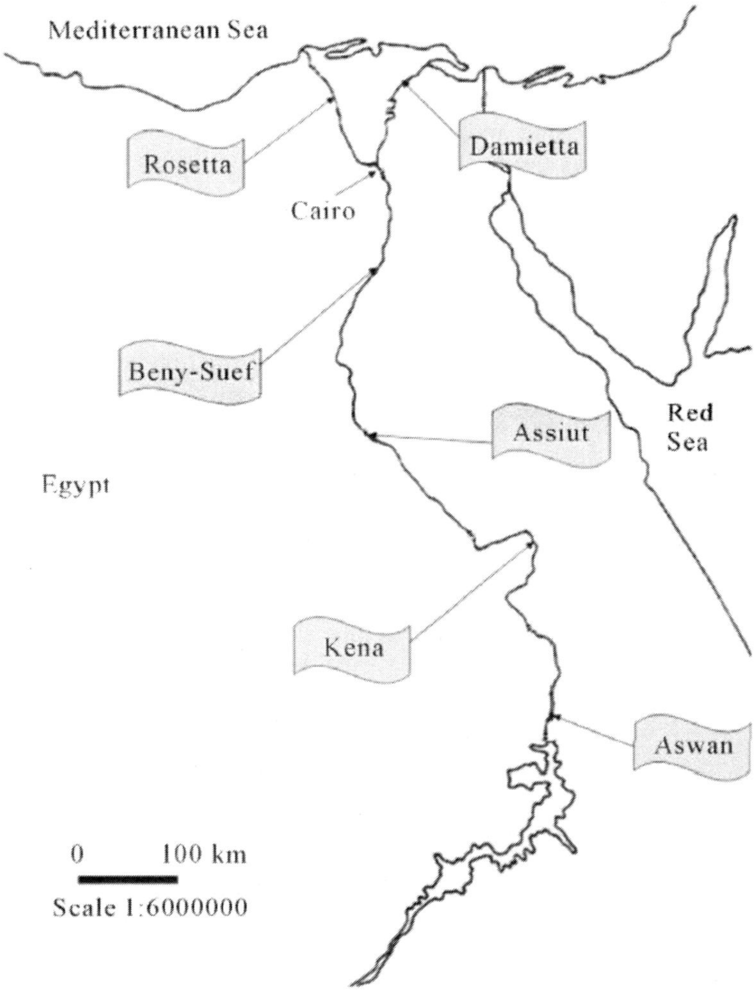

Figure 1. Map showing the sampling sites along the whole course of the Egyptian river Nile from its start at Aswan to its estuaries at Damietta and Rosetta sites.

(every two months) from the main course of the river Nile during the period from July 2009 to Jun 2010. Water samples (108 samples) were collected by polyvinyl chloride Van Dorn water sampler (5 L capacity) at two meter depths. Sampling and preservation of water samples were done according to the traditional manual methods [24]. Nile tilapia (Oreochromis niloticus niloticus) was caught by gill net from the selected sites (72 specimens; 12 fish from each sampling site). The body weight of the collected specimens ranged from 240 - 290 g. The target organs (liver, gills, gonads and muscles) were excised for subsequent analysis.

2.2. Water Quality Assessment

Water quality criteria including the pH, electrical conductivity (mScm-1) and temperature (°C) were measured by using water checker U-10 Horiba Ltd. The other water criteria [chemical oxygen demand (COD), total organic carbon (TOC), total solids (TS), ammonia (NH_3), nitrate (NO_3), chloride (CL) florid (F), orthophosphate (O-PO_4), sulphate (SO_4) and phenolics (Phenol)] were measured according to the traditional manual methods [24]. Total copper (Cu), chromium (Cr), cadmium (Cd), zinc (Zn), iron (Fe), lead (Pb), manganese (Mn) and mercury (Hg) were measured after digestion using Graphite Furnace AA (GFAA) spectroscopy. Data on the selected sites are shown in Table 1.

2.3. Bioaccumulation Biomarker

Liver, gills, gonads and muscles were transported in liquid nitrogen container to the laboratory for chemical analysis. They were washed with tap water (previously analyzed for Pb and Cd) followed by bi-distilled water, then oven-dried to constant weight at 105°C. The dried fish was crushed and powdered in an agate mortar, then, they were kept in polyethylene bottles for analysis. One gram portions of fish tissues were digested by means of a microwave after addition of nitric acid and hydrogen peroxide. The results were calculated in milligram per kilogram wet weight (mg/kg wet wt). Chemicals concentration were analyzed according to German industrial standard, DIN 38406-6, (DEV, E6) with an Atomic Absorption Spectrometer using flame and graphite furnace technique.

2.4. Biochemical Biomarker

The activities of Glucose-6-phosphate dehydrogenase (G6PDH) and lactate dehydrogenase (LDH) were measured according to a modified protocol based on [25,26]. Liver and muscles samples were pulverized under liquid nitrogen and about 100 mg of ground tissue powder was added to 5 vol. of buffer (50 mM Tris, pH 7.4, 1 mM EDTA and 2 mM $MgCl_2$). Tissue was homogenized briefly with an Ultra-Turrax (temperature was maintained at 4°C during homogenization). The homogenate was centrifuged for 15 min at 10,000 × g and 4°C and supernatants were used for the enzyme activity assays using spectrophotometer (Micro Lab 200 Vital Scientific, Dieren, The Netherlands) at a wavelength of 340 nm and at 37°C using kits, Stanbio LDH (UV-Rate) procedure no. 0940 USA for the quantitative determination of LDH and RANDOX Laboratories Ltd., PD410, UK BT294QY, for the quantitative determination of G6PDH. The catalytic activities of the selected enzymes were calculated in accordance with the recommendations of the French Society of Clinical Biology and were expressed as U/g tissues [22].

Table 1. Physical and chemical parameters of water samples collected from different sites along the whole course of the river Nile, Egypt.

Sites / Parameter (unit)	Aswan	Kena	Assiut	Beny-Suef	Damietta	Rosetta	Permissible limit*
	Mean ± SD	Mean ± SD	Mean ± SD	Mean ± SD	Mean ± SD	Mean ± SD	
PH (Unit)	7.86 ± 0.25	8.01 ± 0.41	8.15 ± 0.18	8.27 ± 0.27	8.40 ± 0.44	8.23 ± 0.45	7 - 8.5
Conductivity (Ms/cm)	0.26 ± 0.03	0.27 ± 0.07	0.29 ± 0.09	0.34 ± 0.07	0.38 ± 0.01	0.58 ± 0.12	-
Temperature (°C)	22.68 ± 2.22	23.72 ± 3.15	23.33 ± 4.67	23.64 ± 4.24	25.29 ± 5.66	24.53 ± 4.44	Over 5°C
Chemical oxygen demand (ppm)	10.58 ± 3.62	9.17 ± 3.04	10.63 ± 2.18	7.87 ± 2.15	8.59 ± 2.51	18.00 ± 10.38	10
Total organic carbon (ppm)	5.65 ± 2.88	5.89 ± 1.93	5.74 ± 0.92	4.93 ± 2.58	5.21 ± 2.63	8.61 ± 6.06	-
Total solid (ppm)	198.88 ± 14.01	212.67 ± 23.7	227.75 ± 16.3	259.5 ± 44.33	305.25 ± 55.95	411.25 ± 85.66	500
Ammonia (ppm)	0.11 ± 0.16	0.009 ± 0.001	0.019 ± 0.01	0.01 ± 0.009	0.04 ± 0.05	0.15 ± 0.09	0.5
Nitrate (ppm)	0.81 ± 0.40	0.76 ± 0.39	0.51 ± 0.20	0.72 ± 0.63	1.13 ± 1.13	2.05 ± 2.27	45
Chlorides (ppm)	7.05 ± 1.4	8.56 ± 1.89	10.03 ± 2.58	15.28 ± 5.10	22.42 ± 4.93	40.55 ± 6.72	-
Florid (ppm)	0.28 ± 0.15	0.32 ± 0.13	0.39 ± 0.16	0.31 ± 0.13	0.30 ± 0.09	0.37 ± 0.07	0.5
Ortho phosphate (ppm)	0.01 ± 0.02	0.04 ± 0.04	0.09 ± 0.09	0.03 ± 0.03	0.13 ± 0.07	0.21 ± 0.18	-
Sulphate (ppm)	34.17 ± 13.30	45.33 ± 15.37	47.96 ± 14.11	45.25 ± 15.81	51.00 ± 12.76	68.13 ± 11.42	200
Phenol (ppm)	0.02 ± 0.02	0.02 ± 0.02	0.02 ± 0.016	0.02 ± 0.009	0.03 ± 0.02	0.04 ± 0.02	0.02
Cu (ppm)	0.03 ± 0.03	0.03 ± 0.024	0.03 ± 0.03	0.03 ± 0.03	0.03 ± 0.03	0.05 ± 0.03	1
Cr (ppm)	0.003 ± 0.002	0.007 ± 0.01	0.006 ± 0.01	0.006 ± 0.01	0.05 ± 0.06	0.09 ± 0.15	0.05
Cd (ppm)	0.004 ± 0.004	0.002 ± 0.002	0.006 ± 0.01	0.002 ± 0.002	0.016 ± 0.02	0.01 ± 0.02	0.1
Zn (ppm)	0.21 ± 0.17	0.12 ± 0.10	0.31 ± 0.46	0.34 ± 0.45	0.46 ± 0.68	0.7 ± 0.96	1
Fe (ppm)	0.19 ± 0.17	0.22 ± 0.19	0.34 ± 0.32	0.46 ± 0.39	0.412 ± 0.27	0.5 ± 0.44	1
Pb (ppm)	0.01 ± 0.02	0.02 ± 0.01	0.02 ± 0.01	0.02 ± 0.01	0.03 ± 0.04	0.06 ± 0.078	0.05
Mn (ppm)	0.03 ± 0.03	0.06 ± 0.06	0.05 ± 0.02	0.06 ± 0.06	0.07 ± 0.08	0.1 ± 0.15	0.5
Hg (ppm)	0.0009 ± 0.001	0.0004 ± 0.001	0.0005 ± 0.001	0.0009 ± 0.001	0.002 ± 0.001	0.003 ± 0.001	0.001

*Canadian environmental quality guidelines.

2.5. Tissue Biomarker

Nile tilapias sampled from the selected sites were sacrificed by decapitation. Liver and gill were dissected out, fixed in Bouin's solution and then processed by conventional method [27], sectioned at 4 - 7 μm and stained with Haematoxylin-eosine. The sections were observed under light microscopy. The histological changes in the liver and gill were examined [14] in the randomly selected sections from each fish and photographed using Leica Wild MPS48 microscope and CCD video camera (Sony, AVT-Horn).

2.6. Statistical Analysis

All values from chemical analyses and enzyme activities are presented as mean ± SD. Data obtained from the experiment were subjected to one way analysis of variance (ANOVA) test using the Statistical Package for the Social Sciences [28]. The correlation coefficients between heavy metals of water and tissues samples were calculated by the application of Pearson correlation analysis [29] in order to indicate the nature and the sources of the polluting substances. In all cases, $p < 0.05$ was the accepted significance level [28].

3. RESULTS

3.1. Water Analysis

The results of means and SD of the detected physical and chemical parameters for water samples are given in Table 1. Most of the detected physicochemical parameters exhibited their highest values in the water of river Nile downstream. For nearly all the detected parameters significant differences ($p < 0.05$) were observed between the selected sites. According to the present results the pH seems to be constant all over the river Nile. All the pH values were in the alkaline side (7.8 to 8.4). Mean values of conductivity, COD, TOC, NH_3, NO_3, TS, SO_4, Cl, orthophosphate and phenolics were higher in the water of Damietta and Rosetta sites compared to other sites. The value of COD and the concentrations of phenolics, lead, chromium, cadmium and mercury were higher than the permissible limits in Rosetta and/or Damietta sites (Table 1).

3.2. Bioaccumulation Biomarker

The mean concentrations of the selected heavy metals in liver, gill, gonad and muscle samples of Nile tilapia are presented in Table 2. Detected heavy metals concentrations varied widely between different sites and tissues. The highest concentrations of all the heavy metals were recorded in tissues samples collected from downstream river Nile (Damietta and/or Rosetta) and the lowest ones were detected in the samples collected from upstream areas (Aswan and/or Kena; Table 2). For nearly all the detected heavy metals significant differences ($p < 0.05$) were recorded between the selected sites. Liver concentrated higher levels of Cu, Cr, Cd, Zn, Fe and Pb than the other organs for all the six sites (Table 2).

The distributions of all these heavy metals were in the order of liver > gill > gonad > muscle. The highest concentration of Mn was recorded in the gills of fishes collected from nearly all the sites (Table 2). The distribution of Mn was in the order of gill > liver > gonad > muscle nearly in all the sites (excluded Aswan and Damietta). The mean concentration of Hg was very low in all the tissues. According to the present result the Hg accumulated mainly in the liver followed by gills of fishes collected from all the sites except Kena and Assiut. In the latter two sites, the highest Hg concentrations were recorded in the gill tissues. Gonad and muscle were found to accumulate small amounts of all the detected heavy metals. Significant ($p < 0.05$) high correlations (from 0.951 to 0.988) were observed between the concentrations of the selected heavy metals in the water and tissues of Nile tilapia.

3.3. Biochemical Biomarkers

The activities of the selected two metabolic enzymes (G6PDH and LDH) in the liver and muscles of Nile tilapia collected from the whole course of the Egyptian river Nile are presented in Table 3.

For both enzymes significant differences ($p < 0.05$) were detected between the selected sites. The activity of G6PDH in the liver and muscles of Nile tilapia decreased significantly ($p < 0.05$) from Aswan toward Damietta and Rosetta sites (Table 3). The highest G6PDH activity was recorded in the liver (57.7 U/g) and muscles (61.8 U/g) of fish collected from Aswan (Table 3). On the other hand the lowest G6PDH activity was recorded in the liver and muscle tissues collected from Rosetta (Table 3). The activity of LDH in the muscles and liver of Nile tilapia increased significantly ($p < 0.05$) from upstream toward downstream river Nile (Table 3). The lowest LDH activity was recorded in the liver of fish sampled from Kena and in the muscles of fish collected from Aswan (Table 3). The highest values of LDH activity was detected in the liver and muscles of Nile tilapia sampled from Damietta. The activities of G6PDH and LDH in the muscles were always higher than that in the liver of fish collected from the same site (Table 3).

Table 2. Heavy metals concentrations in different tissues of Nile tilapia Oreochromis niloticus niloticus collected from different sites along the whole course of the river Nile, Egypt.

Parameters	Organs \ Localities	Aswan	Kena	Assiut	Beny-Suef	Damietta	Rosetta
		Mean ± SD	Mean ± SD	Mean ± SD	Mean ± SD	Mean ± SD	Mean ± SD
Cu	Liver (L)	16.39 ± 3.66	35.83 ± 7.06	25.75 ± 17.44	20.04 ± 9.07	34.30 ± 26.59	23.53 ± 4.56
	Gills (GL)	3.13 ± 0.91	4.38 ± 4.01	17.51 ± 2.94	8.59 ± 7.88	23.96 ± 10.28	12.54 ± 1.01
	Gonads (G)	2.45 ± 3.06	3.29 ± 1.84	12.2 ± 6.59	5.08 ± 6.39	5.58 ± 2.21	27.27 ± 22.44
	Muscles (M)	0.796 ± 0.67	0.47 ± 0.32	4.53 ± 2.71	2.43 ± 1.86	5.75 ± 4.43	10.69 ± 4.59
Pattern		L > GL > G > M	L > GL > G > M	L > GL > G > M	L > GL > G > M	L > GL > G > M	L > G > GL > M
Cr	Liver (L)	6.26 ± 1.98	6.63 ± 2.42	6.39 ± 1.63	6.27 ± 3.14	10.56 ± 2.83	13.74 ± 3.82
	Gills (GL)	5.74 ± 3.29	6.59 ± 1.28	5.13 ± 1.96	5.69 ± 3.34	6.05 ± 4.63	13.11 ± 2.02
	Gonads (G)	5.19 ± 1.35	5.52 ± 2.03	4.78 ± 1.07	4.53 ± 2.14	4.01 ± 3.67	11.65 ± 1.16
	Muscles (M)	3.93 ± 2.43	4.78 ± 1.79	3.71 ± 1.23	4.45 ± 2.1	2.64 ± 1.76	7.87 ± 7.67
Pattern		L > GL > G > M	L > GL > G > M	L > GL > G > M	L > GL > G > M	L > GL > G > M	L > GL > G > M
Cd	Liver (L)	0.94 ± 0.25	1.19 ± 0.63	1.03 ± 0.21	1.11 ± 0.57	3.32 ± 1.32	1.7 ± 1.17
	Gills (GL)	0.80 ± 0.39	0.75 ± 0.48	0.94 ± 0.13	1.05 ± 0.41	1.68 ± 0.67	1.18 ± 0.59
	Gonads (G)	0.79 ± 0.47	0.62 ± 0.17	0.92 ± 0.37	0.98 ± 0.45	1.52 ± 0.47	1.07 ± 0.62
	Muscles (M)	0.67 ± 0.64	0.58 ± 0.29	0.89 ± 0.34	0.71 ± 0.25	0.89 ± 0.43	1.02 ± 0.42
Pattern		L > GL > G > M	L > GL > G > M	L > GL > G > M	L > GL > G > M	L > GL > G > M	L > GL > G > M
Zn	Liver (L)	78.77 ± 7.29	67.67 ± 14.17	67.30 ± 7.84	53.65 ± 9.63	121.82 ± 5.8	78.66 ± 10.18
	Gills (GL)	49.37 ± 6.84	53.98 ± 7..5	62.71 ± 5.02	52.57 ± 7.82	107.28 ± 5.15	59.08 ± 10.50
	Gonads (G)	48.63 ± 8.46	43.93 ± 8.24	48.61 ± 5.41	50.15 ± 8.04	75.86 ± 6.39	49.66 ± 2.89
	Muscles (M)	41.84 ± 7.62	42.87 ± 10.49	43.37 ± 6.89	46.85 ± 7.13	58.43 ± 5.43	54.69 ± 5.76
Pattern		L > GL > G > M	L > GL > G > M	L > GL > G > M	L > GL > G > M	L > GL > G > M	L > GL > G > M
Fe	Liver (L)	64.71 ± 7.67	68.78 ± 12.29	74.87 ± 11.61	52.86 ± 110.79	93.58 ± 8.52	87.88 ± 6.57
	Gills (GL)	59.13 ± 6.13	57.48 ± 11.68	47.76 ± 7.26	49.66 ± 9.82	82.8 ± 11.94	63.61 ± 16.97
	Gonads (G)	42.49 ± 5.47	43.78 ± 13.82	46.76 ± 12.54	32.38 ± 7.01	34.36 ± 10.42	57.49 ± 8.45
	Muscles (M)	26.08 ± 6.44	34.94 ± 7.37	30.39 ± 8.34	28.56 ± 7.63	30.23 ± 12.79	29.09 ± 8.59
Pattern		L > GL > G > M	L > GL > G > M	L > GL > G > M	L > GL > G > M	L > GL > G > M	L > GL > G > M
pb	Liver (L)	7.54 ± 0.75	4.05 ± 2.53	7.21 ± 6.29	7.18 ± 5.39	11.31 ± 6.87	29.77 ± 6.26
	Gills (GL)	5.42 ± 2.80	3.66 ± 1.75	6.04 ± 2.17	6.54 ± 3.89	8.01 ± 4.05	24.29 ± 3.36
	Gonads (G)	4.21 ± 2.29	2.94 ± 0.47	5.67 ± 1.49	5.76 ± 4.70	6.9 ± 2.45	18.9 ± 3.61
	Muscles (M)	3.99 ± 2.64	2.01 ± 0.86	4.32 ± 1.38	4.79 ± 3.68	5.36 ± 2.66	14.44 ± 2.43
Pattern		L > GL > G > M	L > GL > G > M	L > GL > G > M	L > GL > G > M	L > GL > G > M	L > GL > G > M
Mn	Liver (L)	6.34 ± 1.79	7.72 ± 1.97	8.89 ± 1.7	8.92 ± 5.09	13.97 ± 2.1	15.77 ± 2.23
	Gills (GL)	4.89 ± 2.55	9.32 ± 1.54	11.27 ± 5.54	9.79 ± 3.54	17.06 ± 3.03	10.39 ± 1.66
	Gonads (G)	4.73 ± 1.79	5.54 ± 1.50	8.46 ± 6.29	6.53 ± 2.54	16.77 ± 4.12	11.54 ± 3.99
	Muscles (M)	3.42 ± 1.66	2.62 ± 2.73	2.72 ± 1.67	3.15 ± 1.9	3.52 ± 0.71	7.02 ± 1.48
Pattern		L > GL > G > M	GL > L > G > M	GL > L > G > M	GL > L > G > M	GL > L > G > M	L > GL > G > M
Hg	Liver (L)	0.005 ± 0.01	0.0009 ± 0.001	0.002 ± 0.003	0.06 ± 0.02	0.05 ± 0.04	0.04 ± 0.02
	Gills (GL)	0.001 ± 0.002	0.004 ± 0.01	0.003 ± 0.01	0.05 ± 0.02	0.04 ± 0.03	0.03 ± 0.02
	Gonads (G)	0.001 ± 0.001	0.0005 ± 0.001	0.002 ± 0.004	0.04 ± 0.03	0.01 ± 0.02	0.012 ± 0.01
	Muscles (M)	0.0005 ± 0.001	0.0009 ± 0.002	0.0006 ± 0.001	0.05 ± 0.04	0.02 ± 0.02	0.013 ± 0.01
Pattern		L > GL > G > M	GL > M > L > G	GL > G > L > M	L > GL > M > G	L > GL > M > G	L > GL > M > G

Table 3. Activities of Glucose-6-Phosphate Dehydrogenase (G6PDH) and Lactate Dehydrogenase (LDH) in liver and muscle tissues of Nile tilapia Oreochromis niloticus niloticus collected from different sites along the whole course of the river Nile, Egypt.

Sites	G6PDH (U/g) Mean ± SD		LDH (U/g) Mean ± SD	
	Liver	Muscles	Liver	Muscles
Aswan	57.7 ± 5.9	61.8 ± 4.8	1177.3 ± 42.9	1190.9 ± 85.4
Kena	42.1 ± 3.6	59.8 ± 3.5	1136.4 ± 40.3	1210.8 ± 44.3
Assiut	39.2 ± 11.1*	50.0 ± 5.4*	1191.7 ± 34.4	1232.0 ± 40.3*
Beny-Suef	40.3 ± 1.5*	42.4 ± 2.2*	1172.9 ± 40.6	1227.7 ± 17.7
Damietta	38.2 ± 1.6*	37.5 ± 3.2*	1558.1 ± 57.8*	1587.2 ± 37.1*
Rosetta	28.2 ± 5.1*	37.2 ± 1.4*	1491.8 ± 58.6*	1550.05 ± 37.7*

*Significant comparing to Aswan site at 0.05 levels.

3.4. Tissues Biomarkers

Livers of the healthy Nile tilapia displayed a normal appearance macroscopically and had a normal histological appearance as well. Hepatocytes are polygonal in shape, arranged in several cellular layers and surrounded by sinusoids (S), which could be seen to converge into the central vein (V). Pancreatic (P) tissue was present in association with venous vessels (Figures 2(a), (b)). Remarkable structural changes were detected in the liver of Nile tilapia collected nearly from the whole course of the Egyptian river Nile with moderate to totally damaged liver in the heavily polluted sites. The histopathological changes found in the liver of the examined fish included irregular arrangements of hepatocytes, blood congestion (bc) in blood vessel, vacuolation (vac) and necrosis (nec) (Figures 2(c), (d)). Leukocyte infiltration (inf) and hemorrhage (hem) were also detected in the hepatic tissues of some fish (Figure 3(a)). In the fishes collected from Damietta and Rosetta, patchy degeneration (pd) elements around the pancreatic cells (Figure 2(c)) and parenchyma cells (Figures 3(b), (c)) were observed with progressive increase of fibro-connective tissue (fct; Figure 3(b)). Also, dilatation of sinusoids (ds), dilatation of blood vessels (db) (Figure 3(b)) and extensive vascular congestion (Figure 3(c)) were observed. Acute and extensive vacuolation (vac) and focal areas of necrosis (nec) were observed (Figures 3(c), (d)) in the liver of fish collected from downstream areas.

Gill of healthy Nile tilapia displayed normal appearance (Figure 4(a)). In the core was a cartilaginous supporting rod (car) and blood vessels (bv) with traces of sinusoidal blood spaces. The primary lamella (P) was rounded at the apices while the projecting secondary lamellae (S) were clearly interspaced (Figure 4(a)).

Histological alterations in the gill of fish collected from the river Nile included fusion of adjacent secondary gill lamellae (fl) and epithelial lifting (el) (Figure 4(b)). There was also shrinkage of cartilaginous supporting mass resulting in a decrease in the size of the gills (Figures 4(b)-(d)) and curling of primary lamellae (cu) was also evident (Figure 4(c)). At Rosetta and Damietta, the epithelium was disrupted owing to sever degeneration (deg) and necrotic (nec) changes in gill filaments (Figure 4(d)). The increase in intracellular vacuolation (Figure 4(d)) signals onset of oedematous changes (oed). Frequently, alterations such as dilation and congestion in blood vessel (bc) were detected (Figure 4(d)). The gills show further shrinkage making the cartilaginous core absolutely obscured.

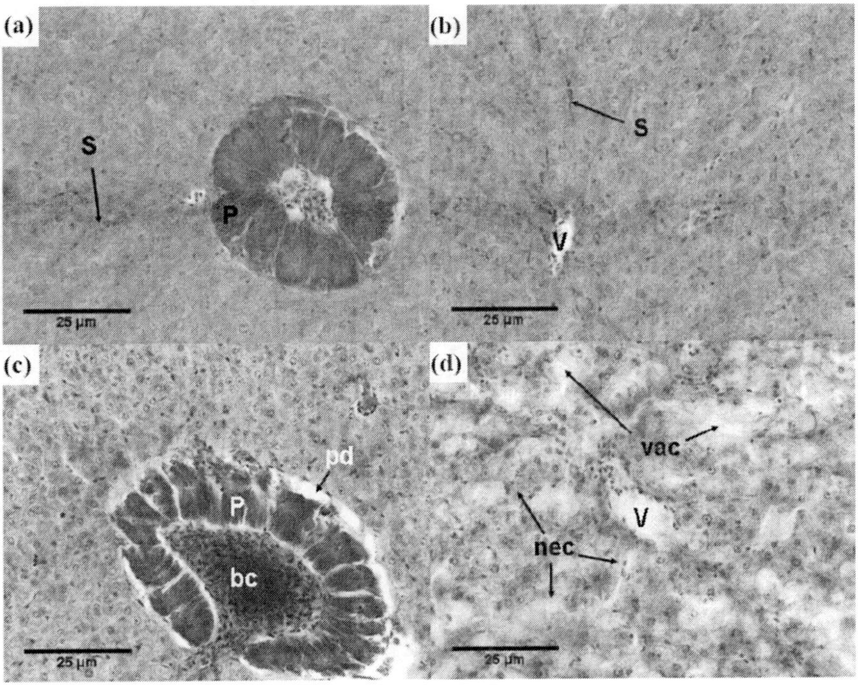

Figure 2. Photomicrograph of liver tissue showing (a), (b) Normal liver of Nile tilapia collected from Kena; (c), (d) Histopathological changes in the liver of Nile tilapia collected from Damietta and Rosetta sites. (S) sinusoid, (V) central vein, (P) pancreas, (nec) necrosis, (vac) vacuolation, (pd) patchy degeneration, (bc) blood congestion. Stained with H & E.

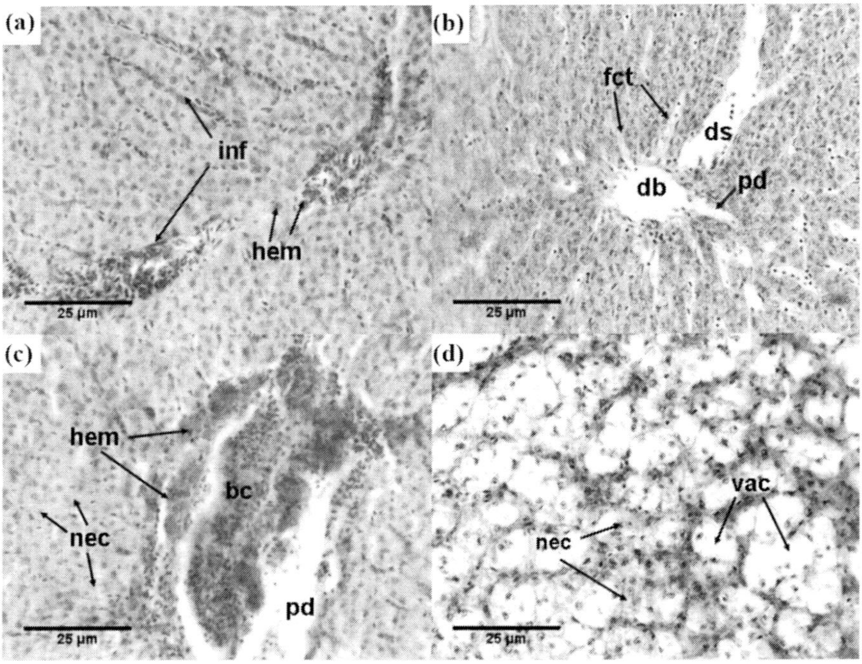

Figure 3. Photomicrograph of liver tissue showing histopathology in tissue of Nile tilapia collected from Damietta and Rosetta sites. (a) Liver tissues with leucocytes infiltration (inf) and haemorrhage (hem); (b) Showing progressive increase of fibro-connective tissue (fct), dilation of sinusoids (ds) and blood vessels (db); (c) Showing extensive haemorrhage (hem), blood congestion (bc), patchy degeneration (pd) and necrosis (nec); (d) Showing extensive vacuolation (vac) and necrosis (nec). Stained with H & E.

4. DISCUSSION

While most of the Nile's water quality is within acceptable levels, there are several hot spots mostly found in downstream areas of the river Nile. Most of the detected physical and chemical parameters were significantly higher in the water samples collected from downstream areas (Rosetta and Damietta sites) compared to sites located upstream. Detailed investigation about the level of physical and chemical parameters in water and sediment of the selected sites have been presented previously [1]. From that study we concluded that mean values of conductivity, chemical oxygen demand

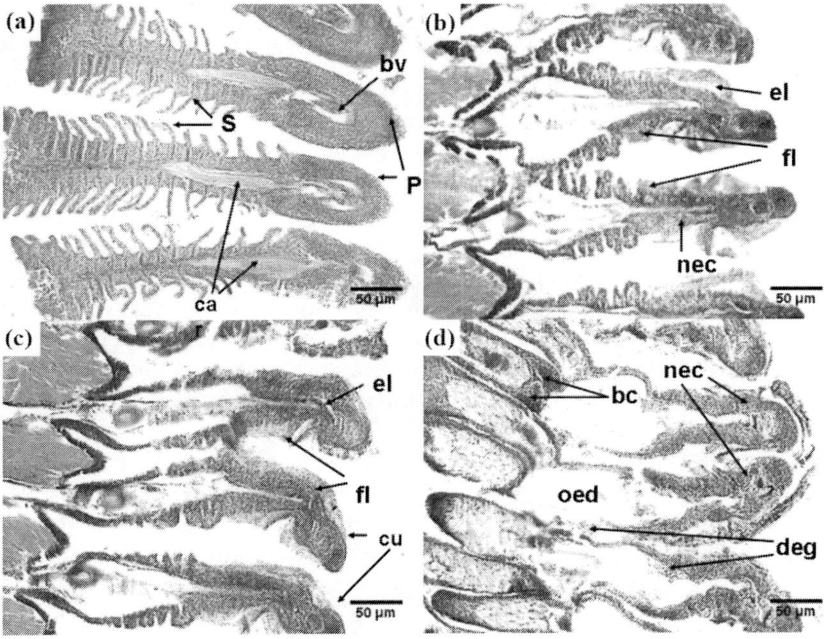

Figure 4. Photomicrograph of gills of Nile tilapia showing (a) Gill of healthy fish, collected from Aswan site, containing normal structures; (b) Histopathological changes in the gill of Nile tilapia collected from Damietta and Rosetta sites showing fusion of adjacent secondary lamellae (fl), epithelial lifting (el) and necrosis (nec); (c) Gill tissues with curling of primary lamellae (cu); (d) Gill epithelium was disrupted owing to sever degeneration (deg) and necrosis (nec) with signals onset of oedematous changes (oed). (P) Primary lamella, (S) secondary lamella, (ca) cartilage (bv) blood vessel. Stained with H & E.

(COD), total organic carbon (TOC), ammonia (NH_3), nitrate (NO_3), total solid (TS), sulphate (SO_4), chloride (Cl), orthophosphate and phenolics were recorded to be higher in the water of Damietta and Rosetta sites compared to the other sites. Also heavy metals levels increased in the water and sediment significantly from upstream to downstream areas. Such increase proves the presence of large quantities of organic and inorganic pollutants in the sites located downstream river Nile [1]. Physicochemical analysis gave useful information on the levels of contamination along the whole course of the Egyptian river Nile but, generally, did not give information on the effects of the contaminants on biological systems. So, it was essential to use biomarker responses in fishes. Here I present data on bioaccumulation, biochemical and tissues biomarkers in Nile tilapia (Oreochromis niloticus niloticus) as early warning indicators of river Nile pollution.

Bioaccumulation is the ability of fish to concentrate an element from water to a level higher than that of its environment. Therefore, bioaccumulation of metals in fish can be considered as an index of metal pollution in the aquatic bodies [30,31]. Fish surviving at highly polluted areas accumulate higher levels of heavy metals than those surviving at less polluted areas [32]. In the present work, Nile tilapia sampled from downstream river Nile accumulated higher levels of all the detected heavy metals than those collected from upstream sites. Mean levels of the selected heavy metals in the tissues of Nile tilapia followed the same pattern as in water, in being higher at Damietta and Rosetta. This indicated a close correlation between the concentrations of metals in water and in fish tissues. Heavy metal residues in the tissues of Nile tilapia exhibited different patterns of accumulation and distribution among the selected tissues and localities. Effects of high concentrations of heavy metals on fish are not well established; however, there is evidence that high concentrations in fish can experience toxicity [33]. Some heavy metals as copper can combine with other contaminants to produce an additive toxic effect on fish [34]. Chromium bioaccumulation in fish has been reported to cause impaired respiratory and osmoregulatory functions through structural damage to gill epithelium [9]. Cadmium and lead are highly toxic non-essential heavy metals and they do not have a role in biological processes in living organisms [35]. Thus even in low concentration, Cd and Pb could be harmful to fishes. Lead was found to cause pathological changes in tissue and organs. Manganese functions as an essential constituent for bone structure, reproduction and normal functioning of the enzymes system [35]. It is toxic only when present in higher amount, but at low level is considered as micronutrient [35].

According to the current results, the highest concentrations of all the detected heavy metals (excluded Mn) were recorded in the liver tissues followed by gills. The distributions of all these heavy metals were in the order of liver > gill > gonad > muscle. The high concentration of these metals in the liver is related to the fact that the liver plays an important role in accumulation. High accumulation of metals in liver may be due to their strong binding with cystine residues of metallothionein (MT), which serves as a detoxification mechanism [34]. MT has high affinities for heavy metals and in doing so, concentrate and regulate these metals in the liver [36]. The lower concentration of nearly all the detected metals in gills (compared to liver) was possibly due to lower binding affinity of metals on the gill surface. Such low accumulation may also be due to development of some defensive mechanism such as excessive mucous secretion and clogging of gills. Gonad and muscle were found to accumulate small amounts of all the heavy metals and might have received them through circulation. It is suggested that the low accumulation of metals in gonad and muscle may be due to lack of binding affinity of these metals with the proteins of gonad and muscle. This is particularly important because muscles contribute the greatest mass of the flesh that is consumed as food. Heavy metal concentrations reported in the present study is consistent and within the range reported in a previous study that evaluated metal contamination of African catfish sampled from the rive Nile [1].

Biochemical biomarkers are among the first measurable responses to the presence of chemical pollutants and thus are early, sensitive indicators of possible damage at higher levels of organization [37]. The alteration in enzyme activity has been advocated to provide an early warning of potentially damaging changes in stressed fish. Especially LDH and G6PDH have been employed for diagnosing liver and muscle damages caused by pollutants in fish [38] and could be useful biomarkers of water pollution. Remarkable alterations in the activity of both enzymes were detected in the present study. These alterations go in parallel with the elevation in the levels of water chemical parameters detected in the downstream river Nile as a result of pollution stress in these areas. The activity of G6PDH was significantly decreased from upstream to downstream river Nile recording the lowest values at Rosetta and/or Damietta water. Such reduction in G6PDH activity in the liver and muscles of Nile tilapia was previously recorded in some fishes exposed to toxicants [22,39]. A reduction in G6PDH activity in the liver was also observed in bullhead (Cottus gobio) exposed to paper mill effluents [40]. Thus the observed decrease in activity of G6PDH in the present work may reflect metabolic imbalance after pollution stress in direct relation to the limited availability of NADPH [22]. This probably reflects an adaptation to oxidative conditions to which fish have been exposed. GSH is used by antioxidant defense mechanisms and its production requires NADPH to be synthesized in the pentose phosphate metabolic pathway in which G6PDH participate. For this reason, G6PDH is considered as antioxidant enzyme [41]. The decreased activities of G6PDH with higher concentrations of metals in downstream river Nile are most likely related to the displacement of magnesium cofactor by cadmium, copper or other metals such that the enzyme lost sensitivity to magnesium activation [16].

LDH commonly reflects the metabolic capacity of tissues after long-term exposure to contaminated water bodies [22,42]. The activity of LDH, which is a cytoplasmic enzyme, showed a marked elevation in the tissues of Nile tilapia collected from downstream river Nile compared to other sites. LDH activity is generally associated with cellular metabolic activity and is a pivotal enzyme between the glycolytic pathway and the tricarboxylic acid cycle [43]. Thus, elevation of LDH may suggest a bias towards the anaerobic glycolytic pathway. By other meaning, changes in the lactate dehydrogenase activity may indicate the facility with which the Nile tilapia can shift to anaerobic metabolism under stressed conditions. Anaerobic capacity was most impacted in the tissues of fish collected from heavily polluted sites with the LDH activity in the tissues of fish from downstream river Nile higher than that of fish sampled from upstream river Nile. Liver and muscles of Nile tilapia seem to be able to react strongly to the presence of higher chemical pollution by increasing anaerobic capacity, possibly facilitating elimination of offending chemicals [44]. LDH activity in muscle was always higher than that in liver. This biochemical response also complies with low partial pressure and oxygen concentrations in the muscle, prompting increased anaerobic capacity.

Histopathological biomarkers can be used as indicators for the effects of various anthropogenic pollutants on organisms and are a reflection of the overall health of the entire population in the ecosystem [45]. The severity of damage depends on the toxic potentiality of a particular toxicant accumulated in the tissue. Therefore, exposure to polluted water may adversely affect various organs in fish which ultimately could lead to overall toxic impact on organs like gill and liver [46]. Different kind of toxicity effects were noticed in various fish exposed to different pollutants [18,22,47]. Liver as the main organ of metabolism comes into contact with xenobiotics absorbed from the aquatic environment and liver lesions are often associated with exposure to aquatic pollutants [48]. One of the most important functions of liver is to clean pollutants from the blood coming from the intestine, so it is considered as indicator of aquatic environmental pollution. The monitoring of histological changes in fish liver is a highly sensitive and accurate way to assess the effects of xenobiotic compounds in field [18]. The fish that was sampled from upstream river Nile showed healthy liver structure. In the present study irregular arrangements of hepatocytes, leukocyte infiltration, necrosis and vacuolation of hepatocytes was observed in the livers of Nile tilapia collected from the whole course of the river Nile even though their livers were normal in appearance macroscopically. The most prominent histological alterations observed in the fish that showed abnormal livers macroscopically were dilation of blood vessels and sinusoids, severe congestion in sinusoids and blood vessels, increased vacuolation of hepatocytes and focal areas of necrosis. Vacuolation of hepatocyte is a nonspecific response of fish due to toxic conditions (Roberts 1978). The vacuolization of hepatocytes might indicate an imbalance between the rate of synthesis of substances in the parenchyma cells and the rate of their release into the circulation [49]. Severe congestion in sinusoids and small blood vessels makes the blood flow from the hepatic portal vein and hepatic artery into the central vein rather difficult. This may be responsible for the cellular degeneration and necrosis in the livers of some Nile tilapia collected from downstream river

Nile. Observed liver abnormalities reflect the biological impacts of complex mixture of pollutants present in this downstream areas of the river Nile.

Gills are the first target of waterborne pollutants due to the constant contact with water, as well as the main place for some heavy metals uptake [50]. It is well known that changes in fish gill are among the most commonly recognized responses to aquatic pollutants [50]. Histological study of the gills showed a typical structural organization of the lamella in the fishes collected from upstream river Nile. However, fish sampled from heavily polluted sites (downstream areas) showed several histological alterations, namely fusion of gill lamellae, epithelial lifting, necrosis, curling of primary lamellae, shrinkage of cartilaginous supporting mass, sever degeneration in gill filaments and dilation and congestion in blood vessel. The cellular damages observed in the gills in term of epithelial lifting and necrosis can adversely affect the gas exchange and ionic regulation. As a consequence of the increased distance between water and blood due to epithelial lifting, the oxygen uptake is impaired. However, fishes have the capacity to increase their ventilation rate, to compensate low oxygen uptake. Fusion of some secondary lamellae is an example of defense mechanisms that reduce the branchial superficial area in contact with the external environment [18]. The observed edematous changes in gill filaments is probably due to increased capillary permeability [22,47]. Edema could serve as a mechanism of defense, because separation of the epithelium of the lamellae increases the distance across which waterborne pollutants must diffuse to reach the bloodstream [18]. These mechanisms also increase the diffusion barrier to the pollutants [51]. Similar lesions have been previously reported by some authors in some fish species exposed to different kind of pollution [22,52-54]. This means, such alterations are non-specific and may be induced by different types of contaminants.

5. CONCLUSION

It is evident from this study that, liver was the site of maximum accumulation for the elements while muscle was the overall site of least metal accumulation. Altered activities of G6PDH and LDH can provide a useful biomarker for environmental managers in investigating the exposure of fish to contaminated waters. Histopathological lesions provide a reliable, easily quantifiable index of low-level toxic stress to a broad range of environmental pollutants. These results provide evidence that bioaccumulation, biochemical and tissue biomakers can be sensitive indicators of exposure to mixed organic pollutants in surface waters.

6. ACKNOWLEDGEMENTS

Alaa G. M. Osman, (Project ID 448). Most of this study were conducted during 2010-2012 in the framework of my Alexander von Humboldt foundation postdoctoral fellowship (3.4-AGY/1134716 STP), hosted by Leibniz-Institute of Freshwater Ecology and Inland Fisheries (IGB), Berlin, Germany. My special thanks go to Prof. Dr. Werner Kloas who offered me a warm welcome and gave me the opportunity to use the research facilities in his department. I would like to thank Mr. A. Gad El-Rab, A. Moustafa, M. Nassar, R. Said, Department of Zoology (Al-Azhar University, Egypt) for their support during sampling. I gratefully acknowledge Dr. Alaa ElDin Sayed, Department of zoology (Assiut University, Egypt) for his help during enzyme analysis.

REFERENCES

1. G. M. Osman and W. Kloas, "Water Quality and Heavy Metal Monitoring in Water, Sediments and Tissues of the African Catfish Clarias gariepinus (Burchell, 1822) from the River Nile, Egypt," Journal of Environmental Protection, Vol. 1, 2010, pp. 389-400. HUdoi:10.4236/jep.2010.14045
2. G. M. Osman, A. M. Abd El Reheem, M. A. Moustafa, U. M. Mahmoud, K. Y. Abuel-Fadl and W. Kloas, "In Situ Evaluation of the Genotoxic Potential of the River Nile: I. Micronucleus and Nuclear Lesion Tests of Erythrocytes of Oreochromis niloticus niloticus (Linnaeus, 1758) and Clarias gariepinus (Burchell, 1822)," Toxicological and Environmental Chemistry, Vol. 93, No. 5, 2011, pp. 1002-1017. HUdoi:10.1080/02772248.2011.564496
3. R. Tejeda-Vera, E. Lopez-Lopez and J. E. Sedeno-Diaz, "Biomarkers and Bioindicators of the Health Condition of Ameca splendens and Goodea atripinnis (Pisces: Goodeaidae) in the Ameca River, Mexico," Environment International, Vol. 33, No. 4, 2007, pp. 521-531. HUdoi:10.1016/j.envint.2006.11.018
4. H. I. Falfushynska and O. B. Stolyar, "Responses of Biochemical Markers in Carp Cyprinus carpio from Two Field Sites in Western Ukraine," Ecotoxicology and Environmental Safety, Vol. 72, No. 3, 2009, pp. 729-736. HUdoi:10.1016/j.ecoenv.2008.04.006
5. G. Ambedkar and M. Muniyan, "Bioaccumulation of Some Heavy Metals in the Selected Five Freshwater Fish from Kollidam River, Tamilnadu, India," Advances in Applied Science Research, Vol. 2, No. 5, 2011, pp. 221- 225.
6. A. Otitoloju, "Evaluation of the Joint-Action Toxicity of Binary Mixtures of Heavy Metals against the Mangrove Periwinkle Tympanotonus fuscatus var radula (L.)," Ecotoxicology and Environmental Safety, Vol. 53, No. 3, 2002, pp. 404-415. HUdoi:10.1016/S0147-6513(02)00032-5
7. N. Barak and C. Mason, "Mercury, Cadmium and Lead in Eels and Roach: The Effects of Size, Season and Locality on Metal Concentrations in Flesh and Liver," Science of the Total Environment, Vol. 92, 1990, pp. 249-256. HUdoi:10.1016/0048-9697(90)90334-Q
8. Haluzova, H. Modra, J. Blahova, M. Havelkova, Z. Siroka and Z. Svobodova, "Biochemical Markers of Contamination in Fish Toxicity Tests," Interdisciplinary Toxicology, Vol. 4, No. 2, 2011, pp. 85-89.

9. G. Heath, "Water Pollution and Fish Physiology," Lewis Publs., Boca Raton, 1996.
10. P. P. Pandolfi, F. Sonati, R. Rivi, P. Mason, F. Grosveld and L. Luzzatto, "Targeted Disruption of the Housekeeping Gene Encoding Glucose 6-Phosphate Dehydrogenase (G6PD): G6PD Is Dispensable for Pentose Synthesis But Essential for Defense against Oxidative Stress," The EMBO Journal, Vol. 14, No. 21, 1995, pp. 5209-5215.
11. P. C. Das, S. Ayyappan, B. K. Das and J. K. Jena, "Nitrite Toxicity in Indian Major Carps: Sublethal Effect on Selected Enzymes in Fingerlings of Catla catla, Labeo rohita and Cirrhinus mrigala," Comparative Biochemistry and Physiology. Toxicology & Pharmacology: CBP, Vol. 138, No. 1, 2004, pp. 3-10.
12. P. C. Das, S. Ayyappan, J. K. Jena and B. K. Das, "Acute Toxicity of Ammonia and Its Sub-Lethal Effects on Selected Haematological and Enzymatic Parameters of Mrigal, Cirrhinus mrigala (Hamilton)," Aquaculture Research, Vol. 35, No. 2, 2004, pp. 134-143. HUdoi:10.1111/j.1365-2109.2004.00994.x
13. S. M. Long, K. J. Ryder and D. A. Holdway, "The Use of Respiratory Enzymes as Biomarkers of Petroleum Hydrocarbon Exposure in Mytilus edulis Planulatus," Ecotoxicology and Environmental Safety, Vol. 55, No. 3, 2003, pp. 261-270. HUdoi:10.1016/S0147-6513(02)00137-9
14. D. Bernet, H. Schmidt, W. Meier, P. Burkhardt-Holm and T. Wahli, "Histopathology in Fish: Proposal for a Protocol to Assess Aquatic Pollution," Journal of Fish Diseases, Vol. 22, No. 1, 1999, pp. 25-34. HUdoi:10.1046/j.1365-2761.1999.00134.x
15. S. J. Teh, S. M. Adams and D. E. Hintonm, "Histopathologic Biomarkers in Feral Freshwater Fish Populations Exposed to Different Types of Contaminant Stress," Aquatic Toxicology, Vol. 37, 1997, pp. 51-70. HUdoi:10.1016/S0166-445X(96)00808-9
16. S. J. The, S. L. Clark, C. L. Brown, S. N. Luoma and D. E. Hinton, "Enzymatic and Histopathologic Biomarkers as Indicators of Contaminant Exposure and Effect in Asian Clam (Potamocorbula amurensis)," Biomarkers, Vol. 4, No. 6, 1999, pp. 497-509. HUdoi:10.1080/135475099230660
17. H. Moeller, "A Critical Review on the Role of Pollution as a Cause of Fish Diseases," In: A. E. Ellis, Ed., Fish and Shellfish Pathology, European Association of Fish Pathology, Academic Press, London, 1985, pp. 169-182.
18. Figueiredo-Fernandes, J. V. Ferreira-Cardoso, S. Garcia-Santos, S. M. Monteiro, J. Carrola, P. Matos and A. Fontainhas-Fernandes, "Histopathological changes in Liver and Gill Epithelium of Nile Tilapia, Oreochromis niloticus, Exposed to Waterborne Copper," Pesquisa Veterinaria Brasileira, Vol. 27, No. 3, 2007, pp. 103-109. HUdoi:10.1590/S0100-736X2007000300004
19. Figueiredo-Fernandes, A. Fontainhas-Fernandes, R. Monteiro, M. A. Reis-Henriques and E. Rocha, "Effects of the Fungicide Mancozeb on Liver Structure of Nile Tilapia, Oreochromis niloticus: Assessment and Quantification of Induced Cytological Changes Using Qualitative Histopathology and the Stereological Point-Sampled Intercept Method," Bulletin of environmental contamination and toxicology, Vol. 76, No. 2, 2006, pp. 249-255. HUdoi:10.1007/s00128-006-0914-1
20. G. M. Osman, E. Ali, M. Hashem, M. Mostafa and I. Mekkawy, "Genotoxicity of Two Pathogenic Strains of Zoosporic Fungi (Achlya klebsiana and Aphanomyces laevis) on Erythrocytes of Nile Tilapia Oreochromis niloticus niloticus," Ecotoxicology and environmental safety, Vol. 73, No. 1, 2010, pp. 24-31. HUdoi:10.1016/j.ecoenv.2009.08.021
21. K. M. Gadagbui, M. Addy and A. Goksoyr, "Species Characteristics of Hepatic Biotransformation Enzymes in two Tropical Freshwater Teleosts, Tilapia (Oreochromis niloticus) and Mudfish (Clarias anguillaris)," Comparative Biochemistry and Physiology C-Pharmacology Toxicology & Endocrinology, Vol. 114, No. 3, 1996, pp. 201- 211.
22. G. M. Osman, A. M. Abd El Reheem, K. Y. AbuelFadl and A. G. GadEl-Rab, "Enzymatic and Histopathologic Biomarkers as Indicators of Aquatic Pollution in Fishes,"

Natural Science, Vol. 2, No. 11, 2010, pp. 1302- 1311. HUdoi:10.4236/ns.2010.211158

23. G. Osman, K. Y. Abuel-Fadl and W. Kloas, "In Situ Evaluation of the Genotoxic Potential of the River Nile: II. Detection of DNA Strand-Breakage and Apoptosis in Oreochromis niloticus niloticus (Linnaeus, 1758) and Clarias gariepinus (Burchell, 1822)," Mutation Research, Vol. 747, 2012, pp. 14-21. HUdoi:10.1016/j.mrgentox.2012.02.013
24. APHA, "Standard Methods for the Examination of Water & Wastewater," Amer Public Health Assn, 2005.
25. Hardewig, H. O. Pörtner and P. van Dijk, "How Does the Cold Stenothermal Gadoid Lota lota Survive High Water Temperatures during Summer?" Journal of Comparative Physiology. B, Biochemical, Systemic and Environmental Physiology, Vol. 174, No. 2, 2004, pp. 149- 156. HUdoi:10.1007/s00360-003-0399-8
26. G. M. Osman, I. A. A. Mekkawy, J. Verreth and F. Kirschbaum, "Effects of Lead Nitrate on the Activity of Metabolic Enzymes during Early Developmental Stages of the African Catfish Clarias gariepinus (Burchell, 1822)," Fish Physiology and Biochemistry, Vol. 33, 2007, pp. 1-13. HUdoi:10.1007/s10695-006-9111-8
27. J. D. Bancroft and A. Stevens, "Theory and Practice of Histological Techniques," 4th Edition, Churchill Livingstone, New York, 1996.
28. SPSS, "SPSS-Inc. for Windows Release," Vol. 10, Chicago, 1998.
29. Statistica, "StatSoft-Inc. for Windows Release, USA," 2007.
30. M. Javed, "Heavy Metal Contamination of Freshwater Fish and Bed Sediments in the River Ravi Stretch and related Tributaries," Pakistan Journal of Biological Sciences, Vol. 8, No. 10, 2005, pp. 1337-1341. HUdoi:10.3923/pjbs.2005.1337.1341
31. H. Karadede-Akin and E. Unlü, "Heavy Metal Concentrations in Water, Sediment, Fish and Some Benthic Organisms from Tigris River, Turkey," Environmental Monitoring and Assessment, Vol. 131, No. 1-3, 2007, pp. 323-337. HUdoi:10.1007/s10661-006-9478-0
32. M. Bahnasawy, A. A. Khidr and N. Dheina, "Seasonal Variations of Heavy Metals Concentrations in Mullet, Mugil Cephalus and Liza Ramada (Mugilidae) from Lake Manzala, Egypt," Applied Sciences Research, Vol. 5, No. 7, 2009, pp. 845-852.
33. L. A. Woodward, M. Mulvey and M. C. Newman, "Mercury Contamination and Population-Level Responses in Chironomids: Can Allozyme Polymorphism Indicate Exposure?" Environmental Toxicology and Chemistry, Vol. 15, No. 8, 1996, pp. 1309-1316. HUdoi:10.1002/etc.5620150808
34. M. Yacoub, "Study on Some Heavy Metals Accumulated in Some Organs of Three River Nile Fishes from Cairo and Kalubia Governorates," African Journal of Biological Science, Vol. 3, 2007, pp. 9-21.
35. El-Naggar, S. Mahmoud and S. Tayel, "Bioaccumulation of Some Heavy Metals and Histopathological Alterations in Liver of Oreochromis niloticus in Relation to Water Quality at Different Localities along the River Nile, Egypt," World Journal of Fish and Marine Sciences, Vol. 1, No. 2, 2009, pp. 105-114.
36. J. A. McCarter and M. Roch, "Hepatic Metallothionein and Resistance to Copper in Juvenile Coho Salmon," Comparative Biochemistry and Physiology. C, Comparative Pharmacology and Toxicology, Vol. 74, No. 1, 1983, pp. 133-137. HUdoi:10.1016/0742-8413(83)90164-0
37. L. Vigano, A. Arillo, F. Melodia, P. Arlati and C. Monti, "Biomarker Responses in Cyprinids of the Middle Stretch of the River Po, Italy," Environmental Toxicology and Chemistry, Vol. 17, No. 3, 1998, pp. 404-411. HUdoi:10.1002/etc.5620170309
38. V. Lenartova, K. Holovska, J. R. Pedrajas, E. MartinezLara, J. Peinado, J. LopezBarea, I. Rosival and P. Kosuth, "Antioxidant and Detoxifying Fish Enzymes as Biomarkers of River Pollution," Biomarkers, Vol. 2, No. 4, 1997, pp. 247-252. HUdoi:10.1080/135475097231625
39. Korsgaard, "Metabolic Changes Associated with 17alpha-thinylestradiol Exposure in the Pregnant Teleost Zoarces Viviparous," Electric Journal of Ichthyology, Vol. 1, 2005, pp. 10-20.

40. F. Bucher, R. Hofer, G. Krumschnabel and C. Doblander, "Doblander in the Prooxidant-Antioxidant Balances in the Liver of Bullhead (Cottus gobio) Exposed to Treated Paper Mill Effluents," Chemosphere, Vol. 27, 1993, pp. 1329-1338. HUdoi:10.1016/0045-6535(93)90227-V

41. R. Reiter, L. Tang, J. J. Garcia and A. MunozHoyos, "Pharmacological Actions of Melatonin in Oxygen Radical Pathophysiology," Life Sciences, Vol. 60, No. 25, 1997, pp. 2255-2271. HUdoi:10.1016/S0024-3205(97)00030-1

42. M. M. Gagnon and D. A. Holdway, "Metabolic Enzyme Activities in Fish Gills as Biomarkers of Exposure to Petroleum Hydrocarbons," Ecotoxicology and Environmental Safety, Vol. 44, No. 1, 1999, pp. 92-99. HUdoi:10.1006/eesa.1999.1804

43. R. Chourpagar and G. K. Kulkarni, "Toxic Effect of Copper Sulphate on Lactate Dehydrogenase Activity in a Freshwater Crab, Barytelphusa cunicularis (Westwood)," World Journal of Zoology, Vol. 4, No. 3, 2009, pp. 180- 183.

44. Cohen, M. M. Gagnon and D. Nugegoda, "Alterations of Metabolic Enzymes in Australian Bass, Macquaria Novemaculeata, after Exposure to Petroleum Hydrocarbons," Archives of Environmental Contamination and Toxicology, Vol. 49, No. 2, 2005, pp. 200-205. HUdoi:10.1007/s00244-004-0174-1

45. Z. Khoshnood, S. Khodabandeh, S. Mosafer and R. Khoshnood, "Effects of Cortisol on Gill Chloride Cells in Persian Sturgeon, Acipenser persicus, Fry," Yakhteh, Vol. 11, No. 4, 2010, pp. 424-431.

46. Parvathi, P. Sivakumari and C. Sarasu, "Effect of Chromium on Histological Alteration of Gill, Liver and Kindey of Freshwater Teleost, Cyprinus caprio (L.)," Journal of Fisheries International, Vol. 6, No. 1, 2011, pp. 1-5. HUdoi:10.3923/jfish.2011.1.5

47. E. Olojo, K. Olurin, G. Mbaka and A. Oluwemimo, "Histopathology of the Gill and Liver Tissues of the African Catfish Clarias gariepinus Exposed to Lead," African Journal of Biotechnology, Vol. 4, No. 1, 2005, pp. 117-122.

48. Fernandes, A. Fontainhas-Fernandes, E. Rocha and M. A. Salgado, "Monitoring Pollution in Esmoriz-Paramos Lagoon, Portugal: Liver Histological and Biochemical Effects in Liza saliens," Environmental Monitoring and Assessment, Vol. 145, No. 1-3, 2008, pp. 315-322. HUdoi:10.1007/s10661-007-0041-4

49. W. H. Gingerich, "Hepatic Toxicology of Fishes," In: L. J. Weber, Ed., Aquatic Toxicology, Raven Press, New York, 1982, pp. 55-105.

50. P. Laurent and S. F. Perry, "Environmental-Effects on Fish Gill Morphology," Physiological Zoology, Vol. 64, No. 1, 1991, pp. 4-25.

51. van Heerden, A. Vosloo and M. Nikinmaa, "Effects of Short-Term Copper Exposure on Gill Structure, Metallothionein and Hypoxia-Inducible Factor-1alpha (HIF-1 alpha) Levels in Rainbow Trout (Oncorhynchus mykiss)," Aquatic Toxicology, Vol. 69, No. 3, 2004, pp. 271-280. HUdoi:10.1016/j.aquatox.2004.06.002

52. Fanta, F. S. A. Rios, S. Romão, A. C. C. Vianna and S. Freiberger, "Histopathology of the Fish Corydoras paleatus Contaminated with Sublethal Levels of Organophosphorus in Water and Food," Ecotoxicology and environmental safety, Vol. 54, No. 2, 2003, pp. 119-130. HUdoi:10.1016/S0147-6513(02)00044-1

53. B. R. Martinez, M. Y. Nagae, C. T. B. V. Zaia and D. A. M. Zaia, "Acute Morphological and Physiological Effects of Lead in the Neotropical Fish Prochilodus lineatus," Brazilian Journal of Biology = Revista brasleira de biologia, Vol. 64, No. 4, 2004, pp. 797-807.

54. R. Triebskorn, I. Telcean, H. Casper, A. Farkas, C. Sandu, G. Stan, O. Colărescu, T. Dori and H.-R. Köhler, "Monitoring Pollution in River Mureş, Romania, Part II: Metal Accumulation and Histopathology in Fish," Environmental Monitoring and Assessment, Vol. 141, No. 1-3, 2008, pp. 177-188. HUdoi:10.1007/s10661-007-9886-9

CHAPTER 14

Dynamic Assessment of Comprehensive Water Quality Considering the Release of Sediment Pollution

Tianxiang Wang, Shiguo Xu * and Jianwei Liu

Institution of water and water environment, Dalian University of Technology, Dalian 116023, China

ABSTRACT

Comprehensive assessment of water quality is an important technological measure for water environmental management and protection. Previous assessment methods tend to ignore the influences of sediment pollutant release and dynamic change of the water boundary. In view of this, this paper explores a new method for comprehensive water quality assessment. Laboratory simulation experiments are conducted to analyze the influences of sediment pollutant release on water quality, and the results are taken as increments, coupled with original samples, to constitute a new set of evaluation samples. Dynamic and comprehensive water quality assessment methods are created based on a principal component analysis (PCA)/analytic hierarchy process (AHP)–variable fuzzy pattern recognition (VFPR) model and adopted to evaluate water quality. A geographic information system (GIS) is applied to visually display the results of water quality assessment and the change of the water boundary. This study takes Biliuhe Reservoir as an engineering example. The results show the change process of the water boundary, during which the water level is reduced from 63.10 m to 54.15 m. The reservoir water quality is fine, of which the water quality level (GB3838-2002) is between level 2 and level 3, and closer to level 2 taking no account of sediment pollutant release. The water quality of Biliuhe Reservoir, overall, is worse in summer and better in winter during the monitoring period. Meanwhile, the water quality shows the tendency of being better from upstream to downstream, and the water quality in the surface layer is better than that in the bottom layer. However, water quality is much closer, or even inferior, to level 3 when considering the release of nitrogen and phosphorus in sediments, and up to 42.7% of the original assessment results of the samples undergo changes. It is concluded that the proposed method is comparatively reasonable as it avoids neglecting sediment pollutant release in the water quality assessment, and the presentation of the evaluation results and change of the water boundary is intuitive with the application of GIS.

Keywords: dynamic; water quality assessment; sediment pollution; water boundary

1. INTRODUCTION

Water quality assessment is the foundation of water environmental protection and management [1]. Noteworthy, primary or secondary pollutants in the basin would enter the reservoirs, lakes, rivers, or oceans due to rainfall erosion and river transportation, and then gradually accumulate at the bottom and result in sediment pollution [2]. Sediment pollutants have characteristics of coexistence of aqueous-biological-sedimentary phases [3], and pollutants will release from the sediments under the influence of external conditions, such as dissolved oxygen, hydrodynamics, pH, temperature, and the pollutant concentration in the overlaying water [4,5,6]. In addition, sediment pollution has dual characters of being a source and a sink [7]. In general, the flux direction of liquid pollutants is upward, and the intensity is associated with the concentration of the overlying water [5,8,9], while the flux direction of solid pollutants is the opposite, especially sand, minerals, biological residual bodies, and solid waste [10]. The release of sediment pollutant is a complicated process which is affected by multiple factors which change constantly [11]. Sediment pollutant release is essentially a process where pollutants diffuse from a region with higher concentration to an adjoining region with lower concentration [12]. According to the EPA (1998), 10% of the sediment pollution release could threaten fish, and animals and persons feeding on fish [13]. The nutrients released from sediments in Chaohu Lake in Anhui Province, China, accounts for 21% of exogenous pollutants. Up to 90% of the phosphorus and 80% of the nitrogen accumulates in sediments in the

Dianchi Lake in Yunnan Province, China [14]. Reservoirs also face the problem of sediment pollution, such as Yuqiao Reservoir and Xinlicheng Reservoir [15,16]. Thus, sediment pollution is a factor that cannot be ignored in water quality assessment. In addition, the reservoir water level undergoes seasonal and inter-annual fluctuation, which leads to the dynamic change of the water boundary [17]. Thus, the relative positions of water monitoring points change accordingly. A monitoring point that is in the middle level in the vertical direction during flood season may turn into a surface layer in the dry season. The neglect of this change may lead to miscalculations in the water quality assessment. The spatial and temporal variability due to water level fluctuation increases the difficulty of the water quality assessment.

According to the characteristics of uncertainty, randomness, and variability of water quality, researchers have put forward a series of assessment methods, which have been effectively applied in water quality assessments [18,19,20]. Sinha et al. used the comprehensive index method to assess water quality of Hooghly River [21]. Chu et al. explored the use of an artificial neural network to assess water quality [22]. Yan et al. developed a dynamic variable fuzzy set assessment model to judge the water quality [23]. However, the sediment pollution and the water boundary change are ignored. The objectives of this paper are to explore a water quality assessment method that can reflect the influence of sediment pollution and present the evaluation results and water boundary visually.

2. MATERIALS AND METHODS

2.1. Study Area and Sampling

Biliuhe Reservoir (122°29′24.11″ N; 39°49′12.52″ E) provides up to 80% of the water demand of Dalin City, Northeast China. Biliuhe Reservoir has three main tributaries, including the Biliuhe River, Gelihe River, and Bajiahe River (Figure 1). The watershed area is 2814 km^2 and the forest coverage rate is over 80%. The annual rainfall is 742.8 mm, which concentrates in summer (66.9%), and the annual runoff is 6.6 × 108 m^3. The maximum water storage of the reservoir is 9.34 × 108 m^3. Nitrogen and phosphorus pollution are the main water quality issues of the reservoir. This paper combines the characteristics of the reservoir and the “Specification for Water Environment Monitoring, SL 219-2013” to design a sampling plan.

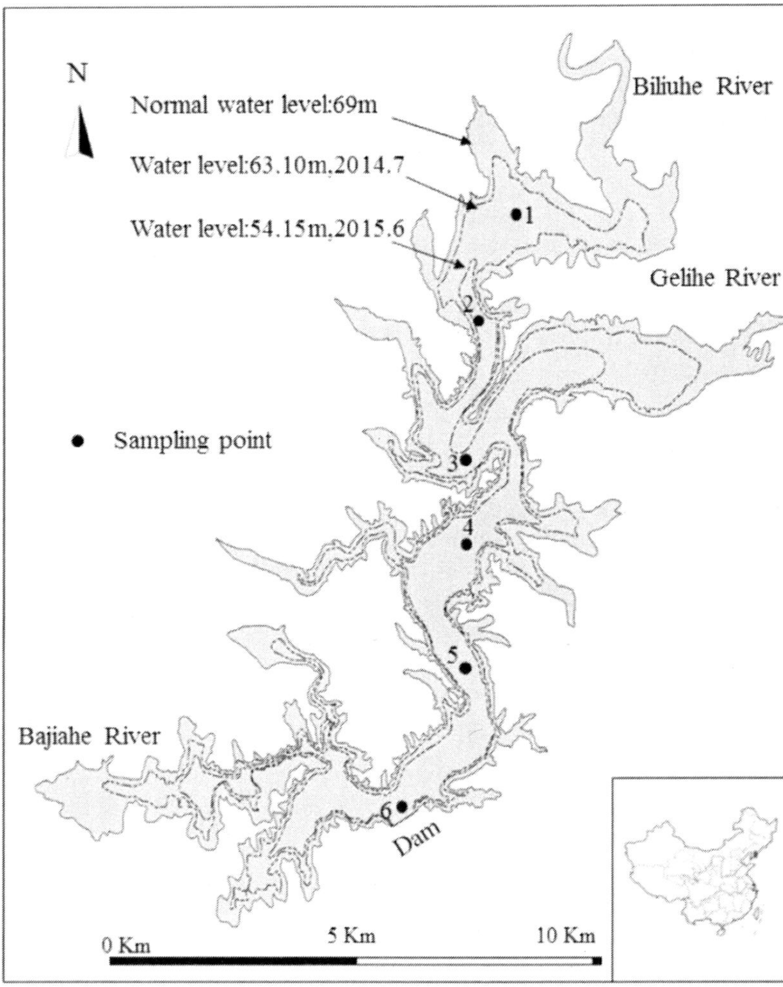

Figure 1. The horizontal sampling point map of Biliuhe Reservoir.

The location map of sampling points is shown in Figure 1 and Figure 2. Samples are collected from July 2014 to June 2015, and December, January, and March belong to the freeze-thawing period which cannot be sampled. In order to simplify the problem, this study considered only the monitoring points in the main stem of Biliuhe Reservoir (Figure 1 and Figure 2). During the monitoring periods, the water level decreases from 63.10 m to 54.15 m and the difference of the water level is 8.95 m. Correspondingly, the water boundary changes. Water samples are collected in accordance with specific conditions: collect one water sample if the water depth is less than 5 m;

collect two water samples separately from surface and bottom layer when the depth is between 5 and 10 m; and collect samples every 5 m when the depth is over 10 m. Sediment samples were collected in October 2014, including points 1–6. As the water level fluctuates, the relative positions of the sampling sites change accordingly. To solve this problem, a dynamic sampling method is developed as follows: (1) Cover as many planned sampling sites as possible; (2) add the number of sampling points when the water level is at a high depth; (3) retreat or cancel the relevant sampling points when the water level is at a low depth. A Garmin GPS (GARMIN Corporation, Shengyang, China) is used to mark the sampling point and collect the location information. The water samples are sampled with Niskin bottles (HYDRO-BIOS), the sediment samples are collected by a Van Veen sampler (KC-Denmark). Monitoring indicators of water samples cover dissolved oxygen (DO), total phosphorus (TP), total nitrogen (TN), ammonia nitrogen, chemical oxygen demand (CODMn), pH, iron, and manganese. Monitoring indicators of sediment samples include total nitrogen and total phosphorus. The water samples were carried out in accordance with the "Water and wastewater monitoring and analysis method (Fourth Edition), Beijing: China Environmental Science Press, 2002". The sediment samples of total nitrogen and total phosphorus were measured by Kjeldahl determination and the Standard measurements and testing (SMT) protocol method [2,5]. The monitoring results were expressed with average values and the errors of the parallel experiments were less than 5%.

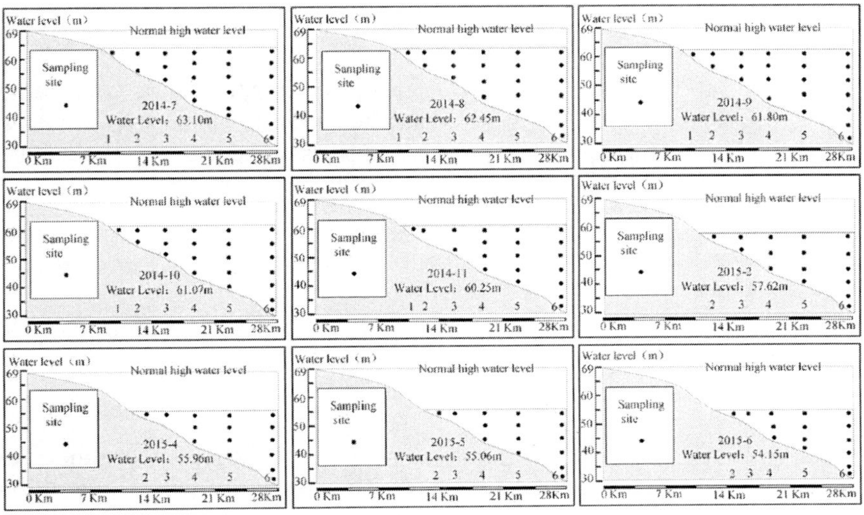

Figure 2. The vertical sampling points map in Biliuhe reservoir.

2.2. Experiment on Sediment Pollutant Release

This paper studies the release capacity of sediment pollutants under different concentrations of overlying water through a simulation experiment. As eutrophication is the main problem for reservoirs, this paper only considers the release of nitrogen and phosphorus in sediments. Sediment samples are put in a cool and dry place to let them naturally air dry. Then the samples are screened through a 100 mesh sieve (0.15 um) and preserved at −20 °C. Sediment samples (0.2 g each) are placed in a 20 mL centrifuge tube and 20 mL standard solution is added. The phosphorous standard solution of 0, 0.1, 0.2, 1, 2, 4, 6, 10, 15 and 20 mg/L is made by potassium dihydrogen phosphate (KH2PO4, GR) and nitrogen standard solution of 0, 0.1, 0.5, 2, 4, 6, 8, 10, 15 and 20 mg/L is made by potassium nitrate (KNO3,GR). The centrifuge tubes are shaken (180–200 rpm) at an indoor temperature (20 ± 1 °C) for 24 h. After that, samples are then centrifuged (3000 rpm) for 10 min and the filtrates are collected through a 0.45 um fiberglass filter membrane. Next, the sorption-release characters of nitrogen and phosphorus can be calculated by measuring the equilibrium concentration of the collected filtrate. All experiments are carried out in triplicate and averaged, with relative errors less than 5%. Based on the experiment results, the release amount of nitrogen and phosphorus in sediments under different concentrations of each sampling site are calculated [5,6]:

$$q_m = \frac{(C_e - C_0) \times V}{M} \tag{1}$$

where C_e is the equilibrium concentration, mg/L; C_0 is the initial concentration, mg/L; $q_m < 0$ is the adsorption quantity, mg/kg; $q_m > 0$ is the release quantity, mg/kg; M is the weight of the sediment, g; V is the volume of adding standard solution, mL.

Furthermore, the amount of sediment release of different initial concentration can calculate by Equation (2) based on the interpolation method:

$$q_k = q_a + \frac{(q_b - q_a)}{(c_b - c_a)} \times (c_k - c_a) \tag{2}$$

where qk is the adsorption/release quantity when the initial concentration is c_k, mg/kg; q_b is the adsorption/release quantity when the initial concentration is c_b, mg/kg; q_a is the adsorption/release quantity when the initial concentration is c_a, mg/kg; the c_a, c_b belong to the standard solution of the experiment and c_k is between c_a and c_b, mg/L.

Next, the water load increment is obtained:

$$\Delta x(k) = \frac{Q \times q_k}{h_w \times 100} \tag{3}$$

where $\Delta x(k)$ is the increment of the contaminant when the initial concentration is k, mg/L; Q is the sediment content per unit area, g /cm^2; $q_k < 0$ is the adsorption quantity, $q_k > 0$ is the release quantity calculated by the interpolation method, mg/kg; and hw is the average water depth, m.

Finally, the new samples are generated:

$$X_{ij} = x_{ij} + \Delta xij \tag{4}$$

where X_{ij} is defined as the evaluation characteristic value considering the influence of sediment pollutant release; X_{ij} is the evaluation characteristic value of indicator j and sample i, mg/L; x_{ij} is the actual monitoring value of indicator j and sample i, mg/L; Δx_{ij} is the increment value considering the sediment pollutant release of indicator j and sample i, which is obtained by the laboratory simulation experiment, mg/L.

2.3. Dynamic Display of Water Quality Elements (Assessment Results)

To fully understand the situation of water quality, stratified monitoring is normally needed. Reservoir water depth changes dynamically due to rainfall and reservoir operation [24], and the position of sampling sites changes accordingly.

As can be seen in Figure 2, the water level during monitoring period fluctuates between 63.10 m and 54.15 m. Obviously, when the water level is 63.10 m, the distribution of sampling sites is different from that of 54.15 m. Water samples collected at the same depth of the water at different times may belong to different water layers. Geostatistics models in GIS, combined with the image editing capability, can visually export the dynamic status of the topography and the corresponding water quality element data graphically [25,26]. The process is as follows: First, extract the data of the horizontal and vertical topography boundaries of each sampling point based on the topographic data (digital elevation model, DEM). Then, import the position data of the sampling points and corresponding water quality data to the boundary documents. Afterwards, geostatistics is used to conduct the interpolation of water quality elements (assessment results) of the sampling points. Finally, export the water quality distribution diagram of each sampling, respectively, to dynamically and visually display the water quality assessment results. Based on the statistics and pollutant diffusion mechanism, the closer the distance of the points, the more similar the pollutant concentrations and water quality status are. The inverse distance weighting method in GIS is a interpolation method based on the above principles, and the original data will not be changed. Therefore, in this paper, the inverse distance weighting method is selected. Since the interpolation results are used only for display, the interpolation error is not considered in this paper. The detailed steps are as follows:

(1) Generate the DEM map of the study area based on the 1:5000 topographic map

(2) Extract the dynamic boundaries according to the water level

(3) Import the positions of the sampling points

(4) Import the water quality elements (assessment results) of each sample

(5) Generate the distribution of water quality, overall, by the inverse distance weighting method

(6) Export the dynamic distribution diagram of the water quality after inserting a legend

2.4. Assessment Model of Water Quality

The variable fuzzy pattern recognition (VFPR) model has strong adaptability to the dynamic and fuzzification characteristics of water quality assessments, which is applied in various kinds of evaluation issues [27,28]. This paper uses the VFPR model to comprehensively assess water quality.

Assume the set of samples of the water quality is X = (x_{ij}), and the set of standards of the water quality is expressed as Y = (y_{hj}), where I = 1, 2, ..., m, j = 1, 2, ... , n, h = 1, 2, ... , c. To eliminate the influences of different dimensions and inverse indices, the samples (x_{ij}) and standards (y_{hj}) are normalized to rij and shj:

$$r_{ij} = \begin{cases} 0 & x_{ij}\pounds y_{cj}(\text{positive index}), x_{ij}y_{cj}(\text{inverse index}) \\ \frac{y_{cj}-x_{ij}}{y_{cj}-y_{1j}} & \text{positive index or inverse index} \\ 1 & x_{ij}y_{1j}(\text{positive index}), x_{ij}\pounds y_{1j}(\text{inverse index}) \end{cases} \quad (5)$$

$$s_{hj} = \begin{cases} 0 & y_{hj} = y_{cj}, \text{ positive index or inverse index} \\ \frac{y_{cj}-y_{hj}}{y_{cj}-y_{1j}} & \text{positive index or inverse index} \\ 1 & y_{hj} = y_{1j}, \text{ positive index or inverse index} \end{cases} \quad (6)$$

where x_{ij} is the value of indicator j of sample i, where i is the number of samples and j is the number of indicators; y_{hj} is the value that defines standard h of indicator j, where h = 1, 2 ..., c, c represents the highest level of the standard; r_{ij} and s_{hj} are the results of the normalization of the samples (x_{ij}) and standards (y_{hj}).

Next, uhi, the synthetic relative membership degree for sample j belonging to standard h, is obtained:

$$u_{hi} = \begin{cases} 0 & , \quad 1 \le h \le a_i \text{ or } c \ge h \ge b_i \\ \dfrac{1}{\sum\limits_{k=a_i}^{b_i} \left[\dfrac{\sum\limits_{j=1}^{m} [w_j |r_{ij} - s_{hj}|]^p}{\sum\limits_{j=1}^{m} [w_j |r_{ij} - s_{kj}|]^p} \right]^{\frac{a}{p}}} & , \quad a_i \le h \le b_i \end{cases} \tag{7}$$

where u_{hi} is the synthetic relative membership degree for sample i belonging to standard h; h is the standard level of water quality; w_j is the weight of indicator j; a_i is the minimum level of sample i and bi is the maximum level of sample i. a is the optimization criteria parameter, a = 1 (linear), a = 2 (nonlinear); and p is the distance parameter, p = 1 (Hamming distance), p = 2 (Euclidean distance).

When u_{hi} is obtained, the traditional fuzzy assessment model generally takes the maximum membership degree as the final level of the sample, neglecting the information of other membership degrees. This paper regards the level characteristic value as the final assessment result, and the VFPR model considers all possibilities of relevant membership degrees of samples to each standard, which reflects the dynamic change of water quality levels. The model is as follows:

$$H = \sum_{h=1}^{c} u_{hi} h \tag{8}$$

where H is the level characteristic value; and h is the level of standard (h = 1, 2... c).

2.5. Steps of Dynamic Assessment of Comprehensive Water Quality Considering the Release of Sediment Pollution

This paper combines the VFPR model and GIS to dynamically assess water quality and creates a visual display of the results by combining with GIS. The details are as follows:

In the first step, the indicator system of the water quality assessment should be developed on the principles of scientificity, systematicness, feasibility, and practicability.

In the second step, laboratory simulation experiments are conducted to evaluate the influence of sediment release under conditions of different concentrations of the overlying water, so as to obtain water quality assessment samples considering sediment release.

In the third step, Equations (5) and (6) are used to normalize the indicators and standards.

In the fourth step, an appropriate method for weighting indicators should be selected. The common methods are equal weight (EW), principal component analysis (PCA), the analytic hierarchy process (AHP) method, the entropy weight method, and synthetic weight.

In the fifth step, synthetic weight w and the normalization results of r_{ij} and s_{hj} are put into Equations (7). Thus, four results of the synthetic relative membership degree uhj are calculated by changing the model parameters ($a = 1$ or $a = 2$; $p = 1$ or $p = 2$).

In the sixth step, Equation (8) is used to calculate the level characteristic value of the sample i based on the fifth step, and then the average value of the four models is taken as the final assessment result.

In the seventh step, a visible method is applied to visually display the assessment results.

3. RESULTS AND DISCUSSION

3.1. Analysis of the Reasonability of Assessment Model

This paper selects eight indicators, i.e., dissolved oxygen (DO, X1), total phosphorus (X2), total nitrogen (X3), ammonia nitrogen (X4), chemical oxygen demand (COD_{Mn}, X5), pH (X6), ions (X7), and manganese (X8), to assess water quality. The indicator system covers eutrophication, heavy metal, and comprehensive pollution.

The results of water quality assessment should reflect the actual status of the water quality. To verify the reasonability of the assessment method, eight virtual water quality samples are developed based on the National Surface Water Quality Standard of China (GB3838-2002) and are assessed using different assessment methods and weights.

The created virtual water quality samples can be seen in Table 1. All of the indicator values of sample 1 are better than level 1. Sample 2 is the same as sample 1 except the dissolved oxygen belongs to level 2. Similarly, sample 3 is also the same as sample 1, except the total phosphorus is level 2. Conversely, the indicator values of sample 8 are all worse than level 5, and for samples 4, 5, 6 and 7, the indicator values are just the average values of the neighboring levels from level 1 to level 5, respectively; for example, the indicator values of sample 4 are the average values of level 1 and level 2.

Table 1. Virtual samples and water quality standards (GB3838-2002).

		Indicators							
		X1 (mg/L)	X2 (mg/L)	X3 (mg/L)	X4 (mg/L)	X5 (mg/L)	X6 (-)	X7 (mg/L)	X8 (mg/L)
Virtual water samples	1	7.6	0.005	0.1	0.1	1	7	0.05	0.005
	2	6.2	0.005	0.1	0.1	1	7	0.05	0.005
	3	7.6	0.02	0.1	0.1	1	7	0.05	0.005
	4	6.75	0.0175	0.35	0.325	3	7.5	0.15	0.03
	5	5.5	0.0375	0.75	0.75	5	8.5	0.25	0.075
	6	4	0.075	1.25	1.25	8	9.5	0.4	0.15
	7	2.5	0.15	1.75	1.75	12.5	12	0.75	0.3
	8	1	0.4	3	3	16	14	2	0.5
Water quality standards	1	7.5	0.01	0.2	0.15	2	6–9	<0.3	<0.1
	2	6	0.025	0.5	0.5	4			
	3	5	0.05	1	1	6			
	4	3	0.1	1.5	1.5	10	>9 or <6	>0.3	>0.1
	5	2	0.2	2	2	15			

By reference to the literature, this paper uses different methods to determine the weights of indicators which can be seen in Table 2, Table 3 and Table 4, of which the judgment matrix of the analytic hierarchy process (AHP) method is referred to in [28,29,30,31] and the procedure of the PCA method can be seen in [32].

As can be seen in Table 2, the first PC explains 95.829% (>85%) of total variance, and the component matrix of first PC is $-0.965X1 + 0.960X2 + 0.994X3 + 0.982X4 + 0.982X5 + 0.980X6 + 0.960X7 + 0.995X8$. Next, the normalization method is used to calculate the weight of factors based on the component matrix of the first PC (Table 4).

Table 2. The eigenvalues and contribution rates of PCA.

Component	Initial Eigenvalues		
	Total	% of Variance	Cumulative %
1	7.666	95.829	95.829
2	0.248	3.101	98.930
3	0.063	0.790	99.721
4	0.017	0.206	99.927
5	0.003	0.043	99.970
6	0.002	0.026	99.996
7	0.000	0.003	99.999
8	4.423×10^{-5}	0.001	100.000

Table 3. The judgment matrix of AHP.

	Judgment Matrix							
	X1	X2	X3	X4	X5	X6	X7	X8
X1	1	1	2	4	4	4	3	3
X2	1	1	2	4	4	4	3	3
X3	0.5	0.5	1	2	2	2	2	2
X4	0.25	0.25	0.5	1	1	1	0.5	0.5
X5	0.25	0.25	0.5	1	1	1	0.5	0.5
X6	0.25	0.25	0.5	1	1	1	0.5	0.5
X7	0.33	0.33	0.5	2	2	2	1	1
X8	0.33	0.33	0.5	2	2	2	1	1
The consistency ratio is 0.0078 < 0.1								

Table 4. Weights of indicators determined by different methods.

Serial Number	Indicators	Weight			
		AHP	PCA	AHP-PCA	EW
1	DO, X1	0.251	0.124	0.249	0.125
2	Total phosphorus, X2	0.251	0.123	0.247	0.125
3	Total nitrogen, X3	0.137	0.127	0.140	0.125
4	Ammonia nitrogen, X4	0.057	0.127	0.058	0.125
5	COD_{Mn}, X5	0.057	0.125	0.057	0.125
6	Ph, X6	0.057	0.125	0.057	0.125
7	Iron, X7	0.095	0.123	0.094	0.125
8	Manganese, X8	0.095	0.127	0.097	0.125

The Table 3 shows the judgment matrix of the AHP, and then the eigenvector of the largest eigenvalue could be used to calculate the weight of factors by the normalization method (Table 4)

Obviously, different methods determine the weights of the indicators from different aspects. Appropriate weights can be determined by comparison and optimization.

Then the virtual samples are assessed with different methods and different weights, and the assessment results can be seen in Table 5. WQI is a comprehensive index method based on equal weight, EW-VFPR is a variable fuzzy pattern recognition (VFPR) model based on equal weight. AHP-VFPR is a VFPR model based on the analytic hierarchy process (AHP) weight. PCA-VFPR is a VFPR model based on principal component analysis (PCA) weighting. PCA/AHP-VFPR is a VFPR model based on the synthesis weight of PCA and AHP. PCA/AHP-FCA is a fuzzy synthetic evaluation [33] (FCA) model based on the synthesis weight of PCA and AHP. I-PCA/AHP-FCA is an improved FCA model using Equation (4), based on the synthesis weight of PCA and AHP. As is shown in Table 3, the assessment results of the WQI method and PCA/AHP-FCA method are inattentive with the actual situation, because some indicators in samples 2 and 3 are actually inferior to level 1. The WQI method is based on equal weight, so it considers all other indicators' indices and weakens each single indicator's impact, which leads to unsatisfactory results; the PCA/AHP-FCA method determines the result according to the maximum membership degree selection principle, which neglects the information of other membership degrees and, finally, results in inaccuracy to some extent. For example, the membership degree of sample 3 is 0.8 to level 1 and 0.2 to level 2. However, according to the maximum membership degree selection principle, the result of sample 3 is level 1, which neglects the information of level 2. The assessment result of VFPR model based on the synthesis weight is in accordance with the actual case. The model is able to reflect the variability of the indicators when the indicators are changed. Previous studies show that the synthesis weight not only includes the information of the original data, but includes the researchers' knowledge about specific problems, which makes the results more reliable. The assessment result of the I-PCA/AHP-FCA method is similar to the PCA/AHP-VFPR method, but the I-PCA/AHP-FCA method would lose some information of the comprehensive samples.

Above all, the PCA/AHP-VFPR model considers the successive information of relevant membership degrees of each single indicator to the water standard, and the successive information of the relevant membership degrees of the comprehensive water quality to water levels. The PCA/AHP-VFPR model is a dynamic and fuzzy method, which is in line with the water quality properties, so it is appropriate for dynamic assessment of the comprehensive water quality.

Table 5. Assessment results of different methods.

Samples	Methods						
	WQI	EW-VFPR	PCA-VFPR	AHP-VFPR	PCA/AHP-VFPR	I-PCA/AHP-FCA	PCA/AHP-FCA
1	1.00	1.00	1.00	1.00	1.00	1.00	1.00
2	1.00	1.23	1.22	1.49	1.48	1.21	1.00
3	1.00	1.05	1.04	1.08	1.07	1.16	1.00
4	2.00	1.50	1.50	1.50	1.50	1.50	1~2
5	3.00	2.50	2.50	2.50	2.50	2.50	2~3
6	4.00	3.50	3.50	3.50	3.50	3.50	3~4
7	5.00	4.50	4.50	4.50	4.50	4.50	4~5
8	5.00	5.00	5.00	5.00	5.00	5.00	5.00

3.2. Analysis of Release Ability of Nitrogen and Phosphorus in Sediment

To simplify, temporal variation of sediment and complicated changes of hydraulics are not considered. According to the field investigation of Biliuhe Reservoir, the average deposit depth is 40 centimeters, the dry soil density is 1.55 g/cm^3, and the sediment moisture content measured is about 45.5%. Based on the results above, the sediment content of per unit area (Q) is calculated to be 34.4 g/cm^2. Given that the average water depth (h_w) of Biliuhe Reservoir is 12.8 m, by inputting the release amount (q_k) (mg/kg) of each point into the calculation equation we can obtain the increment of the sediment release into the water. Further, the amount of sediment release and water load increment is calculated by Equations (1)–(4). Adding the water load increment to the original monitoring data creates new water quality data that considers the sediment release. When those above data are introduced into the assessment model, the assessment result of the comprehensive water quality considering the release of sediment pollution is acquired.

For simplification, we take the average concentration of nitrogen and phosphorus of the monitoring points during monitoring period as the concentration in the overlaying water, i.e., the concentration of nitrogen and phosphorus in the overlaying water is 2.29 mg/L and 0.027 mg/L separately. Then the release amount (q_k) of the sites are calculated by Equation (2). Based on the above-mentioned data, the increment of nitrogen and phosphorus of each monitoring point is obtained, which can be seen in Table 6.

Table 6 shows the release ability of nitrogen and phosphorus in sediment is greater under shaking, leading to an increase of the water pollution load. It must be declared that the shaking in the experiment does not agree with the actual water environment, and the influence of the solid matter's settlement is not considered, either. However, the result could be regarded as the potential release ability of nitrogen and phosphorus in the sediment, and is still valuable in analyzing the influences of sediment release on water quality assessment, indeed.

Table 6. The increment of sediment release on water quality.

Sampling Sites	TN		TP	
	Q (mg/kg)	△x (mg/L)	Q (mg/kg)	△x (mg/L)
1	41.97	1.13	17.28	0.47
2	67.32	1.81	11.85	0.32
3	88.00	2.37	10.13	0.27
4	56.22	1.52	17.68	0.48
5	45.68	1.23	8.22	0.22
6	52.25	1.41	9.67	0.26
Average values	58.57	1.58	12.47	0.34

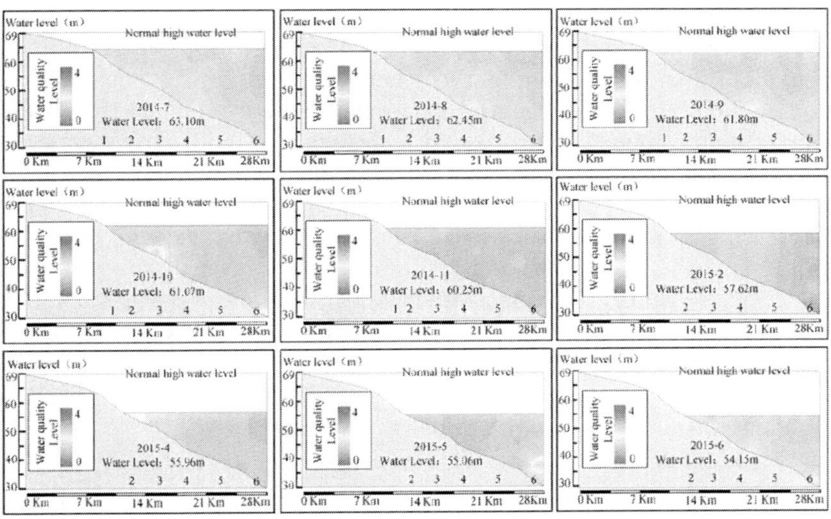

Figure 3. The vertical change of water quality of Biliuhe Reservoir.

This paper adds the increment of total nitrogen and total phosphorus with the corresponding original data to obtain new evaluation samples, and then analyzes the water quality comprehensive status, considering the influences of sediment release.

3.3. Results of Comprehensive Water Quality Assessment

We input the original monitoring samples into the assessment model, and use GIS to export the dynamic assessment results, which are shown in Figure 3. The comprehensive water quality index varies from 1.97 to 3.03. The average water quality index is 2.36 and the variable coefficient is 7.15.

It can be easily seen in Figure 3 that the water level is reduced from 63.10 m to 54.15 m and the water boundary is clear. The water quality is fine (2.36), overall, and worse in summer and better in winter, overall. The monthly average water quality level in July, August, and September in 2014 are 2.39, 2.35 and 2.37, respectively. The monthly average water quality level in February in 2015 is 2.27. The change of the comprehensive water quality level is in accordance with the actual reservoir environmental change. During summer, heavy rainfall would import a large amount of pollutants, which decrease the water quality. Meanwhile, the thermocline would impede the dissolved oxygen exchange, which contributes to the anaerobic environment at the bottom and sediment pollution release [34,35]. In winter, small amounts of runoff carry relatively fewer pollutants into the reservoir, which is in quiescent state, overall, and low temperature can also promote the increase of dissolved oxygen [36]. Meanwhile, the water quality shows the tendency of becoming better from upstream to downstream (dam) where the average level of the water quality is 2.31, and is better than others. The process of hydraulic transport agrees with this trend [37]. The water quality in the surface layer is better than that in the bottom layer because of sediment pollution, which influences water quality by releasing contaminants and consuming dissolved oxygen [38]. The comprehensive water quality indices in the surface layer of Points 3–6 are 2.34, 2.33, 2.27 and 2.27, respectively, while in the bottom layer the indices are 2.62, 2.63, 2.42 and 2.42. This case also occurred in July 2014, September 2014, and June 2015, which are mainly in relation with the change of dissolved oxygen. Thus, the influence of sediment pollution should be considered in the water quality assessment.

3.4. Assessment Results of Comprehensive Water Quality Considering Release of Sediment Pollution

We input the evaluation samples considering sediment release into the assessment model, and use GIS to export the dynamic assessment results.

By comparing Figure 3 and Figure 4, we can find that the average assessment results considering the release of nitrogen and phosphorus in sediment is 3.03 and the variable coefficient is 3.60. The comprehensive water quality level significantly declines. The comprehensive water quality level of 42.7% of the sampling points descends to levels inferior to level 3, which means the results of the assessment also change. Therefore, sediment pollution release can significantly change the status of water quality. Actually, sediment pollutants have dual characters of being sources and sinks of pollutants and change with external conditions [10]. Thus, it is necessary to combine the actual external conditions with the water quality comprehensive assessment to accurately assess the water quality status in the next.

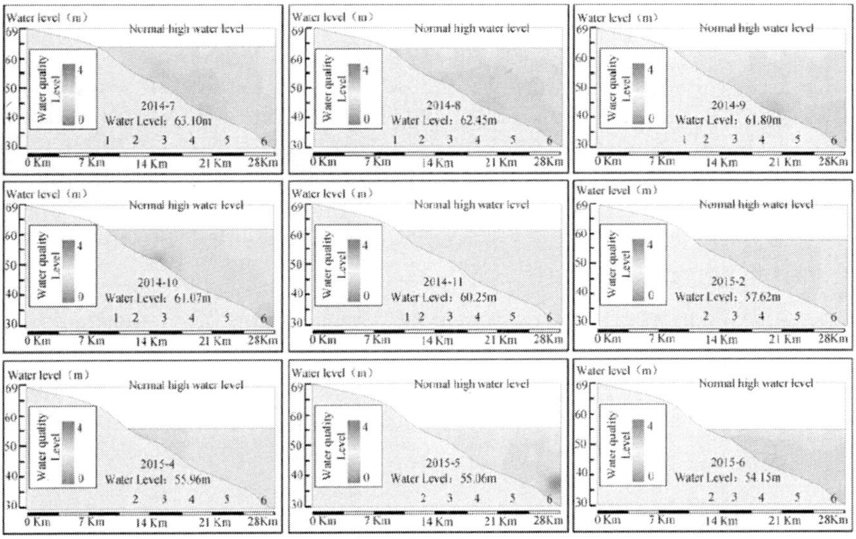

Figure 4. The vertical change of comprehensive water quality considering sediment release of Biliuhe Reservoir.

4. CONCLUSIONS

This paper analyzes the influence of the release of nitrogen and phosphorus in sediments on water quality through a physical simulation experiment, and explores a method to dynamically assess comprehensive water quality based on the PCA/AHP-VFPR model. GIS is applied to visually display the assessment results of water quality and reflect the change of the water boundary. Taking Biliuhe Reservoir as an example, the results show that the water quality is good, which is between level 2 and level 3, and much closer to level 2; the water boundary can also be visual demonstrated, in which the water level is reduced from 63.10 m to 54.15 m. The water quality of Biliuhe Reservoir, overall, is worse in summer and better in winter during the monitoring period. Meanwhile, water quality shows the tendency of becoming better from the entrance to the dam site, and the water quality in the surface layer is better than that in the bottom layer. The release of nitrogen and phosphorus in sediments could prominently affect water quality. Water quality is closer to, or worse than, level 3 considering the effects, and 42.7% of the samples are changed to being inferior to level 3.

This paper preliminarily studies the influence of sediment release on water quality assessment, but the effects of the microenvironment in water are ignored, which requires further study. This paper believes that the proposed method contributes to the study of the dynamic change of water quality, and it could also provide a reference for water resource protection and similar studies.

Acknowledgments: The authors would like to thank the anonymous reviewers. This work was supported by the National Natural Science Foundation of China (No. 51327004; No. 51279022; No. 51679026), the project of the comprehensive investigation of water quality in Biliuhe Reservoir, and the National Key Research and Development Program of China (2016YFC0400903). The datasets are provided by the Biliuhe Reservoir Management Bureau of Dalian.

Author Contributions: Tianxiang Wang and Shiguo Xu conceived and designed the experiments; Tianxiang Wang performed the experiments; Jianwei Liu and Shiguo Xu analyzed the data; Shiguo Xu contributed reagents/materials/analysis tools; and Tianxiang Wang wrote the paper.

REFERENCES

1. Tan, X.; Sheldon, F.; Bunn, S.E.; Zhang, Q.F. Using diatom indices for water quality assessment in a subtropical river, China. Environ. Sci. Pollut. Res. 2013, 20, 4164–4175.
2. Xu, S.G.; Wang, T.X. Review of research on accumulation process and effect of internal pollution of reservoir. Adv. Sci. Technol. Water Resour. 2015, 35, 162–167.
3. Jin, X.C.; Wang, S.R.; Jiang, X. Preliminary study of the three dimension model of the lake water sediment interface. Res. Environ. Sci. 2004, 17, 1–6. (In Chinese)
4. Cheng, H.M.; Hua, Z.L. Effects of hydrodynamic disturbances and resuspension characteristics on the release of tetrabromobisphenol A from sediment. Environ. Pollut. 2016, 219, 785–793.
5. Zhang, Y.; He, F.; Liu, Z.S.; Liu, B.Y.; Zhou, Q.H.; Wu, Z.B. Release characteristics of sediment phosphorus in all fractions of West Lake, Hang Zhou, China. Ecol. Eng. 2016, 95, 645–651.
6. Puttonen, I.; Kohonen, T.; Mattila, J. Factors controlling phosphorus release from sediments in coastal archipelago areas. Mar. Pollut. Bull. 2016, 108, 77–86.
7. Hongthanat, N.; Kovar, J.L.; Thompson, M.L. Phosphorus source-sink relationships of stream sediments in the Rathbun Lake watershed in southern Iowa, USA. Environ Monit. Assess. 2016, 188, 453.
8. Wang, H.; Zhou, Y.Y.; Wang, X. Transport dynamics of Cr and Zn between deposited sediment and overlying water. Clean Soil Air Water 2016, 44, 1453–1460.

9. Yang, Q.Z.; Zhao, H.Z.; Zhao, N.N.; Ni, J.R.; Gu, X.J. Enhanced phosphorus flux from overlying water to sediment in a bioelectrochemical system. Bioresour. Technol. 2016, 216, 182–187.
10. Wang, H.; Zhao, Y.J.; Wang, X.; Liang, D.F. Fluctuations in Cd release from surface-deposited sediment in a river-connected lake following dredging. J. Geochem. Explor. 2017, 172, 184–194.
11. Wu, Y.H.; Wen, Y.J.; Zhou, J.X. Phosphorus Release from Lake Sediments: Effects of pH, Temperature and Dissolved Oxygen. Ksce J. Civ. Eng. 2014, 18, 323–329.
12. Liu, C.; Zhong, J.C.; Wang, J.J.; Zhang, L.; Fan, C.X. Fifteen-year study of environmental dredging effect on variation of nitrogen and phosphorus exchange across the sediment-water interface of an urban lake. Environ. Pollut. 2016, 219, 639–648.
13. National Service Center for Environmental Publications (USEPA). EPA's Contaminated Sediment Management Strategy; National Service Center for Environmental Publications: Washington, DC, USA, 1998.
14. Zhang, X.H. Water Environment Restoration Engineering Principle and Its Application; Chemical Industry Press: Beijing, China, 2002. (In Chinese)
15. Wu, G.D.; Wei, Z.; Su, R.X. Distribution, accumulation and mobility of mercury in superficial sediment samples from Tianjin, northern China. J. Environ. Monitor. 2011, 13, 2488–2495.
16. Zheng, B.H.; Qin, Y.W.; Zhang, L.; Ma, Y.Q.; Zhao, Y.M.; Wen, Q. Sixty-year sedimentary records of polymetallic contamination (Cu, Zn, Cd, Pb, As) in the Dahuofang Reservoir in Northeast China. Environ. Earth Sci. 2016, 75, 486.
17. Ji, D.B.; Wells, S.A.; Yang, Z.J.; Liu, D.F.; Huang, Y.L.; Ma, J.; Berger, C.J. Impacts of water level rise on algal bloom prevention in the tributary of Three Gorges Reservoir, China. Ecol. Eng. 2017, 98, 70–81.
18. Lamastra, L.; Balderacchi, M.; Di, G.A.; Monchiero, M.; Trevisan, M. A novel fuzzy expert system to assess the sustainability of the viticulture at the wine-estate scale. Sci. Total Environ. 2016, 572, 724–733.
19. He, T.; Zhang, L.G.; Zeng, Y.; Zuo, C.Y.; Li, J. Water Quality Comprehensive Index Method of Eltrix River in Xin Jiang Province using SPSS. Procedia Earth Plantetary Sci. 2012, 5, 314–321.
20. Lopez-Doval, J.C.; Montagner, C.C.; de Alburquerque, A.F.; Moschini-Carlos, V.; Umbuzeiro, G.; Pompeo, M. Nutrients, emerging pollutants and pesticides in a tropical urban reservoir: Spatial distributions and risk assessment. Sci. Total Environ. 2017, 575, 1307–1324.
21. Sinha, K.; Das, P. Assessment of water quality index using cluster analysis and artificial neural network modeling: A case study of Hooghly river basin, west Bengal, India. Desalin. Water Treat. 2015, 54, 28–36.
22. Chu, H.B.; Lu, W.X.; Zhang, L. Application of artificial neural network in environmental water quality assessment. J. Agric. Sci. Technol. 2013, 2, 343–356.
23. Yan, F.; Liu, L.; Zhang, Y.; Chen, M.S.; Chen, N. The research of dynamic variable fuzzy set assessment model in water quality evaluation. Water Resour. Manag. 2016, 30, 63–78.
24. Logez, M.; Roy, R.; Tissot, L.; Argillier, C. Effects of water-level fluctuations on the environmental characteristics and fish-environment relationships in the littoral zone of a reservoir. Fund. Appl. Limnol. 2016, 189, 37–49.
25. Easton, Z.M. Defining Spatial Heterogeneity of Hillslope Infiltration Characteristics Using Geostatistics, Error Modeling, and Autocorrelation Analysis. J. Irrig. Drain-ASCE 2013, 139, 718–727.
26. Deshmukh, K.K.; Aher, S.P. Assessment of the Impact of Municipal Solid Waste on Groundwater Quality near the Sangamner City using GIS Approach. Water Resour. Manag. 2016, 30, 2425–2443.
27. Peng, H.; Zhou, H.C.; Li, M. Assessing water renewal of the northern coastal zone in China using a variable fuzzy pattern recognition model. J. Hydroinform. 2010, 12, 339–350.

28. Xu, S.G.; Wang, T.X.; Hu, S.D. Dynamic Assessment of Water Quality Based on a Variable Fuzzy Pattern Recognition Model. Int. J. Environ. Res. Public Health 2015, 12, 2230–2248.
29. Chakraborty, S.; Kumar, R.N. Assessment of groundwater quality at a MSW landfill site using standard and AHP based water quality index: A case study from Ranchi, Jharkhand, India. Environ. Monit. Assess. 2016, 188, 335.
30. Bozdag, A. Combining AHP with GIS for assessment of irrigation water quality in Cumra irrigation district (Konya), Central Anatolia, Turkey. Environ. Earth Sci. 2015, 73, 8217–8236.
31. Ou, C.P.; St-Hilaire, A.; Ouarda, T.B.M.J.; Conly, F.M.; Armstrong, N.; Khalil, B.; Proulx-Mclnnis, S. Coupling geostatistical approaches with PCA and fuzzy optimal model (FOM) for the integrated assessment of sampling locations of water quality monitoring networks (WQMNs). J. Environ. Monit. 2012, 14, 3118–3128.
32. Dalal, S.G.; Shirodkar, P.V.; Jagtap, T.G.; Naik, B.G.; Rao, G.S. Evaluation of significant sources influencing the variation of water quality of Kandla creek, Gulf of Katchchh, using PCA. Environ. Monit. Assess. 2010, 163, 49–56.
33. Lu, X.W.; Li, L.Y.; Lei, K.; Wang, L.J.; Zhai, Y.X.; Zhai, M. Water quality assessment of Wei River, China using fuzzy synthetic evaluation. Environ. Earth Sci. 2010, 60, 1693–1699.
34. Flores, P.E.D.; Geronimo, F.K.F.; Alihan, J.C.P.; Kim, L.H. Transport of nonpoint source pollutants and stormwater runoff in a hybrid rain garden system. J. Wetl. Res. 2016, 18, 481–487.
35. Nemirovskaya, I.A. Variations in different compounds in Volga water, suspension, and bottom sediments in the summer of 2009. Water Resour. 2012, 39, 533–545.
36. Stefanovic, D.L.; Stefan, H.G. Two-dimensional temperature and dissolved oxygen dynamics in the littoral region of an ice-covered lake. Cold Reg. Sci. Technol. 2002, 34, 159–178.
37. Reagan, M.T.; Moridis, G.J.; Keen, N.D.; Johnson, J.N. Numerical simulation of the environmental impact of hydraulic fracturing of tight/shale gas reservoirs on near-surface groundwater: Background, base cases, shallow reservoirs, short-term gas, and water transport. Water Resour. Res. 2015, 51, 2543–2573.
38. Doig, L.E.; North, R.L.; Hudson, J.J.; Hewlett, C.; Lindenschmidt, K.E.; Liber, K. Phosphorus release from sediments in a river-valley reservoir in the northern Great Plains of North America. Hydrobiologia 2017, 787, 323–339.

CHAPTER 15

Impact of Industrial Water Pollution on Rice Production in Vietnam

Huynh Viet Khai[1] and Mitsuyasu Yabe[2]

[1]*Can Tho University, Vietnam*

[2]*Kyushu University, Japan*

1. INTRODUCTION

Vietnam has achieved the average GDP growth rate of 6.71% per year. The industrial sector has mainly contributed economic development in Vietnam, with annual growth of 12% during the period of 200-2009. In line with its industrialization and modernization policies, Vietnam has rapidly changed economic structure from agriculture base to industrial economy. The industrial and construction sector only contributed 26 percent of national GDP in 1986, but it rapidly increases to 40.3 percent in 2009.

Economic development has brought many benefits to Vietnam. Income, public transportation and, in general, quality of life have gradually improved while the percentage of people below the poverty threshold has reduced. However, there have also been many negative consequences of rapid industrialization, particularly on agriculture and ecosystem health, because of the exploitation of natural resources and pollution. The two biggest cities in Vietnam, Ha Noi and Ho Chi Minh, have been ranked as the worst cities in Asia for dust pollution (The World Bank, 2008). Within Vietnam, Ho Chi Minh, the largest city, is at the top of the national pollution list (The World Bank, 2007). This pollution, into the air, water and land, is released by various, large industries. For instance, footwear manufacturing releases 11% of the air pollution load, 10% of the land pollution load and 6% of the water pollution load, while the plastic products manufacturing industry produces 10, 13 and 9% of the air, land and water pollution load, respectively. The main pollution sources do not necessarily come from the largest industries. The cement industry, which only has 12 factories and employs 0.5% of the provincial workforce, releases 24% of the air pollution load (ICEM, 2007). Similarly, the 160 paper factories employ only 0.8% of the provincial workers but contribute 14% of the water pollution load.

According to the Department of Science, Technology, and Environment of Tay Ninh, since almost all industrial zones have not installed wastewater treatment systems in Vietnam, the existence of industrial wastewater contamination appears almost everywhere. Wastewater from thousands of industrial facilities in 30 industrial areas and from small factories and businesses in the basin is the main source of pollution in the Dong Nai river of Ho Chi Minh City [1] - . The untreated wastewater contaminating oil from Hai Au concrete factory has been released directly into paddy fields approximately 200m^3/day and 1,500 m^3/day for Phuoc Long textile firm. However, it is difficult to know the actual damage and loss due to the contamination of untreated wastewater from industrial activities in Vietnam (Quang, 2001).

There have been a number of empirical agricultural studies concerning environmental problems, such as soil degradation, wind and water erosion in the world; however, few have specifically examined the impact of industrial pollution. Bai (1988) conducted field experiments in wheat lands irrigated with wastewater from the Liangshui River, the Tonghui River and the Wanquan River. He reported that wastewater irrigation caused a reduction in wheat yield by 8–17.1%. Similar studies in the Geobeidian area of the Tonghui River and the Yizhuang area of the Lianghe River reported that yields of wheat and rice cultivated in unpolluted soils in the sewage-irrigated area decrease by about 10% of the yields obtained in clean water-irrigated areas. In the sewage-irrigated area with polluted soils, yields of wheat and rice grown reduce by 40.6% and 39% of those in clean irrigation areas.

Chang *et al.* (2001) analyzed the impact of industrial pollution on agriculture, human health and industrial activities in Chongqing. To determine the effect of sewage-irrigation, they proposed expressing yield reductions as a function of the comprehensive water pollution index. Using this approach, reductions in yield due to sewage irrigation were about 10% for wheat and 30% for rice and vegetables. To evaluate the effects of polluted water irrigation, Lindhjem (2007) compared crop quality and quantity between a wastewater-irrigated area and a clean water-irrigated area. The total loss of corn and wheat production was estimated to be RMB

360 per mu, of which RMB 285 was caused by reduction in quantity, and RMB 75 was the reduction in quality. This paper also cites the study of Song (2004) that used dose-response functions to estimate the reductions in quantity and quality of crops from polluted water irrigation. Water pollution decreased rice production by 20% and quality by about 4%.

A study by The World Bank (2007) also used dose-response functions to calculate the economic losses from crop damage caused by water pollution, in terms of both reductions in crop quantity and quality (excess pollutant levels and substandard nutritional value). The economic cost of wastewater irrigation in China was estimated to be about 7 billion RMB annually for the four major crops (wheat, corn, rice, and vegetables). Reddy and Behera (2006) evaluated the impact of water pollution on rural communities in India, in terms of agricultural production, human heath, and livestock, using the effects on production, replacement costs and human capital approaches. The study estimated that the total loss per household per annum due to water pollution was $282.5, of which $213.2 was from agriculture, $16.3 from livestock and $53 from human health.

There has been some studies in recent years on industrial pollution in Vietnam such as the report written by Thong and Ngoc (2004) presented a descriptive analysis of data collected from 32 industrial estates in southern Vietnam to determine the factors affecting investment on wastewater treatment plants. It performed that water pollution was a serious problem in the big industrial estates of Ho Chi Minh City, Binh Duong, Dong Nai and Ba Ria-Vung Tau Provinces, and that financial constraints and lack of space were the main reasons why many small and medium-sized enterprises did not invest in wastewater treatment systems. Hung *et al.*(2008) studied the effects of trade liberalization on the environment, using data from the Viet Enterprise Survey of 2002 and the World Bank's Industrial Pollution Projection System. They found that trade liberalization led to greater pollution and environmental degradation but that the Vietnamese people have gradually recognized the importance of environmental protection.

However, because of a lack of information on the costs of pollution, national and local authorities in Vietnam have not paid much attention to pollution control measures. In this study, we review the literature on this topic and estimate the damage of rice production due to water pollution. Our findings could help governmental bodies enforce existing water pollution regulations, for example, TCVN 5945 on water pollution standards or Decree 67 on wastewater pollution charges, also help recognize and understand the failure of some of the current environmental policies in Vietnam. Our study could also provide useful information to authorities, such as the Natural Resources and Environment, and industries to manage water pollution and data for cost-benefit analyses of treatment projects in the industrial zones of Vietnam.

2. EVALUATION CONCEPT

The total economic loss of rice production includes three factors. First, a reduction in crop quantity assumes that water pollution decreases rice yield. Second, a reduction in rice quality, which is measured as price, assumes that the lower price of rice in a particular region could reflect reduced rice quality due to water pollution. Third, an increase in input costs assumes that farms may attempt to compensate for the possible productivity losses by implementing activities that are capable of offsetting this possible loss but are more costly to implement. The expectation of the profit loss is summarized by the following formula:

$$\begin{aligned}\pi_p &= \left(\overline{P} - \Delta P\right)\left(\overline{Q} - \Delta Q\right) - \left(\overline{C} + \Delta C\right)\\ &= \overline{P}\,\overline{Q} - \overline{P} \times \Delta Q - \Delta P \times \overline{Q} + \Delta P \times \Delta Q - \overline{C} - \Delta C\\ &= \left(\overline{P}\,\overline{Q} - \overline{C}\right) - \left(\overline{P} \times \Delta Q + \Delta P \times \overline{Q} + \Delta\overline{C}\right) + \Delta P \times \Delta Q\\ &= \quad \pi_n \quad - \quad \text{Profit loss}\end{aligned} \tag{1}$$

$$\begin{aligned}\Rightarrow \text{Profit loss} &= \overline{P} \times \Delta Q + \Delta P \times \overline{Q} + \Delta\overline{C}\\ &= \text{Quantity loss} + \text{Quality loss} + \text{Cost increase}\end{aligned} \tag{2}$$

where πn and πp are the rice profits in the non-polluted and polluted areas. Because $\Delta P \times \Delta Q$ is small compared with the other parts of the equation, it can be ignored and assumed to be 0.

However, it is complicated to estimate quality loss through the proxy of price because there are many other unobservable factors, excepting water pollution, which affect the price of rice. Thus, the study only calculates the three elements affected by water pollution:

- *Quantity loss:* Water pollution causes a decrease in rice yield. The production function approach is used to estimate the loss of rice yield.
- *Cost increase:* Since farms may aim and indeed be able to compensate for the possible productivity losses by implementing activities which are capable of offsetting this possible loss but are more costly to implement. In such circumstances, because it is not productivity which will be impacted, but production costs, cost function approach is applied to assess the impacts of pollution in economic terms.
- *Profit loss:* This is defined as total loss of net economic return estimated by the comparison of profit functions between two selected areas (one is considered as the polluted, other is the non-polluted area). The

difference in rice profits of two regions is considered as total loss of net economic return due to industrial pollution.

3. EMPIRICAL MODEL

We surveyed rice farmers in two areas with the assumption that they had the same natural environment conditions and social characteristics, and only differed with respect to pollution. One area was considered to be the polluted area, receiving wastewater from nearby industrial parks, while the other area was assumed to be the non-polluted area, being distant from sources of industrial pollutants. The productivity loss of rice production caused by water pollution was estimated by the difference in rice yield between the two regions (Translog production function approach). The similar calculation was applied for cost increase and profit loss due to water pollution by applying the methods of Cobb-Douglas cost function and translog profit function respectively.

3.1. Production Function Approach

The production function approach is that industrial activities possibly have a negative impact on the outputs, cost and profit of producers through the effect of environment. Environment affects goods or services existing in the market through the value change of their outputs, for instance, the reduced value of fish caught because of river pollution. The production function approach is often used to estimate the effect of environment change on soil erosion, deforestation, fisheries, the impact of air and water pollution on agriculture and so on (Bateman *et al.*, 2003)

A literature search on the production function approach in rice production Vietnam was conducted to make sure that relevant variables will be included in the farm survey questionnaire and to examine the suitability of existing rice production models for the research. There are a number of studies related to rice production in Vietnam. Kompas (2004) and Linh (2007) used a stochastic production frontier to estimate the technical efficiency of rice production in Vietnam. Do and Bennett (2007) used a production function approach with flood duration and relative location of upstream and downstream farmers variables to estimate the cost of changing wetland management, representing the reduced income of rice production in the Mekong River Delta. The loss of rice productivity was estimated based on the differences in rice yield between upper and lower of the Tram Chim park dyke. The results showed that the rice productivity in the lowering of park dyke decreased 0.06 tons per hectare per annum, which led to the profit loss of VND 0.07 million per hectare per annum.

These three studies used the Cobb-Douglas functional form of the rice production function approach. This study uses a translog functional form and does test for checking the existence of Cobb-Douglass. The model takes the basic form:

$$Y=f(L,K,I,Z,E,F)$$

where Y is the rice yield of a farmer in the studied year (tones/ha), L is the number of labors for rice cultivation (man-days/ha), K is capital input (VND/ha), I is a vector of material inputs as seeds (kg/ha), fertilizers (kg/ha), herbicide (ml/ha) and pesticides (ml/ha), Z is a vector of social-economic characteristics of farmers, and E is a vector of farming conditions, and F is the relative location of farms (polluted site = 1, non-polluted site = 0)

The test for the existence of quantity loss due to water pollution is:

H0:Quantity loss = 0 or Coef. of F = 0

H1:Quantity loss > 0 or Coef. of F < 0 (4)

The reduced yield of rice is defined as the difference in the average rice yield between the non-polluted and polluted site. It is estimated by following equation:

$$\Delta Y = f(\overline{L},\overline{K},\overline{I},\overline{Z},\overline{E},F=0) - f(\overline{L},\overline{K},\overline{I},\overline{Z},\overline{E},F=1) \quad (5)$$

where

ΔY

is the average yield loss caused by water pollution (kg/ha); $\overline{L}$, $\overline{K}$, $\overline{I}$, $\overline{Z}$, $\overline{E}$ are the average of labor, capital input, material inputs, social-economic characteristics, and farming conditions, respectively.

As mentioned earlier, a translog functional form is used in the study. The production functional form in the polluted and non-polluted areas is written as followed (Tim & Battese, 2005):

$$\ln(Y)=\alpha_0+\alpha_1\ln(L)+\alpha_2\ln(K)+\alpha_3\ln(I)+\frac{1}{2}\alpha_{11}\left(\ln(L)\right)^2+\alpha_{12}\ln(L)\ln(K)+\alpha_{13}\ln(L)\ln(I)+ \\ +\frac{1}{2}\alpha_{22}\left(\ln(K)\right)^2+\alpha_{23}\ln(K)\ln(I)+\frac{1}{2}\alpha_{33}\left(\ln(I)\right)^2+\sum_{k=1}^{5}\beta_k Z_k+\sum_{h=1}^{4}\delta_h E_h+\gamma F \quad (6)$$

where Y, L, K, I, F are the same as in the above equations and $Z1$, $Z2$, $Z3$, $Z4$ are the variables of the gender (1 = male, 0 = female), the age (years), the number of school year (years), attending trainings (1 = Yes, 0 = No) of rice households, and $E1$, $E2$, $E3$, $E4$ are the variables of serious diseases happening during the study year (1 = Yes, 0 = No), rice monoculture (1 = yes, 0 = No), soil quality (1 = fertile soil, 0 = other soils), off-farm income ratio.

Some restrictions are used to check the constant returns to scale:

$$\alpha_1 + \alpha_2 + \alpha_3 = 1$$
$$\alpha_{11} + \alpha_{12} + \alpha_{13} = 0$$
$$\alpha_{12} + \alpha_{22} + \alpha_{23} = 0$$
$$\alpha_{13} + \alpha_{23} + \alpha_{33} = 0 \quad (7)$$

Then, the following restriction is applied to test the existence of Cobb-Douglass function:

$$\alpha_{11} = \alpha_{12} = \alpha_{13} = \alpha_{22} = \alpha_{23} = \alpha_{33} = 0 \quad \textbf{(8)}$$

3.2. Replacement Cost (RC)

Replacement cost approach is defined as payment for restoring original environment (unpolluted state) if it has already been damaged. The costs of moving away from the polluted area suffered by the victims of environmental damage or actual spending on safeguards against environmental risks are called replacement costs (Bateman *et al.*, 2003; Winpenny, 1991). In the study written by Reddy and Behera (2006), the replacement cost method is used to estimate the damage costs of pump sets due to water pollution. In this study, farmers in the polluted areas might spend more input costs for the compensation of rice productivity loss because they directly use the polluted water for irrigation. Thus, it is assumed that the costs of farmers in polluted areas are more than those in the non-polluted areas. In this case, the replacement cost is estimated by using the cost function approach. The basic form of cost function is given by:

$$C=C(Ws,Wh,Wf,Wp,Y,Z,E,F) \quad (9)$$

where C is the total cost of a farmer (VND/ha), Ws is the price of seed (VND/kg), Wh is the price of herbicide (VND/100ml), Wf is the price of fertilizers (VND/kg), Wp is the price of pesticides (VND/100ml), Y is the rice yield of a farmer in the studied year (tones/ha), Z is a vector of social-economic characteristics of farmers, and E is a vector of farming conditions, F is the relative location of farms (polluted site = 1, non-polluted site = 0)

The test for the existence of cost increase due to water pollution is:

H_0:Cost increase = 0 or Coef. of F = 0

H_1:Cost increase > 0 or Coef. of F > 0 (10)

The increase in input costs is defined as the difference of the average cost between heavily polluted and less polluted areas. It is estimated by following equation:

$$\Delta C = C(\overline{W}_s, \overline{W}_h, \overline{W}_f, \overline{W}_p, \overline{Y}, \overline{Z}, \overline{E}, F=1) - C(\overline{W}_s, \overline{W}_h, \overline{W}_f, \overline{W}_p, \overline{Y}, \overline{Z}, \overline{E}, F=0) \quad (11)$$

where ΔC is the increase of the average cost per ha because of water pollution (VND/ha);

$\overline{W}_s$, $\overline{W}_h$, $\overline{W}_f$, $\overline{W}_p$, $\overline{Y}$, $\overline{Z}$, $\overline{E}$ are the average price of seed, herbicide, fertilizer, pesticides, social-economic characteristics, and farming conditions, respectively.

The Cobb-Douglas formal function is applied to estimate the cost function in the study (Tim & Battese, 2005):

$$\ln(C) = \varphi_0 + \varphi_1 \ln(W_s) + \varphi_2 \ln(W_h) + \varphi_3 \ln(W_f) + \varphi_4 \ln(W_p) + \varphi_5 \ln(Y) + \sum_{k=1}^{3} \beta_k Z_k + \sum_{h=1}^{3} \delta_h E_h + \gamma F \quad (12)$$

where *C, Ws, Wh, Wf, Wp, F* are the same as in the above equation and *Z1, Z2, Z3*, are, the age (years), the number of school year (years), attending trainings (1 = Yes, 0 = No) of rice households, and *E1, E2, E3*, are serious diseases happening during the year (1 = Yes, 0 = No), rice monoculture (1 = yes, 0 = No), soil quality (1 = fertile soil, 0 = other soils) respectively.

3.3. Profit Function Approach

Net economic return is defined as revenues from rice minus the cost of producing rice. It will be identified by a profit function approach. The profit loss is estimated by the following basic profit function:

$$\pi^* = \pi(W^*, C, Z, E, F) \quad (13)$$

where π^* is normalized profit defined as gross revenue minus variable cost divided by farm-specific output price, W^* is a vector of variable input prices divided by output price, *C* is a vector of fixed factors of the farm, *Z* is a vector of social-economic characteristics of farmers, *E* is a vector of farming conditions, *F* is the relative location of farms (polluted site = 1, non-polluted site = 0).

Hypothesis for the existence of profit loss due to water pollution is:

H_0:π_n=π_p or Profit Loss = 0 or Coef. of F = 0

H_1:π_n>π_p or Profit Loss > 0 or Coef. of F < 0 (14)

The profit loss due to water pollution is defined by the difference in profit between the polluted and non-polluted areas. It is estimated by the equation:

$$\Delta\pi^* = \pi\left(\overline{W}^*,\overline{C},\overline{Z},\overline{E},F=0\right)-\pi\left(\overline{W}^*,\overline{C},\overline{Z},\overline{E},F=1\right) \qquad (15)$$

where $\Delta\pi^*$ is Profit loss in 1000 VND/ha. $\overline{W}^*$, $\overline{C}$, $\overline{Z}$, $\overline{E}$ are the average prices of inputs, the average of the fixed factors, the social-economic characteristics of farmers, the farming conditions, respectively.

We use the translog profit functional form. The formula is given as (Rahman, 2002, Surjit & Carlos, 1981)

$$\ln\pi^* = \alpha_0 + \sum_{j=1}^{4}\alpha_j \ln W_j^* + \frac{1}{2}\sum_{j=1}^{4}\sum_{k=1}^{4}\tau_{jk}\ln W_j^* \ln W_k^* + \sum_{j=1}^{4}\sum_{l=1}^{6}\phi_{jl}\ln W_j^* \ln C_l +$$
$$+\sum_{l=1}^{6}\beta_l \ln C_l + \frac{1}{2}\sum_{l=1}^{6}\sum_{t=1}^{6}\varphi_{lt}\ln C_l \ln C_t + \sum_{m=1}^{3}\varpi_m Z_m + \sum_{n=1}^{4}\eta_n E_n + \gamma F \qquad (16)$$

where π^* is the restricted profit (total revenue minus total cost of variable inputs) normalized by price of output (P); Wj^* is the price of the jth input (Wj) normalized by the output price (P); j is the price of seed (1), the price of herbicides (2), the price of fertilizer (3), the price of pesticide (4); Cl is the quantity of fixed input, where l is total amount of seed used (1), total amount of herbicides used (2), total amount of fertilizer used (3), total amount of pesticides used (4), the number of man-days for rice production (5), the money of machines and services at all stages of rice production (6); $Z1$, $Z2$, $Z3$ are the age (years), the number of school year (years), and attendance at training sessions (1 = Yes, 0 = No) of rice households, respectively; and $E1$, $E2$, $E3$, $E4$ are the variables of serious disease incidence happening during the study year (1 = Yes, 0 = No), rice monoculture (1 = Yes, 0 = No), soil quality (1 = fertile soil, 0 = other soils), and off-farm income ratio, respectively.

Then, the following restriction is applied to test the existence of the Cobb-Douglass function:

$$\tau_{jk}=\phi_{jl}=\varphi_{lt}=0 \qquad (17)$$

4. STUDY SITE AND DATA DESCRIPTION

4.1. Study Site

In the Mekong River Delta, there are approximately 33 industrial parks, which constitute 9.5% of the total industrial parks of the country. Almost all of these 33 parks have no wastewater treatment system. The industrial parks in Can Tho city have released the biggest pollution loads, and the province is ranked in the top 10 most polluted provinces in Vietnam (Table 1). Can Tho is also one of the biggest rice producers in the Mekong River Delta. Because of these reasons, Can Tho was selected as the study site.

Table 1. Top 10 most polluted provinces in Vietnam

Province	Air index	Land index	Water index	Overall
Ho Chi Minh city	1	1	1	1
Hanoi	5	2	2	2
HaiPhong	2	6	4	3
Binh Duong	6	3	3	4
Dong Nai	4	4	5	5
Thai Nguyen	3	5	7	6
PhuTho	7	7	6	7
Da Nang	10	9	8	8
Ba RiaVung Tau	9	8	10	9
Can Tho	8	10	9	10

Note: The pollution loads released to air, land and water were estimated for all 64 provinces in Vietnam, and then pollution indexes were calculated and rankings were made.

Table 2. The industrial zones in Can Tho city

Zones	Size	Main activities	Water treatments
Tra Noc 1	135 ha	Processing, electron, clothes	No [a]
Tra Noc 2	165 ha	Machinery	No [a]
Hung Phu 1	262 ha	Harbor, Store	No
Hung Phu 2	212 ha	Machinery	No
Hong Bang	38.2 ha	Consumer goods	No [a]
Thot Not	150 ha	Processing, clothes, shoes	No [a]

[a] The available decision and acceptation of local authorities to evaluate the impact of environmental pollution.

Source: Resource and Environment Department of Can Tho City (2008)

There are six industrial parks in Can Tho (Table 2), which mainly comprise agricultural and fishery processing industries, clothes and consumer goods manufacturing industries. Almost none of the industrial zones and industrial corporations located near human residences have installed wastewater treatment systems. There has been little management of toxic waste or water pollution by local authorities and business. Tra Noc 1 (built in 1995) and Tra Noc 2 (built in 1999) industrial zones have only recently been acknowledged by the Department of Resources and Environment while Thot Not has been considered by Can Tho authorities to evaluate the impact of environmental pollution (Resource and Environment department of Can Tho city, 2008). As a consequence, Tra Noc 1 and 2 have released large volumes (1000s m^3) of various waste products directly into the river (Tuyen, 2010).

4.2. Data Collection

The study region covers the area within and around Tra Noc 1 and Tra Noc 2 industrial zones, which are two of the greatest polluters in Can Tho. People living in this area have suffered various financial impacts from the pollution: reduced crop yields, the use of cattle and agricultural equipment such as pump sets, contamination of drinking water, and increased incidence of human diseases and deaths directly and indirectly caused by water pollution.

Farmers were randomly selected for interview from two areas (Phuoc Thoi and Thoi An) with similar social and natural conditions (e.g. the same social and farming culture, ethnicity, type of soil). The selection of the polluted and non-polluted area was based on their distance from industrial zones, and on the recommendation or suggestion of local authorities and farmers. Some of the villages in Phuoc Thoi are heavily polluted by wastewater from the TraNoc 1 and 2 industrial zones. The villages in Thoi An are further away from the industrial zones than Phuoc Thoi and deemed to represent a non-polluted area (see Figure 1).

The group of fourteen interviewers and three local guide persons includes ten final year students, four staffs of School of Economics and Business Administration, Can Tho University, one local authority from people's committee, and two local farmers.

The questionnaire composes four main parts. In the first and second parts, the personal and farming information of household such as address, age, gender, training and so on and the situation of environmental pollution were interviewed. The inputs and output of rice production were collected in the three part and income from other activities obtained in the final section of questionnaire.

The household survey took 3 months to complete from January to March 2010 and was divided into two main reporting periods. The first period was called as pilot-survey in January 2010. The aims of this interview were to check and then correct the questionnaire more clearly and concisely, and to help interviewers get used to and understand the content of questionnaire. After the interviewers were trained how to ask by using questionnaire, about 30 farmers were interviewed. The revised questionnaire was used in the second period from February to March 2010. In total, 364 rice farmers, consisting of 214 farmers in the polluted and 150 farmers in the non-polluted area, were interviewed in February and March 2010. Household data were collected on household level information related to production costs and income as well as the social and economic characteristics of the farmers, and their perceived damages and losses due to water pollution.

Table 3 showed the water quality index of the polluted and non-polluted area. The concentrations of Total Suspended Solids (TSS) in the water refer to the concentrations of solid particles that can be trapped by a filter. This can be a problem because high concentrations of TSS can block sunlight from reaching submerged vegetation. This causes a reduction in the photosynthesis rate, and therefore less dissolved oxygen released into the water by plants. If bottom dwelling plants are not exposed to some light, the plants stop producing oxygen and die. Chemical Oxygen Demand (COD) is the amount of oxygen used during the oxidation of organic matter and inorganic chemicals such as ammonia nitrogen (NH3-N). High COD indicates a greater pollution load.

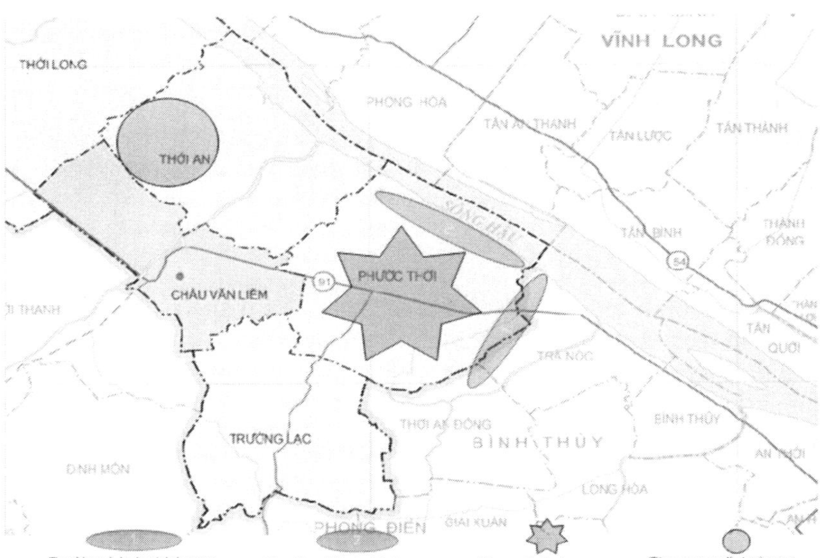

Figure 1. Map of the Study Site

TABLE 3. Water quality of the polluted and non-polluted area

	TSS (mg/l)	COD (mg/l)	NH3-N (mg/l)
The polluted area (PhuocThoi) (1)			
- Sewer mouth	145	720	13.29
- Primary affected water source (2)	50	50	1.23
- Secondary affected water source (3)	60	48	0.63
The non-polluted area (Thoi An)(4)	22	5.1	0.16
Limitation value (TCVN5942,1995)			
- Class A (5)	20	10	0.05
- Class B (6)	80	35	1

Notes:

(1) Measured on January 17th, 2007 (Nga *et al.*, 2008)

(2) The region receives wastewater directly from the industrial park.

(3) The region receives polluted water from the primary affected water source regions.

(4) Measured on January 27th, 2007 (Lang *et al.*, 2009)

(5) Values in Class A are from the surface water used for domestic water supply with appropriate treatments.

(6) Values in Class B are from the surface water used for purposes other than domestic water supply. Water quality criteria for aquatic life are specified in a separate index.

In the polluted area, the concentrations of TSS, COD and NH3-H in the sewer mouth, the primary affected water source and the secondary affected water source regions were mostly much higher than those of the standard water quality (see Table 3). This indicated that our selected pollution area site was heavily polluted. The concentrations of TSS, COD and NH3-N in the sewer mouth region were nearly 2-fold, over 20-fold and 13-fold higher than those of the standard water quality of class B, respectively.

Differences in the water quality index between the polluted and non-polluted area indicate that the water quality in the non-polluted area was much higher than that in the polluted area. However, the concentrations of TSS and NH3-N in the non-polluted were slightly higher than those of the Class A standard. This may be caused by non-point source pollutants, for instance, fertilizer, herbicide and pesticide released by agricultural activities in the region.

Table 4. Description of variables used in rice production models

Variable	Description	Unit
Y	Total yield per hectare	Ton/hectare
P	Price of rice	Thousand

		VND/ton
C	Total cost	Thousand VND/ha
π	Total profit	Thousand VND/ha
C_s	Total amount of seed used	Kg/ha
C_h	Total amount of herbicides used	Equivalent unit of 100 ml/ha
C_f	Total amount of fertilizer used	Kg/ha
C_p	Total amount of pesticide used	Equivalent unit of 100 ml/ha
C_l	The number of man-days for rice production	day/ha
C_c	The money of machines and services at all stages of rice production	Thousand VND/ha
W_s	Price of seed	Thousand VND/kg
W_h	Price of herbicide	Thousand VND/100ml
W_f	Price of fertilizer	Thousand VND/kg
W_p	Price of pesticide	Thousand VND/100ml
Age	The age of respondents	Years
Education	The number of school year of respondents	Years
Training	Respondents attending trainings	1= Yes, 0 = No
Mono	Rice monoculture	1= Yes, 0 = No
Diseases	Diseases happening during the study year	1= Yes, 0 = No
Off-farm ratio	The ratio of off-farm income	
Soil	Soil quality	1 = fertile soil, 0 = other soils

Table 4 showed the descriptions of variables in rice production models. The volumes of herbicide and pesticide used have measurement units of equivalent units of 100 ml per hectare per crop, based on farmers' reports and experts' recommendations. This is because farmers use various types of herbicides and pesticides (mixed with water or as a powder), and sometimes mix them together, which means that it is difficult to estimate exact amounts.

Table 5. Descriptive Statistics of Rice Production per hectare per crop

Variables	Non-polluted area	Polluted area	t-value
Y	5.88	4.99	-7.31***
P	4,157.79	4,060.89	-3.06***
C	10,909.44	10,563.61	-0.84
π	13,623.91	9,759.35	-8.37***
C_s	224.41	206.42	-2.37**
C_h	10.43	11.70	1.33
C_f	475.87	463.76	-0.53
C_p	77.26	70.23	-1.24
C_i	29.03	32.95	1.39
C_c	3283.02	3436.20	1.06
W_s	5.46	5.21	-1.34
W_h	32.79	32.47	0.21
W_f	9.42	9.54	0.92
W_p	24.86	21.70	-1.85*
Age	48.04	48.99	0.81
Education	6.33	6.07	-0.87
Training	0.49	0.35	-2.72***$^{\varphi}$
Mono	0.60	0.58	-0.39$^{\varphi}$
Diseases	0.40	0.42	-0.39$^{\varphi}$
Off-farm ratio	0.20	0.37	4.72***$^{\varphi}$
Soil	0.63	0.75	2.35**$^{\varphi}$

Notes: ***, **, * indicate statistical significance at the 0.01, 0.05 and 0.1 level respectively

$^{\varphi}$ Z-test for the equality of two proportions

Table 5 showed the descriptive statistics of the main variables in the rice production model for the polluted and non-polluted areas. Although soil quality in the non-polluted area was significantly ($P < 0.05$) lower than that in the polluted area, rice productivity and profit in the non-polluted area was significantly ($P < 0.01$) higher than those in the polluted area. The price of rice in the polluted area was significantly ($P < 0.01$) lower than that in the non-polluted area. This indicated that water pollution might have reduced crop quality, and in turn its price. The difference in the off-farm income ratio between the two areas suggests that farmers are aware of the reduced profit from rice cultivation in polluted soil, and therefore have a tendency to find additional work in nearby industrial parks to supplement their income.

Other variables measured did not significantly differ between the two regions (Table 5), except the percentage of respondents attending training. The results also showed that, on average, farmers were 48 years old, have had 6 years of education and 60 % of them grew rice in a monoculture.

5. ESTIMATED RESULTS

5.1. Impact of Water Pollution on Rice Productivity

Table 6 showed the Ordinary Least Squares (OLS) result of rice production function in translog form. The variables estimated in the model were statistically significant at 1 percent level. The estimated R-square was equal to 0.64, revealing the 64 percent change of rice yield possibly explained by independent variables in the model.

The study also examined the null hypothesis in (7) that there was a proportional output change when inputs in the model were varied or farms produce rice with constant returns to scale. The restricted least squares regression with the null hypothesis of constant returns to scale was estimated. The computed F statistic was 37.09 more than the critical value F (7, 327) of 2.69 at 1 percent level of significance [1]. Thus, the null hypothesis was rejected and the study concluded that technology did not exhibit constant returns to scale.

The second test was applied to check the Cobb-Douglass formal existence of the production function. The restricted function was estimated with the null hypothesis of jointed parameters in (8) equal to 0. The computed F statistic of 1.94 was more than the critical F(21,327) of 1.91 at 1 percent level of significance [1]. Thus, the null hypothesis was rejected, meaning that the translog functional form was suitably applied for the data of rice production in the study.

The results of Table 6 showed that there was no multicollinearity in the independent variables of production function because the correlations of these independent variables estimated by using the correlation matrix were less than 70 percent. The null hypothesis homoscedasticity was also accepted by using Breusch-Pagan test because the estimated LM of 49.72 was less than the critical χ^2_{36} of 57.34 at the level of 1 percent [2].

Table 6. The OLS regression of rice production function

Variables	Coef.	t-value	Variables	Coef.	t-value
$\ln(C_s)$	0.696	0.87	$\ln(C_h) \times \ln(C_p)$	0.007	0.7
$\ln(C_h)$	-0.123	-0.8	$\ln(C_h) \times \ln(C_l)$	0.021	1.49
$\ln(C_f)$	0.465	0.84	$\ln(C_h) \times \ln(C_c)$	0.008	0.35
$\ln(C_p)$	-0.157	-0.59	$\ln(C_f) \times \ln(C_p)$	0.060	1.5
$\ln(C_l)$	0.851**	2.55	$\ln(C_f) \times \ln(C_l)$	-0.026	-0.56
$\ln(C_c)$	0.572	1.3	$\ln(C_f) \times \ln(C_c)$	-0.022	-0.27
$\frac{1}{2} \ln(C_s)^2$	0.532***	2.72	$\ln(C_p) \times \ln(C_l)$	-0.031	-1.41
$\frac{1}{2} \ln(C_h)^2$	0.000	0.03	$\ln(C_p) \times \ln(C_c)$	0.029	0.71
$\frac{1}{2} \ln(C_f)^2$	0.037	1.13	$\ln(C_l) \times \ln(C_c)$	-0.023	-0.58
$\frac{1}{2} \ln(C_p)^2$	-0.011	-0.66	*Age*	-0.002**	-2.07
$\frac{1}{2} \ln(C_l)^2$	0.059*	1.81	*Education*	0.004	1.1
$\frac{1}{2} \ln(C_c)^2$	0.075	1.07	*Training*	0.039**	2.07
$\ln(C_s) \times \ln(C_h)$	0.014	0.54	*Disease*	-0.012	-0.67
$\ln(C_s) \times \ln(C_f)$	-0.132	-1.34	*Mono*	0.016	0.7
$\ln(C_s) \times \ln(C_p)$	-0.057	-1.08	*Soil*	0.031	1.64
$\ln(C_s) \times \ln(C_l))$	-0.105*	-1.83	*Off-farm ratio*	-0.054*	-1.93
$\ln(C_s) \times \ln(C_c)$	-0.216*	-1.87	*Pollution*	-0.127***	-6.68
$\ln(C_h) \times \ln(C_f)$	-0.025	-0.7	Constant	-5.615**	-2.26
R-square					0.64
Included observation					364

Notes: ***, **, * indicate statistical significance at the 0.01, 0.05 and 0.1 level respectively

Source: Own estimates; data appendix available from authors.

Table 6 showed that the rice productivity in the polluted was lower than in the non-polluted area because the coefficient of *Pollution* variable was significantly negative at 1 percent level. In addition, the study also revealed that training courses partly contributed an increase in rice yield since the coefficient of *Training* variable was significantly positive.

Moreover, the model also showed that farmer age ($P < 0.05$) and the ratio of off-farm income ($P < 0.1$) explained variation in rice yield. The effect of age might have been caused by declines in the health of older farmers leading to less efficient cultivation. Farmers who earned more off-farm income were associated with less profitable rice cultivation. Our interviews with the farmers in the polluted region suggested that when rice production was no longer profitable, farmers tended to sell their land as construction land or rent their land to farmers from other regions. Local farmers also attempted to secure employment in the nearby industrial parks, from which they could earn more money than compared to rice cultivation. The study also discovered that water pollution made farmers change rice cultivation and crop intensification techniques. Before their income was mainly from rice production with three rice crops per year, now they do rice farming as part-time jobs, only grow one or two crops per year and harvest rice just enough for home consumption. These possibly were the suitable explanations for the negative impact of off-farm income on rice productivity.

The reduced productivity of rice was calculated based on findings from Table 6. After the equation (5) was used to eliminate the effects of other factors, the estimated yield in the non-polluted area was about 5.61 tons and around 4.94 tons for the polluted region. Then, the loss of rice yield due to polluted water irrigation was estimated by subtracting the yield in the polluted from yield in the non-polluted region (equation 5). Using this approach, the estimated result was about 0.67 tons per hectare per crop (5.61 tons – 4.94 tons).

5.2. Increase in Rice Production Cost Due to Water Pollution

Table 7 showed R-square was equal to 0.56, revealing the variation of total rice costs of 56 percent was explained by independent variables in the model. The study also showed that the multicollinearity among the independent variables in cost function did not exist because the results estimated by correlation matrix approach showed that there were no correlations in these independent variables higher than 70 percent. The result of Breusch-Pagan test performed that the estimated LM of 14.96 was less than the critical χ^2_{12} of 26.22 at the level of 1 percent, revealing the absence of heterscedasticity in the estimate of cost function [2)].

The coefficient of *Pollution* variable was statistically significant positive at level of 1 percent, performing rice costs in the polluted region was higher than one in the non-polluted region. Moreover, farmers, who were older, managed their production cost more highly and less efficiently, performed by the positive effect of *Age* variable on total costs at 10 percent level. The significantly positive coefficient of *Rice monoculture* variable ($P < 0.01$) revealed that farmers who grew rice monoculture cost more than ones who cultivated rice rotation or intercropping. Possible explanation is that the cropping system of rice monoculture decreased the fertility of soil.

Like the calculation of yield loss, cost increase due to water pollution was estimated using the coefficients performed in Table 7. After the effect of other factors were eliminated, total cost was estimated about VND 10.37 million for rice production in the polluted area and VND 9.4 per ha per crop for that in the non-polluted area. Cost increase was estimated by subtracting the rice cost in the non-polluted region by the rice cost in the polluted area (equation 11). Using this approach, an increase in cost due to water pollution was calculated around VND 0.97 million per ha per crop (See Table 10).

Table 7. The OLS regression of rice cost function

Variables	Coefficient	t-value
$\ln(W_s)$	0.195***	3.95
$\ln(W_h)$	0.021	0.84
$\ln(W_f)$	0.431***	5.5
$\ln(W_p)$	0.007	0.27
$\ln(Y)$	0.918***	14.79
Age	0.002*	1.9
Education	-0.008	-1.55
Training	-0.058**	-2.02
Diseases	0.045	1.62
Rice monoculture	0.143***	4.72
Soil	-0.032	-1.08
Pollution	0.098***	3.3
Constant	6.140***	22.84
Statistic summary		
R-square	0.56	
Included observation	364	

Notes: ***, **, * indicate statistical significance at the 0.01, 0.05 and 0.1 level respectively

Source: Own estimates; data appendix available from authors.

5.3. Total Loss of Net Economic Return

Table 8 showed the coefficients from the OLS regression of the rice profit model using the translog profit functional form (equation 16). The full model was statistically significant at the 1% level. The estimated R-square revealed that 50% of the variation in the rice profit was explained by the model.

Next, we tested the null hypothesis of the Cobb-Douglass functional form. The restricted function was estimated assuming the null hypothesis that the joint parameters in (17) are 0. The computed F statistic of 1.78 was more than the critical F(55,283) value of 1.57 at the 1 percent level [1)]. The null hypothesis was therefore rejected, which supported the use of the translog functional form in this study. The estimate of profit function also showed the absence of multicollinearity (the correlations of independent variables less than 70 percent) and of heterscedasticity (Breusch-Pagan test showed the critical χ^2_{74} of 105.2 at the level of 1 percent higher than the computed LM of 100.24) [2)].

The coefficient of *Pollution* variable representing the effect of pollution was negative and significant ($P < 0.01$), which confirmed that water pollution reduced the profit of rice cultivation. The reduction in rice profit was calculated using the coefficients presented in Table 8. The estimated profit was approximately VND 9.14 million for rice cultivation in the polluted area and VND 12.34 million for that in the non-polluted area after the influences of other factors were eliminated. The loss of rice profit due to wastewater irrigation was estimated by subtracting the rice profit in the polluted region by the rice profit in the non-polluted region (equation 15). Using this approach, the loss of profit was calculated to be approximately VND 3.2 million per hectare per crop (see Table 10).

Like the results of rice yield loss, this model also performed that farmer age ($P < 0.01$), attending training ($P < 0.01$) and the ratio of off-farm income ($P < 0.1$) explained variation in profit. Moreover, soil quality was also an important factor affecting profit ($P<0.1$).

We also used the same estimate of profit loss due to water pollution to calculate reductions in profit caused by other factors as presented in Table 9. Cultivation in non-fertile soil, instead of fertile soil, could reduce rice profit by 8.24%. Farmers whose main sources of income were from non-agricultural sectors obtained 11.45% less rice profit than those who only had an agricultural income. Participating in trainings was estimated to increase profit by 13.03%. Profit loss caused by water pollution was much higher than the profit loss caused by other factors, which demonstrates that environment pollution has a great significance for rice farmers near industrial parks. Because of this, we suggest that the Vietnamese authorities should place a greater importance on the development and implementation of pollution control policies.

Table 8. The OLS regression of rice profit function

Variables	Coef.	t-value	Variables	Coef.	t-value
$\ln(W_s{*})$	-0.111	-0.02	$\ln(W_p{*})\times\ln(C_c)$	0.162	0.92
$\ln(W_h{*})$	1.211	0.54	$\ln(C_s)$	-3.393	-0.72
$\ln(W_f{*})$	-3.869	-0.71	$\ln(C_h)$	-0.540	-0.34
$\ln(W_p{*})$	-0.698	-0.29	$\ln(C_f)$	3.341	0.99
½ $\ln(W_s{*})^2$	0.322	0.68	$\ln(C_p)$	-2.241	-1.16
½ $\ln(W_h{*})^2$	-0.019	-0.15	$\ln(C_l)$	2.890	1.51
½ $\ln(W_f{*})^2$	-1.245**	-1.99	$\ln(C_c)$	1.432	0.37
½ $\ln(W_p{*})^2$	0.223	1.58	½ $\ln(C_s)^2$	0.338	0.61
$\ln(W_s{*})\times\ln(W_h{*})$	-0.017	-0.09	½ $\ln(C_h)^2$	0.012	0.18
$\ln(W_s{*})\times\ln(W_f{*})$	-0.257	-0.50	½ $\ln(C_f)^2$	-0.117	-0.47
$\ln(W_s{*})\times\ln(W_p{*})$	-0.042	-0.22	½ $\ln(C_p)^2$	0.013	0.18
$\ln(W_h{*})\times\ln(W_f{*})$	0.231	0.91	½ $\ln(C_l)^2$	0.123	1.44
$\ln(W_h{*})\times\ln(W_p{*})$	0.030	0.28	½ $\ln(C_c)^2$	-0.107	-0.52
$\ln(W_f{*})\times\ln(W_p{*})$	0.005	0.02	$\ln(C_s)\times\ln(C_h)$	-0.113	-0.80
$\ln(W_s{*})\times\ln(C_s)$	-0.648*	-1.96	$\ln(C_s)\times\ln(C_f)$	0.500*	1.73
$\ln(W_s{*})\times\ln(C_h)$	0.181	1.32	$\ln(C_s)\times\ln(C_p)$	0.013	0.06
$\ln(W_s{*})\times\ln(C_f)$	0.586**	2.01	$\ln(C_s)\times\ln(C_l))$	-0.250	-1.56
$\ln(W_s{*})\times\ln(C_p)$	-0.165	-0.96	$\ln(C_s)\times\ln(C_c)$	-0.149	-0.48
$\ln(W_s{*})\times\ln(C_l)$	0.079	0.56	$\ln(C_h)\times\ln(C_f)$	-0.130	-1.19
$\ln(W_s{*})\times\ln(C_c)$	0.029	0.08	$\ln(C_h)\times\ln(C_p)$	0.054	0.93
$\ln(W_h{*})\times\ln(C_s)$	-0.314	-1.62	$\ln(C_h)\times\ln(C_l)$	0.148**	2.33
$\ln(W_h{*})\times\ln(C_h)$	-0.064	-0.80	$\ln(C_h)\times\ln(C_c)$	0.012	0.10
$\ln(W_h{*})\times\ln(C_f)$	-0.124	-0.84	$\ln(C_f)\times\ln(C_p)$	-0.137	-1.08
$\ln(W_h{*})\times\ln(C_p)$	0.087	0.89	$\ln(C_f)\times\ln(C_l)$	-0.055	-0.41
$\ln(W_h{*})\times\ln(C_l)$	0.115	1.41	$\ln(C_f)\times\ln(C_c)$	-0.568**	-2.46

$\ln(W_h^*)\times\ln(C_c)$	0.244	1.63	$\ln(C_p)\times\ln(C_l)$	-0.140**	-2.07
$\ln(W_s^*)\times\ln(C_s)$	0.533	1.01	$\ln(C_p)\times\ln(C_c)$	0.192	1.45
$\ln(W_f^*)\times\ln(C_h)$	-0.376*	-1.83	$\ln(C_l)\times\ln(C_c)$	-0.025	-0.22
$\ln(W_f^*)\times\ln(C_f)$	-0.609	-1.63	*Age*	- 0.006***	-2.63
$\ln(W_f^*)\times\ln(C_p)$	-0.104	-0.45	*Education*	0.010	1.22
$\ln(W_f^*)\times\ln(C_l)$	0.106	0.47	*Training*	0.140***	2.90
$\ln(W_f^*)\times\ln(C_c)$	-0.529	-1.26	*Disease*	0.016	0.35
$\ln(W_p^*)\times\ln(C_s)$	0.025	0.12	*Mono*	0.003	0.06
$\ln(W_p^*)\times\ln(C_h)$	-0.010	-0.15	*Soil*	0.086*	1.75
$\ln(W_p^*)\times\ln(C_f)$	-0.164	-1.11	*Off-farm ratio*	-0.126*	-1.73
$\ln(W_p^*)\times\ln(C_p)$	0.107	1.14	*Pollution*	- 0.300***	-5.81
$\ln(W_p^*)\times\ln(C_l)$	-0.098	-1.22	Constant	-20.213	-0.60
R-square	0.50				
Included observation	364				

Table 9. Reduced profit in rice farming and key constraints

Factors	Reduced profit (Thousand VND)	Percentage of reduced profit (%)
Polluted vs. Non-polluted area	3,203	25.95
Non-fertile vs. Fertile soil	874	8.24
Non-training vs. Training	1,465	13.03
The highest off-farm vs. Zero off-farm income ratio	1,229	11.45

Source: Own estimates; data appendix available from authors.

Table 10 summarized the total loss of rice production due to water pollution. The estimated results showed there were about 26 percent of profit loss, including around 12 percent of reduced quantity (yield loss) and 9 percent of cost increase, adversely caused by industrial water pollution. In this study, we also observed that farmers in the polluted area use water irrigation from the highest water tide level to reduce the effects of wastewater on rice production. This was because the farmers thought the water at the high tide level looked less polluted than the waters at other times, despite the fact that the water was always heavily polluted near the industrial parks.

Table 10. Impact of water pollution on rice production

	Amount	Percent
Quantity loss	0.67 tons/ha	12%
Cost increase	0.97 million VND/ha	9%
Total loss of net economic return	3.2 million VND/ha	26%

Source: Own estimates; data appendix available from authors.

Moreover, the use of polluted water also caused the farmers to change their cultivation management. In previous years, three rice crops were produced annually and rice cultivation was the main income source. However, because of pollution, only one or two rice crops is now cultivated in the polluted area each year, and farmers treat rice cultivation as a part-time job, producing rice sufficient only for household consumption.

During our study, we also received reports of skin diseases on the farmers working in the polluted region. For instance, a farmer in the polluted area reported that he had suffered from skin disease 5 days per year, and the treatment cost VND 500,000. The diseases also caused the loss of 2.5 workdays, equivalent to VND 250,000. Therefore, the estimate of total economic loss is underestimated if indirect costs such as the health costs suffered by farmers are not included.

6. CONCLUSIONS AND POLICY IMPLICATION

Local authorities in Vietnam have recently removed or reduced some of the environmental impact requirements to attract industrial investments to their province. Although industrial investments with low environmental standards might increase gross domestic product and create more jobs for local households, they may also bring many problems including water, air and soil pollution. This study provides an example of the negative impacts that arise from pollution by industries.

In this study, we surveyed rice farmers in two areas with the same natural environment conditions, social characteristics (e.g. the same social and farming culture, ethnicity, type of soil), and only differed with respect to pollution. One area was considered to be the polluted area, receiving wastewater from nearby industrial parks, while the other area was assumed to be the non-polluted area, being distant from sources of industrial pollutants. The productivity loss of rice production caused by water pollution was estimated by the difference in rice yield between the two regions. The similar calculation was applied for cost increase and profit loss for using wastewater irrigation. The results showed that the yield loss of rice was about 0.67 tons per hectare per crop, VND 0.97 million for cost increase and totally 26 percent of profit loss due to water pollution. Therefore, since the study includes 214 farmers in the polluted area and these 214 farmers cultivate rice in 148 hectare as a whole, their total cost increase per crop because of water pollution could be estimated about VND 144 million (VND 0.97 * 148ha) and approximately VND 474 million (VND 3.2 million * 148ha) for their total net economic loss.

According to The World Bank (2007), the development of rice roots and seedlings could be influenced by using wastewater for irrigation. Polluted water irrigation causes the reduction of height, leaf area and dry matter. Decrease in leaf surface area leads to the reduction of photosynthesis. These facts have directly impact on rice production. In other words, the impacts of polluted water on rice productivity mainly reduce the number of ears unit area, number of seed per ear and seed weight. The study estimated water pollution caused yield reduction about 12 percent. This result is nearly equal to the reduced yield of 10 percent in the sewage-irrigated area in comparison with clear water-irrigated areas estimated by Bai (2004), but much lower than the rice reduced productivity of 20 percent calculated by Song (2004) in the study of Lindhjem (2007) and 30 percent by Chang *et al.* (2001).

Economic developments that cause damage to natural resources and the environment are unsustainable. We suggest that the Vietnamese government needs to develop policies that ensure sustainable development. Similar to environmental policies in developed countries, the Vietnamese government could consider increasing the current environmental standards and raising environmental taxes. The increase of environmental taxes could not only encourage industries to apply new technologies that reduce environmental pollution, but also generate money to compensate farmers near industrial areas for the damage to their agricultural production and health and to build wastewater treatment facilities in industrial parks. Compensation could be provided directly in cash to the farmers, or indirectly by means such as funding training or activities related to new technologies and the management of agricultural inputs and expenditure. Our study showed that training helped farmers increase their profit, which might partly offset some of the losses caused by environmental pollution.

To reduce polluted water from the industrial parks, an increase in the effectiveness of implementation of Decision 64 and Circular 07 should be recommended. A public disclosure system for the environmental performance of polluters mentioned in Article 104 of the Law on Environmental Protection (dated 2005) and Article 23 of Degree No. 80/2006ND-CP should be considered as one of the best ways to increase the efficiency of Decision 64 and Circular 07.

Article 104 requires polluters to report and publicize the information and data about the environment as follows:

- Reports on the environmental impact assessment, decision on approval for reports on the environmental impact assessment and plan for the implementation of requirements stipulated in the decision on approval for reports on the environmental impact assessment;
- List of and information about sources of wastes, pollutants that seem potentially harmful to people's health and environment;
- Areas where environment is polluted and degraded seriously and extremely seriously, areas in danger of the environmental pollution.
- Report on the environmental situation at the provincial level, report on environmental impact assessment by industries, fields and the national report on the environment
- It is essential to ensure unrestricted access to publicized information
- Agencies publicizing information about the environment have to take responsibility on accuracy, honesty and objectivity of announced information before legal agencies.
- Article 23 provides details and instructions on how to implement Article 104 of the Law on Environmental Protection. These details and instructions include:
- The Ministry of Natural Resources and Environment have responsibility for announcing information and data about the national environment;
- Ministries and ministerial-level agencies, government agencies shoulder responsibility for exposing information and data about the environment in industries and areas under their management;
- Agencies in charge of the environmental protection of People's Committees at all levels bear responsibility for make information

and data about the environment in the area under their management publicly;

- Management board of economic zones, industrial parks, export processing zones, managers of manufacturing and service units accept responsibility for publicizing information and data about the environment in the area under their management;
- Publicity of information and data about the environment is stipulated as follows:
- Information and data about the environment is publicized in form of books, news in newspapers or post on units' websites;
- Information and data about the environment is publicized in form of books, news in newspapers or post on units' websites (if any), reported in people's council meetings, announced on notice boards in residential quarter meetings, or listed in headquarters of units or headquarters of commune, ward, town people's committee where units are in operation.

The requirements of these above public disclosure system illustrate a new and significant approach for environmental authorities to force environmental laws and regulations in strong manner by increasing environmental awareness and permitting the large public to put pressure on polluters to solve current environmental problems. Such public disclosure requirements also create significant pressure on environmental authorities themselves as their own decision failures might also be widely recognized by such requirements. However, the implementation of these requirements in a clear, precise, and systematic manner is strongly needed.

Since water treatment facilities in these industrial parks must be built as soon as possible, the study on their cost effectiveness could be needed and seriously considered to decide whether we should build the water treatment facilities in every individual factory or for the whole industrial parks. Moreover, we suggest that the government should not use high-yield agricultural land for the construction of new industrial parks unless they include the latest pollution treatment technologies. The impact of environmental pollution should continue to be evaluated.

NOTES

[1] Calculated by the formula $F = \frac{(RSS_R - RSS_U)/J}{RSS_U/(N-K)}$, where RSS_R and RSS_U are the restricted and unrestricted sums of squared residuals, J is the number of

restrictions, N is the number of observations, and K is the number of parameters in an unrestricted function.

2) Breusch-Pagan test for heterscedasticity:

$LM=nR^2 \sim X^2_k$

where: n is the number of observations

R^2 is the R-Square of $|\hat{u}_i| = \delta_0 + \delta_1 X_{1i} + \delta_2 X_{2i} + ... + \delta_k X_{ki} + \tilde{v}_i$

k is the number of restricted factors

ACKNOWLEDGMENTS

We would like to express our gratitude to Dr. Benoit Laplante, a long-standing resource person with EEPSEA, and Dr Herminia Francisco, a director of EEPSEA, for their invaluable comments and suggestions in shaping the structure of the study and data analyses. Our warmest thanks go to EESEA for sponsoring the study, to Mr. Joseph Arbiol in our laboratory for reading the draft of the manuscript, and to our colleagues at Can Tho University for assisting in data collection.

NOTES

[1] - The speech of Dr. Trinh Le, the Institute of Tropical Technology and Environmental Protection.

REFERENCES

1. Y Bai, 1988Pollution of irrigation water and its effects, Beijing, China, Beijing Agriculture University Press: Beijing Agriculture University Press.
2. J Bateman, R. T Carson, B Day, M Hanemann, N Hanley, T Hett, M. J Lee, G Loomes, S Mourato, E Ozdemiroglu, D. W Pearce, R Sugden, J Swanson, 2003Economic Valuation With Stated Preference Techniques Edward Elgar Publishing, London.
3. Y Chang, M. S Hans, V Haakon, 2001The Environmental Cost of Water Pollution in Chongqing, China. Environment and Development Economics 6313-333
4. T. N Do, J Bennett, Would wet biodiversity conservation improve social welfare? A case study in Vietnam's Mekong River Delta, International Conference on Sustainable Development: Challanges and Opportunities for GMS,
5. P. T Hung, B. A Tuan, N. T Chinh, 2008The Impact of Trade Liberalization on Industrial Pollution: Empirical Evidence from Vietnam the Economy and Environment Program for Southeast Asia (EEPSEA).

6. ICEM (2007Analysis of Pollution from Manufacturing Sectors in Vietnam International Centre for Environmental Management, Indooroopily, Queenland.
7. H. V Khai, M Yabe, 2011Evaluation of the Impact of Water Pollution on Rice Production in the Mekong Delta, Vietnam. The International Journal of Environmental, Cultural, Economic and Social Sustainability 75
8. H. V Khai, M Yabe, 2012Rice Yield Loss Due to Industrial Pollution in Vietnam. Journal of US-China Public Administration 93
9. T Kompas, 2004Market Reform, Productivity and Efficiency in Vietnamese Rice Production. International and Development Economics Working Papers, Asia Pacific School of Economics and Governmnet, Australian National University, Australia.
10. V. T Lang, K. Q Vinh, N. T. T Truc, 2009Environmental Consequences of and Pollution Control Options for Pond "Tra" Fish Production in Thotnot District, Can Tho city, Vietnam The Economy and Environment Program for Southeast Asia (EEPSEA).
11. H Lindhjem, 2007Emvironmental Economic Impact Assessment in China: Problems and Prospects. Environmnetal Impact Assessment Review 271
12. H. V Linh, 2007Efficiency of Rice Farming Households in Vietnam: A DEA With Bootstrap and Stochastic Frontier Application. University of Minnessota, Minesota, USA.
13. B. T Nga, N. T Giao, P. V Nu, 2008Effects of Waste Water From Tra Noc Industrial Zone on Adjacent Rivers in Can Tho City. Scientific journal of Can Tho University 92
14. M. N Quang, 2001An Evaluation of the Chemical Pollution in Vietnam.Retrieved May 19, 2011, from http://www.mekonginfo.org/mrc_en/doclib.nsf/38bfa13a79f297d2c7256617 0044aaf9/1d952c500be72dc587256b74000703c8?OpenDocument
15. S Rahman, 2002Profit Efficiency Among Bangladeshi Rice Farmers. Food Policy 285-6
16. V. R Reddy, B Behera, 2006Impact of Water Pollution on Rural Communities: An Economic Analysis. Ecological Economics 583
17. Resource and Environment department of Can Tho city (2008Report on the situation of Environment in Can Tho City.
18. S. S Surjit, A. B Carlos, 1981Estimating Farm-Level Input Demand and Wheat Supply in the Indian Punjab Using a Translog Profit Function. American Journal of Agricultural Economics 632
19. The World Bank (2007Cost of Pollution in China- Economic Estimates of Physical Damages.
20. The World Bank (2008Handling Serious Environmnetal Polluters in Vietnam: Review of Implementation and Recommendations.
21. L. Q Thong, N. A Ngoc, 2004Incentives for Wastewater Management in Industrial Estates in Vietnam the Economy and Environment Program for Southeast Asia (EEPSEA).
22. C. D. S. P Tim, G. E Battese, 2005An Introduction to Efficiency and Productivity Analysis, New York, Springer Science.
23. B. C Tuyen, 2010Strategy of Vietnam towards addressing Environment and Climate risks.

CHAPTER 16

The Performance Evaluation of Anaerobic Methods for Palm Oil Mill Effluent (POME) Treatment: A Review

N.H. Abdurahman[1, 2], Y.M. Rosli[1, 2] and N.H. Azhari[1, 2]

[1] *LebuhrayaTun Razak, Gambang, Kuantan, Malaysia*

[2] *Faculty of Chemical and Natural Resources Engineering, University Malaysia Pahang, Malaysia*

1. INTRODUCTION

Palm oil mill effluent (POME) is an important source of inland water pollution when released into local rivers or lakes without treatment. In the process of palm oil milling, POME is generated through sterilization of fresh oil palm fruit bunches, clarification of palm oil and effluent from hydro-cyclone operations [Borja et al.,1996a]. POME is a viscous brown liquid with fine suspended solids at pH ranging between 4 and 5 [Najafpour et al., 2006]. In general appearance, palm oil mill effluent (POME) is a yellowish acidic wastewater with fairly high polluting properties, with average of 25,000 mg/l biochemical oxygen demand (BOD), 55,250 mg/l chemical oxygen demand (COD) and 19,610 mg/l suspended solid (SS). This highly polluting wastewater can cause several pollution problems. Therefore, direct discharge of POME into the environment is not encouraged due to the high values of COD, BOD and SS.

Over the past 20 years, the technique available for the treatment of POME in Malaysia has been biological treatment, consisting of anaerobic, facultative and aerobic pond systems [Chooi,1984], and [N. Ma, 1999]. Anaerobic digestion has been employed by most palm oil mills as their primary treatment of POME [Tay, 1991]. More than 85% of palm oil mills in Malaysia have adopted the ponding system for POME treatment [Ma et al., 1993], while the rest opted for open digesting tanks [Yacop et al., 2005]. These methods are regarded as a conventional POME treatment method involving long retention times and large treatment areas. High-rate anaerobic bioreactors have also been applied in laboratory-scaled POME treatment such as up-flow anaerobic sludge blanket (UASB) reactors [Borja el al., 1994a]; up-flow anaerobic filtration [Borja et al., 1994b]; fluidized bed reactors [Borja et al., 1995a], [Borja et al., 1995b] and up-flow anaerobic sludge fixed-film (UASFF) reactors [Najafpour et al., 2006]. Anaerobic contact digesters Ibrahim et al. (1984) and continuous stirred tank reactors (CSTR) have also been studied for PMOE treatment Chin (1981). Other than anaerobic digestion, POME has also been treated using membrane technology [Ahmad et al., 2006; 2007] and [Fakhru'l-Razi, 1994].

2. ANAEROBIC DIGESTION

Anaerobic digestion is the most suitable method for the treatment of effluents containing high concentration of organic carbon such as POME [Borja et al.,1996a]. Anaerobic digestion is defined as the engineered methanogenic anaerobic decomposition of organic matter. It involves different species of anaerobic microorganisms that degrade organic matter [Cote et al., 2006]. In the anaerobic process, the decomposition of organic and inorganic substrate is carried out in the absence of molecular oxygen. The biological conversion of the organic substrate occurs in the mixtures of primary settled and biological sludge under anaerobic condition followed by hydrolysis, acidogenesis and methanogenesis to convert the intermediate compounds into simpler end products as methane (CH4) and carbon dioxide (CO2) [Gee et al., 1994], [Guerrero et al., 1999], and [Gerardi, 2003]. Therefore, the anaerobic digestion process offers great potential for rapid disintegration of organic matter to produce biogas that can be used to generate electricity and save fossil energy [linke, 2006]. The suggested anaerobic treatment processes for POME include anaerobic suspended growth processes, attached growth anaerobic processes (immobilized cell bioreactors, anaerobic fluidized bed reactors and anaerobic filters), anaerobic blanket processes (up-flow anaerobic sludge blanket reactors and anaerobic baffled reactors), membrane separation anaerobic treatment processes and hybrid anaerobic treatment processes.

2.1. Anaerobic and Alternative Pome Treatment Methods

Currently available alternative methods for POME treatment are: aerobic treatment, membrane treatment systems and the evaporation method. The advantages and disadvantages of anaerobic and alternative treatment methods are shown in Table 1. In terms of energy requirement for POME treatment operation, anaerobic digestion has a greater advantage over the other alternative methods as it does not require energy for aeration. Furthermore, anaerobic POME treatment produces methane gas (CH4) which is a value-added product to digestion that can be utilized in the mill to gain more revenue in terms of CER. For example, the open digesting tank for POME treatment without land application, the capital cost quoted by [Gopal et al., 1986] for a palm oil mill processing 30 tons FFB/h is RM 750,000. Based on the chemical Engineering Plant Cost Index [Ulrich et al., 2004] the capital cost for this system is estimated to be US 370,272 in 2006. Comparing this to the capital cost for a membrane system in POME treatment for a palm oil mill processing 36 tons FFB/h at RM 3,950,000 [Chong, 2007], it is obvious that the former anaerobic treatment has better advantage over other treatment methods in terms of capital cost. The disadvantages of anaerobic treatment are (a) long retention times and (b) long start-up period. However, the problem of long retention times can be rectified by using high-rate anaerobic bioreactors while the long start-up period can be shortened by using granulated seed sludge [McHugh et al., 2003], utilizing seed sludge from same process [Yacob et al., 2006b] or maintaining suitable ph and temperature in the high-rate anaerobic bioreactor for growth of bacteria consortia [Liu et al., 2002].

Table 1. Advantages and disadvantages between anaerobic and alternative treatment methods

Treatment types	Advantages	Disadvantages	Reference
Membrane	Produce consistent and good water quality after treatment, smaller space required for membrane treatment plants, can disinfect treated water	Short membrane life, membrane fouling, expensive compared to conventional treatment	[Ahmad et al.,2006] [Metcalf et al., 2003]
Anaerobic	low energy requirements (no aeration),Producing	Long retention time, large area required For	[Metcalf et al., 2003]

	methane gas as a valuable end product, generated sludge from process could be used for land applications	conventional digesters, slow start-up (granulating reactors)	[Borja et al., 2006a]
Evaporation	solid concentrate from process can be utilized as feed material for fertilizer manufacturing	High energy consumption	[MA et al., 1997]
Aerobic	Shorter retention time, more effective in handling toxic wastes	High energy requirement (aeration), rate of pathogen inactivation is lower in aerobic sludge compared to anaerobic sludge, thus unsuitable for land applications	[Jr et al., 1999] [Doble et al., 2005]

2.2. Anaerobic Treatment Methods

2.2.1. Anaerobic Filtration

Anaerobic digestion has existed as a technology over 100 years. It gradually evolved, from an airtight vessel and a septic tank, to a temperature controlled, completely mixed digester, and finally to a high rate reactor, containing a density of highly active biomass. The microbiology of methane digestion has been examined intensively in the last decade. It has been established that three physiological groups of bacteria, converting hydrogen and carbon dioxide or acetate to methane. In contrast to aerobic degradation, which is mainly a single species phenomenon, anaerobic degradation proceeds as a chain process, in which several sequent organisms are involved. Anaerobic conversion of complex substrates requires the synergistic action of the micro-organisms involved. A factor of utmost importance, in the overall process, is the partial pressure of hydrogen and the thermodynamics linked to it. This fact has been recognized and discussed by researchers [Bryant et al., 1967]; [Boone and

Bryant, 1980]; [McInerney et al. 1979]; [Hickey and Switzenbaum 1988]. Anaerobic filter were favor for wastewater treatment because (a) high substrate removal efficiency (b) it requires a smaller reactor volume which operates on a shorter hydraulic retention times (HRT), [Borja et al., 1994b], (c) the ability to maintain high concentration of biomass in contact with the wastewater without affecting treatment efficiency [Reyes et al., 1999], [Wang et al. (2007)], and (d) tolerance to shock loadings [Reyes et al., 1999], [Van Der Merwe et al., 1993]. Besides, construction and operation of anaerobic filter is less expensive and small amount of suspended solids in the effluent eliminates the need of solid separation or recycle [Russo et al. 1985].

Another factor of fundamental importance has been the identification of new methanogenic species, and the characterization of their physiological behaviour. Of particular interest was the determination of the substrate affinity constants of both hydrogenotrophic and acetotrophic methanogens. While the first exhibit quite high substrate affinities and remove hydrogen down to ppm levels, the second group appears to contain species with only low substrate affinities [Zehnder et al., 1980]; [Huser et al., 1982]. This limited substrate affinity has,, an important consequence for anaerobic wastewater treatment.

A technological advance of utmost importance in anaerobic digestion has been the development of methods to concentrate methanogenic biomass in the reactor, especially in very low solids concentration in the wastewater, 1 - 2%. Such higher concentration of biomass can be achieved using of autoflocculation and gravity settling as, for instance, in the UASB reactor [Lettinga et al. 1983], by attachment to a static carrier (anaerobic filter) [Henze and Harremoes, 1982]; [Van Den Berg and Kennedy 1981]; [Young and McCarty 1969], by attachment to a mobile carrier (fluidized bed) [Binot et Heijnen 1984]; [Bull et al., 1984] or by growth in and on a matrix [Huysman et al., 1983]. All these different methods are in full development

Anaerobic filters have been applied to treat various types of wastewater including soybean processing wastewater [H-Q et al., 2002a], wine vinases [Nebot et al., 1995], [Perez et al., 1998], land fill leachate [Wang et al., 2007], municipal wastewater [Bodkhe, 2008], brewery wastewater [Leal et al., 1998], slaughterhouse wastewater [Ruiz et al., 1997], drug wastewater [Gangagni et al., 2005], and beet sugar water [Farhadian et al., 2007]. However, filter clogging is a major drawback in the continuous operation of anaerobic filters [Bodkhe, 2008], [Jawed et al. 2000], [Parawira et al., 2006]. Clogging of anaerobic filter has only been reported in the treatment of POME at an organic loading rate (OLR) of 20 g COD/l/day [Borja et al., 1995b] and also in the treatment of slaughterhouse wastewater at 6 g COD/l/day. This because the other studies were conducted at lower OLRs which had lower suspended solid content compared to POME. In general, anaerobic filter s are capable of treating wastewaters to obtain good effluent quality with at least 70% of COD removal efficiency with methane gas composition of more than 50%. Table 2 illustrates the COD removal efficiency of some treated wastewater using anaerobic filtration based on highest achievable percentage of methane in the generated biogas. In terms of POME treatment, the highest COD removal efficiency recorded was 94% with 63% of methane at an OLR of 4.5 kg COD/m3/day, while overall COD removal efficiency was up to 90% with an average methane gas composition of 60% [Borja et al., 1994b]. Investigations have been done to improve the efficiency of anaerobic filtration in wastewater treatment. [Yu et al., 2002a] found that operating at an optimal recycle ratio which varies depending on OLR will enhance COD removal. However, methane percentage will be compromised with increase in optimal recycle ratio. Higher retention of biomass in the filter will also lead to a better COD removal efficiency.

Table 2. Operating OLR range; COD removal efficiency in various wastewater treatments using anaerobic filtration based on highest % of methane production

Types of Wastewater	Operating OLR range (Kg $COD/m^3/day$)	COD removal efficiency (%)	Highest methane composition (%)	Reference
Slaughterhouse wastewater	1.0-6.5	79.9 (91.5)	51.1	[Ruiz et al.,1997]
POME	1.2-11.4	94.0 (94.0)	63.0	[Borja et al. 1994b]
Baker's yeast factory effluent	1.8-10.0	69.0 (74.0)	65.0	[Van der et al. 1993]
Distillery wastewaters	0.42-3.4	91.0 (93.0)	63.0	[Russo et al., 1985]
Landfill leachate	0.76-7.63	90.8 (90.8)	N/A	[Wang et al., 2007]

[i] - () - number in bracket denotes highest COD removal efficiency. N/A- data unavailable.

2.2.2. Fluidized Bed Reactor

A fluidized bed reactor (FBR) is a type of reactor device that can be used to carry out a variety of multiphase chemical reactions. Fluidized bed reactor exhibits several advantages that make it useful for treatment of high-strength wastewaters. It has very large surface areas for biomass attachment [Borja et al., 2001], [Toldra et al., 1987] enabling high OLR and short HRTs during operation [Garcia et al., 1998], [Sowmeyan et al., 2008]. Furthermore, fluidized bed has minimal problems of channeling, plugging or gas hold-up [Borja et al., 2001], [Toldra et al. 1987]. Higher up-flow velocity of raw POME is maintained for fluidized bed reactor to enable expansion of the support material bed. Biomass will then attach and grow on the support on material. In this way, biomass can be retained in the reactor. Hickey and [Switzenbaum, 1988] reported on the development of the anaerobic expanded bed process, which was found to convert dilute organic wastes to methane at low temperatures and at high organic and hydraulic loading rates. This process was being evaluated in 1988, on a 10,000 gallons per day pilot scale, consisting of an anaerobic expanded bed followed by post- treatment. [Jeris, 1987] reported on a two year experiment, testing two pilot scale anaerobic fluidized bed reactors, treating primary effluent. One reactor used sand as a carrier, the other granular activated carbon (GAC). Seeding experiments indicated that the GAC developed a biofilm more quickly and had more attached biomass. In addition, better BOD removal was observed with the GAC reactor. He noted that removal efficiencies were essentially independent of organic volumetric loading rates. Over a twelve month period in temperate climates, effluent total BOD5 values were consistently around 40 mg/l.

Investigations have been done on the application of fluidized bed to treat cutting-oil wastewater [Perez et al., 2007]; real textile wastewater [Sen et al., 2003]; slaughterhouse wastewater [Toldra et al., 1987]; wine and distillery wastewater [Garcia et al. 1998], [Sowmeyan et al., 2008]; ice-cream wastewater [Borja et al., 1995a], [Hawkes et al., 1995]; pharmaceutical effluent [Saravanane et al., 2001], and POME [Borja et al., 1995b]. OLR ranges and COD removal efficiencies of various wastewater treatments using fluidized bed is tabulated in Table 3. Based on Table 3, it can be concluded that anaerobic fluidized bed can typically remove at least 65% and up to more than 90% of COD. Inverse flow anaerobic fluidized bed is capable of tolerating higher OLRs compared to up-flow configuration.

Table 3. Operating OLR range; COD removal efficiency of various wastewater treatments using fluidized bed reactor

Types of Wastewater	Operating OLR range (Kg COD/m3/day)	COD removal efficiency (%)	Reactor configuration	Reference
Protein production from extracted				
Sunflower flour effluent	0.6-9.3	80.0-93.3	UF	[Borja et al., 2001]
POME	10.0-40.0	78.0-94.0	UF	[Borja et al., 1995b]
Ice-cream wastewater	3.2-15.6	94.4	UF	[Borja et al.,1995a]
Distillery effluent	6.11-35.09	80.0-92.0	DF	[Sowmeyan et al., 2008]
Brewery wastewater	0.5-70.0	80.0-90.0	DF	[Alvarado et al., 2008]
Real textile wastewater	0.4-5.0	78.0-89.0	UF	[Sen et al., 2003]

UF-upward flow; DF-downward/inverse flow.

The type of support material in the fluidized bed plays an important role to determine the efficiency of the entire treatment system [Garcia et al., 1998], [Sowmeyan et al., 2008] for both inverse flow and up-flow systems. Studies using fluidized bed to treat ice-cream wastewater showed different COD removal efficiencies when different support materials were used. Researcher [Hawkes et al., 1995] found that fluidized bed using granular activated carbon (GAC) gave about 60% COD removal while [Borja et al., 1995a] obtained 94.4% of COD removal using ovoid saponite. Thus suitable support material needs to be selected to obtain high COD removal efficiency in the system.

2.2.3. Up-flow anaerobic sludge blanket (uasb) reactor

Up-flow anaerobic sludge blanket (UASB) technology, normally referred to as UASB reactor, is a form of anaerobic digester that is used in the treatment of wastewater. UASB was developed by [Lettinga et al., 1980] whereby this system has been successful in treating a wide range of industrial effluents including those with inhibitory compounds. The UASB reactor is a methanogenic (methane-producing) digester that evolved from the anaerobic clarigester. A similar but variant technology to UASB is the expanded granular sludge bed (EGSB) digester. The underlying principle of the UASB operation is to have an aerobic sludge which exhibits good settling properties [Lettinga, 1995]. So far, UASB has been applied for the treatment of potato wastewater [Kalyuzhnyi et al., 1998], [Lettinga et al., 1980], [Parawira et al., 2006]; domestic wastewater [Barbosa et al., 1989], [Behling et al., 1997]; slaughterhouse wastewater [Sayed et al., 1984]; POME [Borja et al., 1994c]. UASB has a relatively simple design where sludge from organic matter degradation and biomass settles in the reactor. Organic matter from wastewater that comes in contact with sludge will be digested by the biomass granules. Table 4 shows some performances of wastewater treatment using UASB system. For potato wastewater treatment [Kalyuzhnyi et al., 1998] and [Parawira et al., 2006] both observed foaming and sludge floatation in the UASB reactor when operating at higher OLRs (> 6.1kg COD/m3 day). The ability of UASB to tolerate higher OLR for potato wastewater investigated by [Lettinga et al., 1980] compared due to the fact that the latter two studies were conducted at laboratory scale. In general, UASB is successful in COD removal of more than 60% for most wastewater types except for ice-cream wastewater. Researcher [Hawkes et al., 1995] suggested that the lower COD removal percentage from ice-cream wastewater was due to design faults in the reactor's three phase separator and high contents of milk fat that were hard to degrade.

Table 4. Performance of UASB in various wastewater treatments

Types of Wastewater	Operating OLR range (Kg COD/m3/day)	COD removal efficiency (%)	Methane Composition (%)	Reference
POME single-stage two-stage	1.8-13.9	63.0-81.0	54.0-67.0	[Kalyuzhnyi et al 1998]
(based on methanogenic	1.5-6.1	92.0-98.0	59.0-70.0	[Parawira

reactor)				et al. 2006]
Domestic sewage	3.76	74.0	69.0	[Barbosa et al. 1989]
Ice-cream wastewater	0.5-50	50.0	69.6	[Hawkes et al. 1995]
Sugar – beet	4.0-5.0	95.0	N/A	[Lettinga et al. 1980]
Pharmaceutical wastewater	0.27-2.0	26.0-69.0	N/A	[Stronach et al. 1987]
Slaughter wastewater	7.0-11.0	55.0-85.0	65.0-75.0	[Sayed et al. 1984]
Confectionary wastewater	1.25-2.25	66.0	N/A	[Forster et al. 1983]

[i] - N/A – data unavailable.

POME treatment has been successful with UASB reactor, achieving COD removal efficiency up to 98.4% with the highest operating OLR of 10.63 kg COD/m3 day [Borja et al., 1994c]. However, reactor operated under overload conditions with high volatile fatty acid content became unstable after 15 days. Due to high amount of POME discharge daily from milling process, it is necessary to operate treatment system at higher OLR. UASB reactor is advantageous for its ability to treat wastewater with high suspended solid content [Fang et al. 1994]; [Kalyuzhnyi et al., 1998] that may clog reactors with packing material and also provide higher methane production [Kalyuzhnyi et al., 1996]; [Stronach et al., 1987]. However, this reactor might face long start-up periods if seeded sludge is not granulated.

2.2.4. Anaerobic Contact Digester

The anaerobic contact process is a type of anaerobic digester. Anaerobic digesters are the aerobic equivalents of activated sludge process and are currently used for treating effluents from sugar processing, distilleries, citric acid and yeast production, industries producing canned vegetables, pectin, starch, meat products, etc. This process has been implemented in POME [Ibrahim et al., 1984]; ice-cream wastewater, alcohol distillery wastewater [Vlissidis et al., 1993] and fermented olive mill wastewater treatment [Hamdi et al. 1991]. Concentrated wastewaters are suitable to be treated by anaerobic contact digestion since relatively high quality effluent can be achieved [Jr et al., 1999]. In the study of fermented olive mill wastewater treatment, anaerobic contact was capable of reaching steady state more quickly compared to anaerobic filter; however, more oxygen transfer in the digester (due to mixing) causes this process to be less stable.

2.2.5. Continuous Stirred Tank Reactor (CSTR)

CSTR run at steady state with continuous flow of reactants and products; the feed assumes a uniform composition throughout the reactor, exit stream has the same composition as in the tank. The mechanical agitator provides more area of contact with the biomass thus improving gas production. In POME treatment, CSTR has been applied by a mill under Keck Seng (Malaysia) Berhad in Masai, Johor and it is apparently the only one which has been operating continuously since early 1980's [Tong et al., 2006]. Other applications of CSTR on wastewater treatment include dilute dairy wastewater [Chen et al., 1996]; jam wastewater [Mohan et al., 2008] and coke wastewater [Vazquez et al., 2006].

The CSTR in Kek Seng's Palm oil mill has COD removal efficiency of approximately 83% and CSTR treating dairy wastewater has COD removal efficiency of 60%. In terms of methane composition in generated biogas, it was found to be 62.5% for POME treatment and 22.5-76.9% for dairy wastewater treatment.

2.2.6. Anaerobic Contact Digestion

Presently there are three categories of anaerobic treatment systems. The first category is the conventional anaerobic digester, which includes two basic designs and another that combines the two. The standard rate digester is the most basic treatment system. It mixes the waste is solely by the movement of gas up through the solid matter and into the top of the tank; there is no external mixing. This process is highly inefficient, for it utilizes only 50 percent of the total waste volume, and requires a very long solid retention time (SRT), usually greater than 30 days; this process has been implemented in POME [Ibrahim et al., 1984]; ice-cream wastewater, alcohol distillery wastewater [Vlissidis et al., 1993] and fermented olive mill wastewater treatment [Hamdi et al., 1991]. Concentrated wastewaters are suitable to be treated by anaerobic contact digestion since relatively high quality effluent can be achieved [Jr et al., 1999]. In the study of fermented olive mill wastewater treatment, anaerobic contact was capable of reaching steady state more quickly compared to anaerobic filter; however, more oxygen transfer in the digester (due to mixing) causes this process to be less suitable.

2.2.7. Membrane Separation Anaerobic Treatment Process

Membrane separation has been considered for anaerobic reactors but the technology is still in a development stage. Several studies on membrane anaerobic processes for the treatment of various wastewaters including POME [Fakhru'l et al., 1999] have been performed [Fakhru'l et al., 1994]; [Nagano et al., 1992]; [Pillay et al., 1994]. For example, an ultrafiltration (UF) membrane with a molecular cut-off (MWCO) of 200,000 was used by [76] for biomass/effluent separation in conjunction with an anaerobic process for the treatment of POME. A lower operating pressure (1.5-2 bars) but a higher cross-flow velocity (2.3 m/s) was applied in this study in order to control fouling and to reduce solid deposition on the membrane surfaces. A high COD removal could be obtained in the membrane anaerobic system (MAS), but the permeate displayed a high color content with a low turbidity (less than 10 NTU), including that the color was due to dissolved solids with molecular weights lower than 200,000 g/mol. The particular organics retained in the reactor could be liquefied and decomposed because of the long solid retention time, which was independent of the HRT. The HRT was mainly influenced by the UF membrane flux rates which directly determined the volume of influent that could be fed to the reactor.

2.3. Clean Development Mechanism (CDM)

The Kyoto Protocol is an international agreement linked to the United Nations Framework Convention on Climate Change. The major feature of the Kyoto Protocol is that it sets binding targets for 37 industrialized countries and the European community for reducing greenhouse gas (GHG) emissions.These amount to an average of five per cent against 1990 levels over the five-year period 2008-2012.

The Clean Development Mechanism (CDM), defined in Article 12 of the Protocol, allows a country with an emission-reduction or emission-limitation commitment under the Kyoto Protocol (Annex B Party) to implement an emission-reduction project in developing countries. Such projects can earn saleable certified emission reduction (CER) credits, each equivalent to one tonne of CO2, which can be counted towards meeting Kyoto targets.

The mechanism is seen by many as a trailblazer. It is the first global, environmental investment and credit scheme of its kind, providing a standardized emission offset instrument, CERs. Besides helping to reduce carbon emission to the environment, CDM has the advantage to offer developing countries such as Malaysia to attract foreign investments to sustain renewable energy projects [Menon, 2002]. Thus, palm oil mills could earn carbon credits as revenue by the utilization of methane gas as renewable energy from anaerobic digestion of palm oil mill effluent. There is a lot of attention has been give to develop anaerobic treatment for POME since the implementation of CDM.

2.4. Comparison of Various Anaerobic Treatment Methods In Pome Treatment

Table 5 shows the performance of several of anaerobic digestion or treatment methods under both mesophilic and thermophilic conditions of POME. As can be seen from Table 5, the fluidized bed reactor has the ability to treat POME at very high organic loading rates; OLR with a short retention time, biogas capture is not emphasized using this process. Therefore, it can be concluded that USFF currently gives the best performance in POME treatment, achieving high COD removal efficiency and high OLR methane production at relatively short hydraulic retention time, HRT compared to conventional and other available anaerobic treatment methods.

Table 5. Performance of various anaerobic treatment methods on POME treatment

retention (days)	Operating OLR (Kg COD/m^3/day)	COD removal efficiency (%)	Hydraulic time	Methane composition (%)	Reference
Anaerobic pond 40	1.4		97.8	54.4	[Perez et al., 2001]
Anaerobic digester	2.16	80.7	20	36	[Yacop et al., 2005]
Anaerobic filtration	4.5	94.0	15	63	[Borja et al., 1994b]
Fluidized bed	40.0	78	0.25	N/A	[Borja et al., 1995b]
UASB	10.63	98.4	4	54.2	[Borja et al., 1994c]
UASFF	11.58	97	3	71.9	[Najafpour et al., 2006]

CSTR	3.33	80	18	62.5	[Tong et al., 2006]
Anaerobic contact process[a]	3.44	93.3	4.7	63	[Ibrahim et al., 1984]

[i] - N/A: data unavailable.

[ii] - a In terms of BOD.

Table 6 shows the advantages and disadvantages of each anaerobic treatment method. It can clearly seen that conventional methods are lacking in terms of treatment time, area required for treatment and facilities to capture biogas. However, these methods are more economically viable and have the capacity to tolerate a wider range of OLR. High-rate bioreactors are more effective in biodegradation as shorter retention times are needed, producing higher methane yield while compromising the OLR, capital and operating cost.

Table 6. Advantages and disadvantages of various treatment processes for POME

Treatment processes	advantages	disadvantages	References
Ponding system	Reliable and stable Anaerobically digested POME from the ponds could be used to culture algae. Cheap, simple to construct, low maintenance costs, the energy needed to operate a ponding system is minimal.	large area of land are required, making it unsuitable for factories located in the near urban and other developed areas. no facilities to capture biogas long retention time.	[Chooi et al. 91984]
Anaerobic	Small reactor volume Producing	lower methane emission,	[Borja et

Treatment processes	advantages	disadvantages	References
filtration	high quality effluent, short hydraulic times, able to tolerate shock Loadings, retain high biomass Concentration in the packing	Clogging at high OLRs, High media and support cost Unsuitable for high suspended solid wastewater	al., 1994b]
Fluidized bed	Most compact of all high-rate Processes, very well mixed Conditions in the reactor, large Surface area for biomass Attachment	high power requirements for bed fluidization, high cost of carrier media, not suitable for high suspended solid wastewaters	[Jr et al., 1997]
UASB	Useful for treatment of high suspended solid wastewater Producing high quality effluent No media required (less cost)	Performance dependent on sludge settleability, foaming and sludge floatation at high OLRs, long start up period if granulated, seed sludge is not used	[Lettinga, 1995]
UASFF	Higher OLRs achievable compared to operating UASB or anaerobic filtration alone, problems of	Granulation inhibition at high volatile fatty acid concentration Lower OLRs when treating	[Ayati et al., 2006]

Treatment processes	advantages	disadvantages	References
	clogging eliminated Higher biomass retention, more Stable operation, ability to tolerate Shock loadings, suitable for diluted Wastewater.	suspended solid wastewaters	
CSTR	Provides more contact of wastewater with biomass through mixing, increased gas production compared to conventional method	Less efficient gas production at treatment volume. Less biomass retentio	Hamdi et al., 1991]

2.5. Factors Influencing Anaerobic Digester Performance

Biogas coming from biomethanization or anaerobic digestion represents an attractive strategy for both biomass waste treatment and recycling and is of great interest from an environmental point of view and may benefit society by providing a clean fuel source from renewable energy. This technology is accomplished by a series of biochemical transformations, which can be toughly separated into a first step where hydrolysis, acidification and liquefaction take a place and second step where acetate, hydrogen and carbon dioxide are transformed into biogas with methane content between 60-80%, which cover a large part of energy. Many factors govern the performance of anaerobic digesters where adequate control is required to prevent reactor failure. These factors are operating temperature, pH, mixing, nutrients for bacteria and organic loading rates into the digester.

2.5.1. Operating Temperature

One of the most important factors affecting anaerobic digestion of organic waste is temperature. Anaerobic digestions can be developed at different temperature ranges including mesophilic temperatures (approximately 35ºC) and thermophilic temperatures ranging from 55 ºC to 60 ºC. Conventional anaerobic digestion is carried out at mesophilic temperatures (35–37 ºC), mainly because of the lower energy requirements and better stability of the process. POME is discharged at temperatures around 80-90 oC [Zinatizadeh et al., 2006] which actually makes treatment at both mesophilic and thermophilic temperature feasible especially in tropical countries like Malaysia. Effect of temperature on the performance of anaerobic digestion was investigated by [H-Q et al., 2002a] and found that substrate degradation rate and biogas production rate at 55 oC was higher than operation at 37 oC. Studies have reported that thermophilic digesters are able to tolerate higher OLRs and operate at shorter HRT while producing more biogas [Ahn et al., 2002], [Kim et al., 2006], and [Yilmaz et al., 2008]. However, failure to control temperature increase can result in biomass washout [Lau et al., 1997] with accumulation of volatile fatty acid due to inhibition of methanogenesis. At high temperatures, production of volatile fatty acid is higher compared to mesophilic temperature range [H-Q et al., 2002a].

2.5.2. PH

A pH (potential of Hydrogen) measurement reveals if a solution is acidic or alkaline (also base or basic). If the solution has an equal amount of acidic and alkaline molecules, the pH is considered neutral. The microbial communities in anaerobic digesters are sensitive to pH changes and methanogens are affected to a great extend [Jr et al., 1999]. Several cases of reactor failure reported in studies of wastewater treatment are due to accumulation of high volatile fatty acid concentration, causing a drop in pH which inhibited methanogenesis Parawira et al. (2006), [Patel et al., 2002]. Thus, volatile fatty acid concentration is an important parameter to monitor to guarantee reactor performance [Buyukkamaci et al., 2004]. It was found that digester could tolerate acetic acid concentrations up to 4000 mg/l without inhibition of gas production Stafford (1982). To control the level of volatile fatty acid in the system, alkalinity has to be maintained by recirculation of treated effluent [Najafpour et al., 2006], [Borja et al., 1996a] to the digester or addition of lime and bicarbonate salt [Gerardi, 2003].

2.5.3. Mixing

Distribution of bacteria, substrate, nutrients and temperature equalization by means of adequate mixing, are known to be crucial for the overall anaerobic digester (AD) process [Chapman, 1989]. Several investigations show that improvements in reactor performance can be achieved when changes in mixing intensity are imposed [Angelidaki et al., 2004]. According to [Gerardi, 2003] the main advantages of mixing in AD are: minimization of solids accumulation that may restrict reactor hydraulics, reduction of scum build up, elimination of temperature stratification and maintaining close contact between substrate particles and microbial communities. In a sequential experiment [Stroot et al., 2001] studied the feasibility of co-digestion of municipal solid waste, primary sludge and waste activated sludge (WAS) under mesophilic conditions in laboratory scale continuous stirred tank reactors (CSTRs) under different OLR and mixing conditions.

2.5.4. Organic Loading Rates

Organic loading rate is defined as the application of soluble and particulate organic matter. It's typically expressed on an area basis as pounds of BOD per unit area. Various studies have shown that higher OLRs will reduce COD removal efficiency in wastewater treatment systems [Torkian et al., 2003], [Sanchez et al., 2005]. However, gas production will increase with OLR until a stage when methanogens could not work quick enough to convert acetic acid to methane.

3. CONCLUSIONS

The performance of anaerobic treatment for POME and effects of organic loading rates were thoroughly reviewed. The palm oil industry is an indisputable source of pollution in Malaysia. In order to counteract the negative impact of this source, anaerobic digestion is an advantageous method for POME treatment as it generates valuable and product that can be exchanged into revenue when registered as a clean development mechanism CDM project. Furthermore, research can be done to develop a thermophilic anaerobic bioreactor with minimal control to ease system operation. Moreover, intensity of mixing in the thermophilic range should be investigated to obtain an optimum mixing rate that will keep microbial consortia in close proximity and at the same time improve the system efficiency. Furthermore, operation costs can be reduced through utilization of biogas for heat or electricity energy generation in the plant.

REFERENCES

1. Ahmad, A. L, Chong, M. F, Bhatia, S, & Ismail, S. (2006). Drinking water reclamationfrom palm oil Mill effluent (POME) using membrane technology. *Desalination*, 191,35-44.
2. Ahmad, A. L, Chong, M. F, & Bhatia, S. (2007). Mathematical modeling of multiplesolutes system for reverse osmosis process in palm oil mill effluent (POME) treatment.*Chemical Engineering Journal*, 132, 183-193.
3. Ayati, B, & Ganjidoust, H. (2006). Comparing the efficiency of UAFF and UASB withhybrid reactor in treating wood fiber wastewater. Iranian Journal of EnvironmentalHealthScience Engineering :, 3, 39-44.
4. Alvarado-lassman, A, Rustrian, E, Garcia-alvarado, M. A, Rodriguez, G. C, & Houbron,E. (2008). Brewery wastewater treatment using anaerobic inverse fluidized bedreactors. Bioresource Technology: , 99, 3009-3015.
5. Ahn, J, & Foster, H. . , C. F. (2002). A comparison of mesophilic and thermophilicanaerobic up-flow filters treating paper-pulp-liquors. Process Biochemistry: , 38,257-262.
6. Angelidaki, I, & Sanders, W. (2004). Assessment of the anaerobic biodegradability ofmacropollutants. *Reviews in Environmental Science and Biotechnology*, 3, 117-129.
7. Ma, A. N, Cheah, S. C, & Chow, M. C. (1993). Current status of palm oil processingwastes management. In: Waste Management in Malaysia: Current status and Prospectsfor Bioremediation, , 111-136.
8. Ma, A. N. (1999). Treatment of palm oil mill effluent. Oil Palm and Environment: MalaysianPerspective. Malaysia Oil Palm Growers' Council. (1999), , 277.
9. Borja, R, Gonzalez, E, Raposo, F, Millan, F, & Martin, A. (2001). Performance evaluationof a Mesophilic anaerobic fluidized-bed reactor treating wastewater derivedfrom the production of proteins from extracted sunflower flour. Bioresource Technology: 76, 45-52.
10. Borja, R, Banks, C. J, & Sanchez, E. (2006a). Anaerobic treatment of palm oil mill effluentin a two- stage up-flow anaerobic sludge blanket (UASB) reactor. Journal of Biotechnology, 45, 125-135.
11. Borja, R, & Banks, C. J. (1994c). Anaerobic digestion of palm oil mill effluent using anup-flow Anaerobic sludge blanket reactor. Bimass and Bioenergy: , 6, 381-389.
12. Borja, R, Banks, C. J, & Sanchez, E. (1996a). Anaerobic treatment of palm oil mill effluentin a two- stage up-flow anaerobic sludge blanket (UASB) reactor. Journal of Biotechnology, 45, 125-135.
13. Borja, R, & Banks, C. J. (1994a). Anaerobic digestion of palm oil mill effluent using anup-flow Anaerobic Sludge blanket reactor. Biomass and Bioenergy , 6, 381-389.
14. Borja, R, & Banks, C. J. (1994b). Treatment of palm oil mill effluent by up-flow anaerobicfiltration. Journal of Chemical Technology and Biotechnology , 61, 103-109.
15. Borja, R, & Banks, C. J. (1995a). Response of anaerobic fluidized bed reactor treatingice-cream wastewater to organic, hydraulic, temperature and pH shocks. *Journal ofBiotechnology*, 39, 251-259.
16. Borja, R, & Banks, C. J. (1995b). Comparison of anaerobic filter and an anaerobic fluidizedbed reactor treating palm oil mill effluent. *Process Biochemistry*, 30, 511-521.
17. Bodkhe, S. (2008). Development of an improved anaerobic filter for municipal wastewatertreatment. Bioresource Technology: , 99, 222-226.
18. Behling, E, Diaz, A, Colina, G, Herrera, M, Gutierrez, E, Chacin, E, Fernandez, N, &Forster, C. F. (1997). Domestic wastewater treatment using UASB reactor. BioresourceTechnology: , 61, 239-245.
19. Buyukkamaci, N, & Filibeli, A. (2004). Volatile fatty acid formation in anaerobic hybridreactor. *Process Biochemistry*, 39, 1491-1494.

20. Chen, T, & Shyu, H. , W. -H. (1996). Performance of four types of anaerobic reactorsin treating very dilute dairy wastewater. *Biomass and Bioenergy*, 11, 431-440.
21. Chong, M. F. simulation and design of membrane based palm oil mill effluent(POME) treatment plant from pilot plant studies (2007PhD. Thesis. Univer. siti SainsMalaysia.
22. Chooi, C. F. Ponding system for palm oil mill effluent treatment. PORIM 9, ((1984).
23. Chin, K. K. (1981). Anaerobic treatment kinetics of palm oil sludge. *Water Research*,199 EOF-202 EOF.
24. Cote, C, Daniel, I. M, & Quessy, S. (2006). Reduction of indicator and pathogenic microorganisms by psychrophilic anaerobic digestion in swine slurries. *Bioresource Technology*,
25. 97, 686-691.
26. Chapman, D. (1989). Mixing in anaerobic digesters: state of art. In: P. Cheremisinoff(Ed.) Encyclopedia of environmental control technology, Houston, TX: Gulf, ,325-354.
27. Doble, M, & Kumar, A. (2005). Biotreatment of Industrial Effluents. Elsevier Butterworth-Heinemann, Oxford, United Kingdom, , 19-38.
28. Fakhru, l-R. a. z. i. A. ((1994). Ultrafiltration membrane separation for anaerobicwastewater treatment. Water.Sci. Technol , 30(12), 321-327.
29. Fakhru, l-R. a. z. i, & Noor, A. , M. J. M. M. (1999). Treatment of palm oil mill effluent(POME) with the Membrane anaerobic system (MAS). Water Sci. Technol: , 39,159-163.
30. Farhadian, M, Borghei, M, & Umrania, V. V. Treatment of beet sugar water by UAFBbioprocess. Bioresource Technology ((2007).
31. Forster, C. F, & Wase, D. A. J. (1983). Anaerobic treatment of dilute wastewaters usingan upflow sludge blanket reactor. Environmental Pollution (Series A): , 31, 57-66.
32. Fang, H. H. P, & Chui, H. K. (1994). Comparison of startup performance of four anaerobicreactors for The treatment of high-strength wastewater. ResourcesConservationand Recycling (1994): 11, , 123 EOF.
33. Gangagni RaoA., G. Venkata Naidu., K. Krishna Prasad., N. Chandrasekhar Rao., N.Venkata Mohan., A. Jetty., P.N. Sarma. ((2005). Anaerobic treatment of wastewaterwith high suspended solids from a bulk drug industry using fixed film reactor(AFFR). Bioresource Technology: , 96, 87-93.
34. Gee, P. T, & Chua, N. S. (1994). Current status of palm oil mill effluent by watercoursedischarge. In: PORIM National Palm Oil Milling and Refining TechnologyConference.
35. Guerrero, L, Omil, F, Mendez, R, & Lema, J. M. (1999). Anaerobic hydrolysis andacidogenesis of wastewater from food industries with high content of organic solidsand protein. Water Resource Journal , 33, 3281-3290.
36. Gerardi, M. H. The microbiology of Anaerobic Digesters. Wiley-Interscience, NewJersey ((2003). , 51-57.
37. Gopal, G, & Ma, A. N. (1986). The comparative economics of palm oil mill effluenttreatment and resource recovery systems. National Workshop on Recent Developmentsin Palm Oil Milling Technology & Pollution Control.
38. Garcia-calderon, D, Buffiere, p, Moeltta, R, & Elmaleh, S. (1998). Anaerobic digestionof wine Distillery wastewater in down-flow fluidized bed. *Water Research*, 32,3593-3600.
39. Hamdi, M, & Garcia, J. L. (1991). Comparison between anaerobic filter and anaerobiccontact process for fermented olive mill wastewaters. Bioresource Technology: , 38,23-29.
40. Hawkes, F. R, Donnelly, T, & Anderson, G. K. (1995). Comparative performance ofanaerobic Digesters operating on ice-cream wastewater. *Water Research*, 29, 525-533.

41. Yu, H-Q, Hu, Z-H, Hong, T. Q, & Gu, G. W. (2002a). Performance of an anaerobic filtertreating soybean processing wastewater with and without effluent recycle. *ProcessBiochemistry*, 38, 507-513.
42. Ibrahim, A, Yeoh, B. G, Cheah, S. C, Ma, A. N, Ahmad, S, Chew, T. Y, Raj, R, & Wahid,M. J. A. (1984). Thermophilic anaerobic contact digestion of palm oil mill effluent.Water Science and Technology , 17, 155-165.
43. Jr, G, Leslie, G. T, & Daigger, H. C Lim. ((1999). Biological Wastewater Treatment.second ed. CRC Press. Revised & Expanded.
44. Jawed, M, & Tare, V. (2000). Post-mortem examination and analysis of anaerobic filters.Bioresource Technology: , 72, 75-84.
45. Kim, J. K, Oh, B. R, Chun, Y. N, & Kim, S. W. (2006). Effects of temperature and hydraulicretention time on anaerobic digestion of food waste. *Journal of Bioscience andBioengineering*, 102, 328-332.
46. Kalyuzhnyi, S, & De Los, E. L. Santos., J.R, Martinez. Anaerobic treatment of raw andpreclarified potato-maize wastewater in a UASB reactor. Bioresource Technology((1998).
47. Kalyuzhnyi, S. V, Skylar, V. I, Davlyatshina, M. A, Parshina, S. N, Simankova, M. V,Kostrikina, N. A, & Nozhevnikova, A. N. (1996). Organic removal and microbiologicalfeatures of UASB- reactor under various organic loading rate. s. Bioresource Technology, 55, 47-54.
48. Lettinga, G, Velson, A. F. M, Homba, S. W, De Zeeuw, W, & Klapwiki, A. (1980). Useof up-flow Sludge blanket (USB) reactor concept for biological wastewater treatment,especially for anaerobic treatment. Biotechnology and Bioengineering: , 22, 699-734.
49. Lettinga, G. (1995). Anaerobic digestion and wastewater treatment systems. AntonieVan Leeuwenhoek: , 67, 3-28.
50. Lau, I. W. C, & Fang, H. H. P. (1997). Effect of temperature shock to thermophilicgranules. *Water Research*, 31, 2626-2632.
51. Leal, K, Chachin, E, Gutierez, E, Fernandez, N, & Forster, C. F. (1998). A mesophilicdigestion of brewery wastewater in an unheated anaerobic filter. Bioresources Technology:, 65, 51-55.
52. Liu, W. T, Chan, O. C, & Fang, H. H. P. (2002). Microbial community dynamics duringstart-up of Acidogenic anaerobic reactors. Water Resource: , 36, 3203-3210.
53. Linke, B. (2006). Kinetic study of thermophilic anaerobic digestion of solid wastesfrom potato processing. Biomass and Bioenergy , 30, 892-896.
54. MetcalfEddy. Wastewater Engineering Treatment and Reuse,. fourth ed. McGra Hill ((2003). , 96-97.
55. Mchugh, S, Reilly, C. O, Mahony, T, Colleran, E, & Flaherty, V. O. (2003). Anaerobicgranular Sludge bioreactor technology. *Reviews in Environmental Science and Biotechnology*,2, 225-245.
56. Menon, R. N. Carbon credits and clean development mechanism. Palm Oil EngineeringBulletin ((2002).
57. Mohan, S, & Sunny, N. (2008). Study of biomethanization of waste water from jaindustries. Bioresource Technology: , 99, 210-213.
58. Ma, A. N, Tajima, Y, Asahi, M, & Hanif, J. (1997). Effluent treatment-evaporationmethod. PORIM Engineering News , 44, 7-8.
59. Nagano, A, Arikawa, E, & Kobayashi, H. (1992). The treatment of liquor wastewatercontaining gihg-strength suspended solids by membrane bioreactor system. WaterSci. Technol: , 26, 887-895.
60. Najafpour, G. D, Zinatizadeh, A. A. L, Mohamed, A. R, & Hasnain, M. Isa., H.Nasrollahzadeh.((2006). High-Rate anaerobic digestion of palm oil mill effluent in an upflowanaerobic sludge-fixed film bioreactor. Biochemistry (2006): 41, , 370 EOF.
61. Nebot, E, Romero, L. I, Quiroga, J. M, & Sales, D. (1995). Effect of the feed frequencyon the Performance of anaerobic filters. *Anaerobe*, 1, 113-120.

62. Patel, H, & Madamwar, D. (2002). Effects of temperature and organic loading rateson biomethanation of acidic petrochemical wastewater using an anaerobic upflowfixed-film reactor. Bioresource Technology (2002): 82, , 65 EOF-71 EOF.
63. Pillay, V. L, Townsend, B, & Buckley, C. A. (1994). Improving the performance ofanaerobic digesters at wastewater treatment works: the coupled cross-flow microfiltration/digester process. Water Sci. Technol: , 30, 329-337.
64. Perez, M, Romero, L. I, & Sales, D. Organic matter degradation kinetics in an anaerobicthermophilic fluidized bed bioreactor, Anaerobic ((2001).
65. Perez, M, Romero, L. I, & Sales, D. (1998). Comparative performance of high rateanaerobic thermophilic technologies treating industrial wastewater. *Water Research*,32, 559-564.
66. Perez, M, Rodriguez-cano, R, Romero, L. I, & Sales, D. (2007). Performance of anaerobicthermophilic fluidized bed in the treatment of cutting-oil wastewater. BioresourceTechnology , 98, 3456-3463.
67. Reyes, O, Sanchez, E, Rovirosa, N, Borja, R, Cruz, M, Colmenarejo, M.F, Escobedo, R,Ruiz, M, Rodriguez, X, & Correa, O. . ((1999).). Low-strength wastewater treatmentby a multistage
68. Russo, C, & Jr, G. L. San't Anna., S.E. De Carvalho. ((1985). An Aerobic filter appliedto the treatment of distillery wastewaters. Agricultural Wastes: Anaerobic filterpacked with waste tyre rubber. Bioresource Technology 70, 55-60., 14, 301-313.
69. Ruiz, I, Veiga, M. C, De Santiago, P, & Blazquez, R. (1997). Treatment of slaughterhousewastewater in a USAB reactor and an anaerobic filter. Bioresource Technology:, 60, 251-258.
70. Rbarbosa, R. A, & Sant, G. L. Anna Jr. ((1989). Treatment of raw domestic sewage inan UASB reactor. Water Research: , 23, 1483-1490.
71. Stronach, S. M, Rudd, T, & Lester, J. N. (1987). Start-up of anaerobic bioreactors onhigh strength Industrial wastes. *Biomass*, 13, 173-197.
72. Stroot, P. G, Mcmahon, K. D, Mackie, R. I, & Raskin, L. (2001). Anaerobic co-digestionof municipal solid waste and biosolids under various mixing conditions-I. Digesterperformance. Water Research: , 35, 1804-1816.
73. Saravanane, R, Murthy, D. V. S, & Krishnaiah, K. Treatment of anti-osmotic drugbased pharmaceutical effluent in an upflow anaerobic fluidized bed system. WasteManagement ((2001). , 563 EOF-8 EOF.
74. Sen, S, & Demirer, G. N. (2003). Anaerobic treatment of real textile wastewater with afluidized bed reactor. *Water Research*, 37, 1868-1878.
75. Sanchez, E, Borja, R, Travieso, L, Martin, A, & Colmenarejo, M. F. (2005). Effect of organicloading rate on the stability, operational parameters and performance of a secondaryupflow anaerobic sludge bed reactor treating piggery waste. *BioresourceTechnology*, 96, 335-344.
76. Stafford, D. A. (1982). the effects of mixing and volatile fatty acid concentrations onanaerobic Digester performance. Biomass, 2, 43-55.
77. Sayed, S, De Zeeuw, W, & Lettinga, G. (1984). Anaerobic treatment of slaughterhousewaste using a flocculent sludge UASB reactor. Agricultural Wastes: , 11, 197-226.
78. Sowmeyan, R, & Swaminathan, G. (2008). Evaluation of inverse anaerobic fluidizedbed reactor for Treating high strength organic wastewater. Bioresource Technology: ,99, 3877-3880.
79. Toldra, F, Flors, A, Lequerica, J. L, & Valles, S. (1987). Fluidized bed anaerobic biodegradationof food industry wastewaters. *Biological Wastes*, 21, 55-61.
80. Torkian, A, Eqbali, A, & Hashemian, S. J. (2003). The effect of organic loading rate onthe performance of UASB reactor treating slaughterhouse effluent. Resources Conservation& Recycling: , 40, 1-11.
81. Tong, S. L, & Jaafar, A. B. (2006). POME Biogas capture, upgrading and utilizationPalm Oil Engineering Bulletin: , 78, 11-17.
82. Tay, J. H. Complete reclamation of oil palm wastes. ResourcesConservation and Recycling5, ((1991). , 383 EOF-392 EOF.

83. Ulrich, G. D, & Vasudevan, P. T. (2004). Chemical Engieering Process Design andEconomics: A Practical Guide, second ed. Process Publishing Company.
84. Vlissidis, A, & Zouboulis, A. I. (1993). Thermophilic anaerobic digestion of alcoholdistillery wastewaters. *Bioresource Technology*, 43, 131-140.
85. Vazquez, I, Rodriguez, J, Maranon, E, Castrillon, L, & Fernandez, Y. (2006). Simultaneousremoval of phenol, ammonium and thiocyanate from coke wastewater byaerobic degradation. Journal of Hazardous MaterialsB: , 137, 1773-1780.
86. Van Der Merwe, M, & Britz, T. J. (1993). Anaerobic digestion of baker's yeast factoryefficient using an anaerobic filter and hybrid digester. Bioresource Technology: , 43,169-174.
87. Parawira, W, Murto, M, Zvauya, R, & Mattiasson, B. (2006). Comparative performanceof a UASB reactor and an anaerobic packed-bed reactor when treating potatowaste leachate. *Renewable Energy*, 31, 893-903.
88. Wang, Z, & Banks, C. J. (2007). Treatment of a high-strength substrate-rich alkalineleachate using an Aerobic filter. Waste Management , 27, 359-366.
89. Yilmaz, T, Yuceer, A, & Basibuyuk, M. A comparison of the performance of mesophilicand thermophilic anaerobic filters treating papermill wastewater. BioresourceTechnology ((2008). , 156 EOF-163 EOF.
90. Yacob, S, Hassan, M. A, Shirai, Y, Wakisaka, M, & Subash, S. (2006a). Baseline study of methane Emission from anaerobic ponds of palm oil mill effluent treatment. *Scienceof the Total Environment*, 366, 187-196.
91. Yacop, S, Hassan, M. A, Shirai, Y, Wakisaka, M, & Subash, S. (2005). Baseline study of methane Emission from open digesting tanks of palm oil mill effluent treatment.Chemosphere , 59, 1575-1581.
92. Yacob, S, Shirai, Y, Hassan, M. A, Wakisaka, M, & Subash, S. operation of semi-commercialclosed anaerobic digester for palm oil mill effluent treatment. Process Biochemistry,41, 962-964.
93. Zinatizadeh, A. A. L, Mohamed, A. R, Abdullah, A. Z, Mashitah, M. D, & Hasnain,M. Isa., G.D, Najafpour. ((2006). Process modeling and analysis of palm oil mill effluent treatment in an up-flow anaerobic sludge fixed film bioreactor using response surface methodology (RSM). Water Research, 40, 3193-3208.

CHAPTER 17

Processes of Removing Zinc from Water using Zero-Valent Iron

Tomasz Suponik, Antoni Winiarski, Jacek Szade

Silesian University of Technology, Akademicka 2, 44-100 Gliwice, Poland

University of Silesia, Uniwersytecka 4, 40-007 Katowice, Poland

ABSTRACT

Zero-valent iron has received considerable attention for its potential application in the removal of heavy metals from water. This paper considers the possibility of removal of zinc ions from water by causing precipitates to form on the surface of iron. The chemical states and the atomic concentrations of solids which have formed on the surface of zero-valent iron as well as the type of the deposited polycrystalline substances have been analyzed with the use of X-ray photoelectron spectroscopy (XPS) and X-ray diffraction (XRD), respectively. The BET surface area, the pH at point of zero charge (pH_{PZC}), the ORP of the solutions, and the pH and chemical concentrations in the solutions have also been measured. Furthermore, the paper also considers the possibility of release of zinc from the precipitates to demineralised water in changing physicochemical and chemical conditions. In a wide range of pH values, $Zn_xFe_{3-x}O_4$ (where $x \leq 1$) was the main compound resulting from the removal of zinc in ionic form from water. In neutral and alkaline conditions, the adsorption occurred as an additional process.

Keywords: Zero-valent iron, Zinc, X-ray photoelectron Spectroscopy, Diffraction, Surface charge

1 INTRODUCTION

Zinc is a metal which finds its application in multiple industries. The minerals most commonly exploited for the purposes of zinc production are sphalerite, smithsonite, and calamine. Zinc is essential for human metabolism, but may also pose certain risks. Free zinc ions in solutions are highly toxic to bacteria, plants, invertebrates, and even vertebrate fish (Ronald 1993).

Wastes resulting from the mining of non-ferrous metal ores and hard coal as well as from the manufacturing process of non-ferrous metals accumulated at industrial waste disposal sites contain zinc minerals and, as a result, may affect the quality of ground and surface waters in the southern provinces of Poland. Meanwhile, the Water Framework Directive (Directive 2000/60/EC of the European Parliament and of the Council) issued in October 2000 commits European Union members states to achieve a good qualitative and quantitative status of all water bodies by 2015. Therefore, the problem associated with waste dumps in Silesia, Poland, and their impact on ground and surface waters must be solved as soon as possible. The pH of the groundwater under the waste disposal sites for wastes resulting from the mining of hard coal and non-ferrous metal ores and from the manufacture of non-ferrous metals is slightly acidic and rather neutral or alkaline, respectively.

Permeable reactive barrier technology with the use of zero-valent iron (ZVI) as a reactive material may be applied to remove free zinc ions from groundwater and thus to protect the surface waters. In this technology, the contaminants are removed from the aquifer during the flow of the groundwater through a permeable reactive barrier (PRB) filled with a reactive material. The application of zero-valent iron to remove zinc in cationic form from water has been investigated before (Morrison et al. 2002; Wilkin and McNeil 2003; Kishimoto et al. 2011; Rangsivek and Jekel 2005; Bruzzoniti and Fiore 2014). This research revealed the effectiveness of ZVI in the zinc removal process. The literature, however, includes conflicting reports regarding the mechanism of that process. For example, according to Kishimoto et al. (2011), the mechanism of zinc removal with the use of ZVI powder is as follows: ZVI firstly corrodes and oxidizes into ferric ions by means of dissolved oxygen. The ferric ions then precipitate as iron hydroxide onto the surface of the ZVI powder. Zinc ions adsorb on and/or co-precipitate with the iron hydroxide. The iron hydroxide is then finally oxidized and transformed into iron oxides. Meanwhile, Wilkin and

McNeil (2003) and Oh et al. (2007) claim that the rapid removal of heavy metals using zero-valent iron in water contaminated by acid mine drainage proceeds due to the adsorption onto the surface of iron metal or onto iron corrosion products. In the research presented in a paper by Li and Zhang (2007), the predominant removal mechanism for metal ions with a standard potential very close to or more negative than that of iron is sorption/surface complex formation. At this point, it should be made clear that the type of process obviously depends on the chemical and physicochemical composition of the solution. In general, sorption as a removal mechanism is not preferred because soluble zinc Zn(II) remains in its more soluble oxidation state, and in case of a change in the physicochemical conditions, it may be released back to the environment.

In relation to the above, the matters that could raise interest in the view of ZVI application are the types of products formed on the surface of ZVI as a result of zinc removal from water characterized by low, neutral, and high pH values, the possibility of zinc release from the surface of ZVI to demineralised water upon a change in the physicochemical and chemical conditions, as well as the type of the dominant processes that result in the removal of Zn(II) from water. These issues have been discussed in the study.

According to Eh-pH diagrams of the Zn-O-H system presented by Takeno (2005), zinc in solutions exists mainly in its divalent ionic form up to the pH value of 8.5. Between the pH of 8.5 and ca. 11, Zn(OH)2 and the cationic species such as Zn(OH)+ become increasingly effective. In the pH ranges analyzed in the study, zinc ions were expected to occur in divalent form during the experiments.

2 MATERIALS AND METHOD

In order to remove cationic zinc from solutions at various pH values, zero-valent iron constituted the iron samples. The samples used in the batch tests were made of steel (in accordance with EN 10131:2006, "cold rolled uncoated low carbon and high yield strength steel flat products for cold forming"). The samples were square-shaped, made of cold-rolled steel sheet (0.5 mm thickness), and their dimensions were 5 × 5 mm. Just before their use, the samples were immersed in concentrated nitric acid for ca. 120 s and in demineralised water for ca. 120 s, immediately after that they were used in the batch tests. The tests were conducted in a programmable MULTI BIO RS-24 BIOSAN rotator equipped with plastic tubes filled with zinc solution (so as to eliminate the headspace—a gaseous phase above the solutions; the volume of the solution amounted to 58 cm^3) with the initial pH of 4.5, 7.0, and 9.5 and with iron samples—only one iron sample was introduced to each plastic tube. The pH levels of the solutions were adjusted by slow titration with ultra-pure sulphuric acid solution (0.1 M) or with ultra-pure sodium hydroxide solution (0.1 M). Zinc ion solutions were prepared by adding the desired amounts of metal salt ($ZnSO_4 \cdot 7H_2O$) into bottles and pouring distilled water. The concentration of zinc in the solutions at pH levels of 4.5, 7.0, and 9.5 amounted to 5.57, 7.06, and 6.12 mg/dm^3, respectively. The intention of the authors was to exceed the allowable concentration of zinc by several times within the meaning of legal regulations provided in the Journal of Laws (2014, item 1800). This value in the discharged water is 2 mg/dm^3.

Dissolved oxygen (DO) has a significant impact on the behavior of Zn in the groundwater. The concentration of DO in groundwater (in an unconfined aquifer) located in the vicinity of coal waste disposal sites in the south of Poland amounted to ca. 6 mg/dm^3 (based on own measurements—the samples of water were taken by the authors using a WHG-168 STALTECHNIKA drilling rig. The aquifers were located at a depth of about 10 m below the surface of the ground). In order to reduce the concentration of oxygen in the samples of water and to adjust their condition to the condition of the contaminated aquifer, the solutions were heated before using them in the batch tests (to the temperature of 308 K). This way, the initial concentration of DO amounted to ca. 6 mg/dm^3. The ambient air temperature in the laboratory was ca. 295 K.

After filling, the plastic tubes were sealed with corks and the first batch tests were carried out. Each sample had to undergo 5 min of orbital rotation in a programmable rotator (with a speed range of 20 rpm) after which a 6-s reciprocal motion (with a turning angle of 90°) with a vibration motion followed. This sequence of shaking was repeatedly reiterated. The sequence continued for 24 h (after this time, constant values were achieved). After shaking the samples, the solutions were passed through dense filters and assessed. The pH, oxidation-reduction potential (ORP), DO, and conductivity were measured using Knick PORTAMESS meters. The quantitative analyses of chemicals in solutions were carried out for the following:

- Zn_{total}, with the use of DR5000 Hach Lange UV-Vis Spectrophotometer—Zincon Method; method 8009 of Hach Co.; test results are measured at 620 nm. The concentration of zinc in the solutions was measured in a spectrophotometer as a total value (Zn_{total}). Since there was no zinc speciation (at the beginning) in the solutions other than the Zn(II), the removal of the zinc in the second oxidation state has been assessed in the article.

- Fe(II) with the use of DR5000 Hach Lange UV-Vis Spectrophotometer—1, 10 Phenanthroline Method; method 8146 of Hach Co.; test results are measured at 510 nm.
- Fe_{total} with the use of DR5000 Hach Lange UV-Vis Spectrophotometer—FerroVer Method; method 8008 of Hach Co.; test results measured at 510 nm. The concentration of dissolved Fe(III) was calculated as a difference between Fe_{total} and Fe(II).
- SO_4^{2-}, DR5000 Hach Lange UV-Vis Spectrophotometer—SulfaVer 4 Method; method 8051 of Hach Co.; test results measured at 450 nm.

The measurements were carried out twice. The results, shown in Table 1, were calculated as an arithmetic mean.

Table 1 Physicochemical parameters and concentrations of chemicals in solutions used in the batch tests for the initial pH = 4.5, 7.0, and 9.5

Stage of the batch tests	pH	Cond., uS/cm	ORP, mV	DO, mg/dm^3	Zn, mg/dm^3	Fe(II), mg/dm^3	Fe(III), mg/dm^3	SO_4^{2-}, mg/dm^3
Initial pH=4.5								
Initial values	4.53	137.8	252	6.1	5.57	BDL	BDL	10.2
Values after the first batch tests	5.27	132.4	155	5.0	2.10	0.26	3.22	9.4
Values after the second batch tests	6.83	24.5	124	5.2	0.03	0.15	0.48	1.3
Initial pH=7.0								
Initial values	7.02	110.1	186	5.8	7.06	BDL	BDL	10.2
Values after the first batch tests	6.31	115.4	128	5.2	1.93	0.10	1.04	9.9
Values after the second batch tests	6.95	17.8	121	5.0	0.04	0.06	0.31	0.6
Initial pH=9.5								
Initial values	9.48	98.4	146	6.2	6.12	BDL	BDL	9.3
Values after the first batch tests	6.85	104.7	138	5.5	0.17	0.04	0.10	9.1
Values after the second batch tests	7.08	11.6	120	5.2	0.04	0.02	0.08	BDL

BDL below detection limit

After shaking, all the iron samples were pulled out, washed with demineralised water, wiped gently with a paper towels and dried in an exsiccator. In order to identify the products formed on the surface of the samples and to determine their affinity with zinc in ionic form, the following tests were conducted on iron samples before and after the batch tests:

- Determination of crystalline substances present on the surface of iron samples before and after the batch tests. These measurements were carried out with the use of X-ray diffraction (XRD). For the analysis of solid objects, EMPYREAN X-ray Multipurpose Diffractometer by PANalytical was used. The diffractometer was equipped with PreFIX (pre-aligned, fast interchangeable X-ray) modules allowing for an effortless change in the optical path. The PDF4+ database was used for the identification of chemical compounds. Crystal lattice parameters were measured.

- Identification of elements and determination of atomic concentrations of solids located on the surface of iron samples formed before and after the batch tests. These measurements were carried out with the use of X-ray photoelectron spectroscopy (XPS). This method allowed for the identification of the elements (except for H and He) and their chemical states as well as for the calculation of their atomic concentrations. Measurements were performed using a PHI 5700/660 Multipurpose Electron Spectrometer based on two separate test chambers joined by an UHV transfer system by Physical Electronics using a monochromatized $Al_{K\alpha}$ radiation ($h\nu = 1486.6$ eV). The energy resolution of the spectrometer equipped with a hemispherical energy analyzer was approximately 0.3 eV. The anode was operated at 15 kV and 225 W. Survey and multiplex high-resolution spectra (HRES; multiplex) were measured in ultrahigh vacuum. High-resolution spectra were fitted using mixed Gaussian and Lorentzian functions and Shirley background with the application of MultiPak program. The range in survey mode was from −2 to 1400 eV. The measurement parameters for survey mode and HRES (high-resolution mode) were respectively pass energy 187.85 and 23.50, step 0.800 and 0.100 eV, and time per step 20 and 100 ms. The sizes of the analyzed areas were 1.5 × 2.5 mm (to identify the trends in case different pH values of the solutions were used) and the diameter was 0.8 mm (to compare the main precipitates which appeared on the samples after the batch tests). The binding energy of the XPS lines was normalized to the binding energy of C1s = 285 eV. Lines were standardized to be of the same height.
- Determination of the specific surface area of the iron samples before and after the batch tests. The measurements of multi-point BET surface area were carried out with the use of Micromeritics Gemini 2360 Surface Area Analyzer. The method utilized a flowing gas technique in which the gas used in the analysis (nitrogen) was introduced into a tube containing the iron sample and into a balance tube at the same time. The principle of measurement consisted in the adsorption of nitrogen on the surface of the sample at a constant temperature of liquid nitrogen (77–78 K). An identical internal volume and ambient temperature of both tubes was maintained. The only difference was the pressure of the sample in the sample tube. The measurement range of the analyzer: specific surface from 0.01 m^2/g. In order to remove impurities and moisture, the samples were dried at 383 K before the measurements.

- Determination of the affinity of zinc in ionic form with shells coating the iron sample by measuring the pH at the point of zero charge (pH_{PZC}). The point of zero charge was obtained using batch equilibrium method. This parameter describes the condition (pH value) in which the electrical charge on the surface of ZVI is zero. The testing procedure was as follows: iron samples obtained from the batch tests (only from the first batch tests) were immersed in demineralised water for ca. 120 s and then shaken (separately) in plastic tubes for 24 h with the following: (1) 58 cm^3 of 0.01 M KNO_3 solution and (2) 58 cm^3 of 0.01 M KNO_3 and 10^{-5}M Zn(II) solution, at different initial pH values. A suitable amount of $Zn(NO_3)_2$ was added to the solution to achieve a proper concentration of Zn(II). The applied sequence of shaking was similar to the previous ones (performed with a programmable MULTI BIO RS-24 BIOSAN rotator). The initial pH levels of the solutions were adjusted by slow titration with ultra-pure potassium hydroxide solution (0.1 M) or with ultra-pure nitric acid solution (0.1 M) while keeping the ionic strength constant. The amount of H^+ or OH^- ions adsorbed by iron samples was calculated from the difference between the initial and the final concentrations of H^+ or OH^- ions (Babić et al. 1999). The pH was measured with the use of Knick PORTAMESS 913 meter and the SenTix 41 electrode (Suponik 2015b, in press).

In order to assess the capability of metals to be released from the precipitates found on the surface of iron samples to demineralised water and to identify the products formed on the surface of the samples after this process, tests using a programmable rotator were carried out for a second time, in a manner similar to the one described before. The only difference was that the iron samples used for removing metals from solutions (after finishing the first batch tests) were inserted into 58 cm^3 of demineralised water (DW) poured into plastic tubes. The samples in the second batch tests were shaken in the same way as in the case of the first batch tests. The study conducted after the second test was similar, except for the fact that the pH_{pzc} measurements were omitted.

3 RESULTS AND DISCUSSION

3.1 Study of Water

As presented in Table 1, the concentrations of zinc in water measured after the first tests decreased in a more rapid manner in case of higher pH levels of solutions. This observation was also made in the papers by Kishimoto et al. (2011), Rangsivek and Jekel (2005), and Suponik and Blanko (2014). What is noteworthy here is the fact that after the second batch tests, the zinc from the iron sample was not released to demineralised water. The compounds formed on iron in the first tests were fixed. They were not dependent on previous pH values. The final pH levels after the second tests were similar in case of all iron samples and amounted approximately to 7. The pH of DW amounted to 6.8. After the first tests, also, this parameter had the tendency to approach the value of 7. The parameter increased for low values and decreased for higher ones.

Kishimoto et al. (2011) studied the desorption of zinc from ZVI using sulphuric acid and citric acid solutions. During the backwashing, they observed rapid desorption of zinc when the citric acid solution was used, whereas no desorption was observed when the sulphuric acid was applied. Their conclusion was that the zinc adsorption layer on ZVI was stable in diluted sulphuric acid and became unstable in a reducing agent such as citric acid. This conclusion is consistent with the results obtained in studies using demineralised water.

As the previous work indicates (Suponik 2015a), the lower were the values of pH in the tests, the higher were the initial values of ORP in the water and the faster was the decrease of this potential after the first batch tests. The ORP of pure DW amounted to 159 mV, while in the case of water after the second batch tests, the value reached 120 mV. A slight decrease in oxygen concentration may also be noticed after both of the batch tests, which indicates that the course of redox reactions involves oxygen. The initial concentration of oxygen in DW amounted to 6.4 mg/dm^3.

The oxidation of Fe(0) proceeds faster in low pH (Kowal and Świderska-Bróż 1996; Kishimoto et al. 2011), which is evidenced by the fact that more ions of Fe(II) iron appeared after the first batch tests in water characterized by a lower initial value of pH. This occurred mostly due to processes described by reactions 1 and 2. A similar correlation was observed in case of the second batch tests, though the concentrations of Fe(II) were lower. This was probably caused by the thin shells which covered the ZVI.

$$2Fe^0 + O_2 + 2H_2O^2 \rightarrow 2Fe^{2+} + 4OH^- \quad (1)$$

$$Fe^0 + 2H_2O \rightarrow Fe^{2+} + H_2 + 2OH^- \quad (2)$$

In the presence of dissolved oxygen in water, however, Fe(II) is oxidized to Fe(III), which then may produce either FeOOH, $Fe(OH)_3$, Fe_2O_3, or Fe_3O_4 together with Fe(II)—each in solid form.

$$2Fe^{2+} + {}^1/{}_2O_2 + 2H^+ \rightarrow 2Fe^{3+} + H_2O \quad (3)$$

Due to the presence of oxygen in the solutions (Table 1), the concentrations of Fe(III) in almost all samples were several times higher than the concentrations of Fe(II). As it was reported by Kishimoto et al. (2011) and Rangsivek and Jekel (2005), zinc removal by ZVI was enhanced by the presence of dissolved oxygen.

In accordance with the results presented in Table 1, the sulfates in the first batch tests have been removed from the water to a very small extent. Thus, the sulfates were probably not present on the iron sample in the second tests. The release of the sulfates to the water in these tests was very low—especially at higher pH values.

3.2 XRD

Upon visual inspection, the surfaces of the iron samples applied in the first and second tests were covered with gray and yellow-brown precipitates. The amount of yellow-brown deposits seemed to be higher for lower initial pH values of the solutions. In case of the samples immersed in the solutions of pH amounting to 9.5, spots of metallic iron not covered by precipitates were visible. The amount of gray precipitates was similar in all the samples.

Figure 1 presents the diffraction of X-rays on iron samples before and after the first and second batch tests. Based on Fig. 1a, it may be claimed that iron was found on the surface of iron samples assessed before the batch tests (as it was expected). The network parameters of pure iron are similar (very close) to the parameters of many alloys in which iron is the main component.

Pure iron was also present in samples immersed in solutions characterized by higher initial pH levels, especially at pH = 9.5 (in Fig. 1b, c, the theta angle for Fe(0) amounts to 44.6°, 65.0°, and 82.3°). In both the first and the second batch tests for the initial solutions with the pH = 9.5, the samples exhibited many spots not covered by precipitates.

On the contrary, lepidocrocite γ-FeO(OH) was identified mainly in case of lower pH values, especially when the pH of the solutions was 4.5. According to the XRD graphs plotted for the first and second tests, the theta angle in Fig. 1b, c for γ-FeO(OH) amounts to 14.1°, 27.0°, 36.3°, 38.1°, and 46.8°. γ-FeO(OH) may correspond to the yellow-brown precipitates on

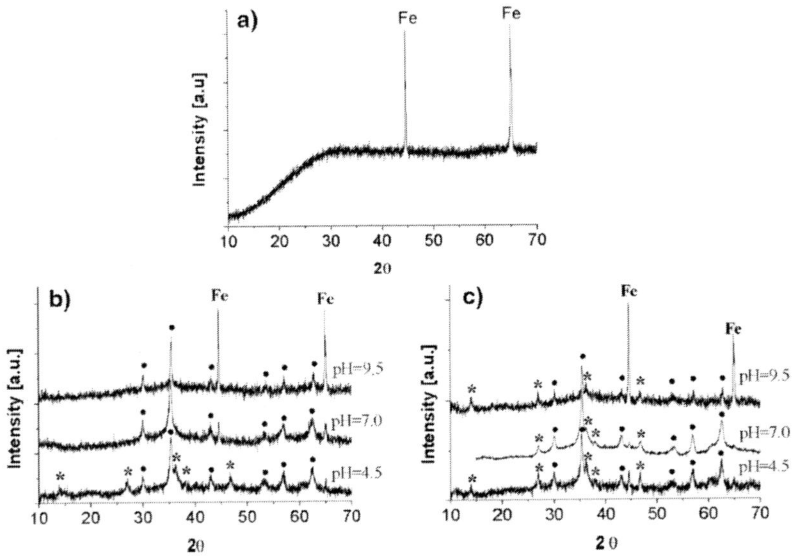

Figure 1 XRD graphs of iron samples: **a** before the batch tests; **b** after the first batch tests for the initial pH of the solution amounting to 4.5, 7.0, and 9.5; and **c** after the second batch tests for the initial pH of the solution amounting to 4.5, 7.0, and 9.5. Fe – Iron in metallic form; * – γ-FeO(OH); – ZnxFe3-xO4

the samples, occurring mainly in lower pH of the solutions that were used. Lepidocrocite is a common phase often forming as a result of iron corrosion process in the presence of dissolved oxygen (Kamolpornwijit et al. 2004). γ-FeO(OH) was also found on the surface of ZVI after the removal of copper and zinc ions from water (Rangsivek and Jekel 2005). The presence of lepidocrocite indicates that the adsorption of zinc may have occurred as an additional process resulting from the uptake of this metal.

According to the XRD graphs (Fig. 1b, c) obtained after the first and second batch tests, zinc was present in the $Zn_xFe_{3-x}O_4$ compound (where $x \leq 1$). Diffraction lines obtained from the XRD tests correspond to the compounds, which can be expressed as $Zn_xFe_{3-x}O_4$, because the diffraction lines of compounds such as $Zn^{2+}Fe^{3+}{}_2O_4$, $Zn^{2+}{}_{0.35}Fe^{2+}{}_{0.65}Fe^{3+}{}_2O_4$, and $Zn^{2+}{}_{0.4}Fe^{2+}{}_{0.6}Fe^{3+}{}_2O_4$ are very similar. Theta angles for $ZnFe_2O_4$ are 30.0°, 35.4°, 43.0°, 53.2°, 56.8°, and 62.4°. It should be added here that the same pattern was exhibited by all the applied samples, irrespective of the initial pH of the solutions. $Zn_xFe_{3-x}O_4$ may correspond to the gray-colored precipitates.

3.3 XPS

Table 2 presents the atomic concentrations of the elements on the iron samples before the study and after the first and second batch tests. Each assessed sample contained carbon and oxygen. It is possible that the oxygen was contained in precipitates in the form of metal oxides and metal hydroxides, while carbon and oxygen in compounds in which the following bonds occur (in very small quantities): -C-O- and -C=O (binding energy: C1s = from 286.51 to 288.64 eV, see Fig. 2). The main contribution to the C1s line is constituted by hydrocarbons deposited on the surface. This always occurs when a sample is stored in atmospheric conditions. The binding energy of the C1s line resulting from the presence of hydrocarbons amounts to 285.0 eV.

Table 2 Atomic concentrations of elements (expressed as percentages) on the iron samples before the study and after the first and second batch tests for the initial pH of the solutions amounting to 4.5, 7.0, and 9.5—on the basis of XPS measurements

Elements	Type of the study												
	Values before the study, in atom %	Values after the first and second batch tests—for the initial pH=4.5, in atom %				Values after the first and second batch tests—for the initial pH=7.0, in atom %				Values after the first and second batch tests—for the initial pH=9.5, in atom %			
	XPS—survey mode	XPS—survey mode		XPS—HRES, multiplex		XPS—survey mode		XPS—HRES, multiplex		XPS—survey mode		XPS—HRES, multiplex	
		After first test	After second test with DW	After first test	After second test with DW	After first test	After second test with DW	After first test	After second test with DW	After first test	After second test with DW	After first test	After second test with DW
C	62.80	16.09	12.10	16.99	12.41	20.50	20.47	21.25	25.03	27.24	30.46	28.13	27.49
O	29.49	55.35	56.09	54.77	57.06	51.94	54.17	52.08	50.36	48.52	46.48	46.90	48.20
S		0.36	0.07	0.59	0.08	2.45		2.73					
Fe	7.71	27.86	31.63	27.23	30.15	20.16	24.28	19.40	23.59	23.54	22.79	24.13	24.05
Zn		0.35	0.11	0.42	0.30	4.95	1.08	4.54	1.02	0.69	0.26	0.84	0.25

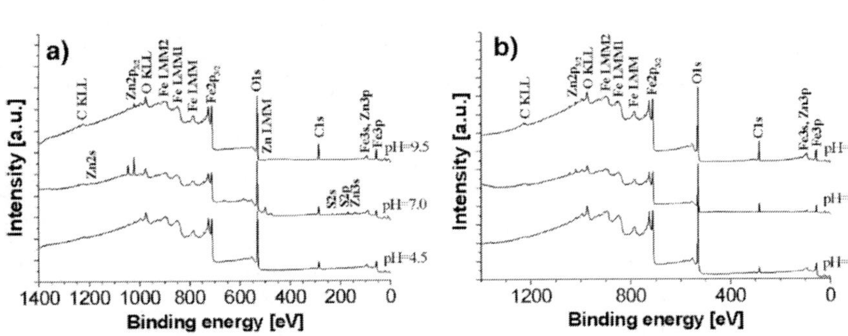

Figure 2 XPS graphs of the iron samples: **a** after the first batch tests for the initial pH of the solution amounting to 4.5, 7.0, and 9.5 and **b** after the second batch tests for the initial pH of the solution amounting to 4.5, 7.0, and 9.5

The oxygen present in the compounds on the surface of the iron samples originated from solutions used in the analysis in the first and second batch tests. The presence of carbon and remaining amounts of oxygen has been explained by the contamination of the samples (by atmospheric air and other factors) during their transport for the measurements. For that reason, the atomic concentrations of oxygen and carbon presented in Table 2 have not been taken into consideration in further analysis.

The results of the tests presented in Table 2 have also shown that zinc and iron formed compounds on the surface of all the samples used in the first and second batch tests. Although the assessments concerned a small area of the samples, it can be concluded that the zinc concentration on iron samples was lower in case of samples introduced to the solution with $pH = 4.5$. This is consistent with the observation that in lower pH levels of solutions, the removal of zinc proceeded in a less effective manner. On the surface of samples immersed in the solution at $pH = 9.5$, no sulfur was observed based on the results of the XPS. This confirms the previous observation that sulfur is not removed from water at higher pH values.

The lower concentrations of zinc on the iron samples after the second batch tests as compared to the first tests results (see Table 2) may indicate that this element is released to demineralised water to a small extent. In the opinion of the authors, however, the decrease in the zinc concentration after the second test for the initial pH of 7 was too big. Most likely, the analysis after the first and the second tests did not include the same locations.

At the same time, while discussing the results of water analysis, it was found that zinc was not released to water after the second batch tests. No indication of zinc in the water may result from the fact that the samples of water were not digested before the measurement in a spectrophotometer. Zinc in the form of other compounds may be transferred to water. In a very a small extent, zinc may also be released into the DW.

HRES spectrums of O1s obtained from the XPS consist of two peaks with similar binding energy for all the analyzed samples. The peaks were as follows: 530.1 ± 0.2 and 531.6 ± 0.3 eV. They correspond to the metal oxides and metal hydroxides, respectively. The peak at 530.1 ± 0.2 eV, which was a result of a metal-oxygen bond, may correspond to many compounds such as the following: γ-Fe_2O_3 (529.8 eV), Fe_3O_4 (530.0 eV), FeO (529.8 eV), α-FeOOH (530.1 eV), or even to ZnO (530.3 eV), while the peak at 531.6 ± 0.3 eV corresponds to γ-FeOOH (531.4 ± 0.2 eV). In case of the first peak, it was difficult to decide which compound corresponded to the binding energy of 530.1 eV. Taking into account the results from the diffractometer, however, it may be established that it corresponded either to Fe_3O_4, in which the iron atoms have been replaced by zinc atoms or pure magnetite. It should be added that in case of the solutions with the initial pH of 7.0 and 9.5, γ-FeOOH was also formed, but in smaller quantities, because the photoelectric lines were weak. In addition, X-ray diffractograms provided no evidence for the presence of lepidocrocite in the crystalline form (Fig. 1b).

Furthermore, in the Fe2p spectra for all the samples, the binding energy amounted to 711.3 ± 0.3 eV. This value confirms the presence of γ-FeOOH on the surface of the samples (Fe2p 711.3 eV—γ-FeOOH).

In case of the iron samples immersed in solutions with the initial pH values of 4.5, 7.0, and 9.5, the Zn2p3/2 emission line in the registered spectra appeared at the binding energy of 1021.5 ± 0.4 eV. This value corresponds to the $ZnFe_2O_4$ compound. The binding energy of this compound amounts to 1021.4 eV. This value has confirmed, according to the data from the XRD tests, the formation of compounds such as $ZnFe_2O_4$.

As provided by Furukawa et al. (2002), Roh et al. (2000), and Rangsivek and Jekel (2005), the secondary minerals which typically form on the surface of ZVI as a result of purification of water contaminated by metals in ionic forms are as follows: γ-FeOOH, Fe_3O_4, Fe_2O_3, α-FeOOH, $CaCO_3$, $Fe(OH)_2$, $Fe_2O_3 \cdot 0.5H_2O$, $FeCO_3$, FeS_2, Fe_3S_4, mackinawite ($(FeNi)_{1+x}S$, where $x = 0$ to 0.11), and green rust ($[Fe_{1-x}^{2+} Fe_x^{3+}(OH)_2]^{x+} [x/n\, A^{n-} \cdot mH_2O]^{x-}$, where x is the Fe^{3+}/Fe_{tot}ratio), which is usually formed under neutral pH conditions. Although according to Wilkin and McNeil (2003), green rust is a primary corrosion product formed on ZVI in sulfate-rich solutions, it has not been observed (by visual inspection) on the surface of the iron sample and there has been no evidence of it in the presented results. This is probably due to the "pure" conditions of purification. As provided by Cornell and Schwertmann (1996), green rust is stable only at low grades of oxide reduction and its oxidation usually leads to the formation of Fe_2O_3 or γ-FeOOH. Roh et al. (2000) also reported that green rust is an intermediate stage and is finally transformed into α-FeOOH, γ-FeOOH, Fe_3O_4, and Fe_2O_3.

The formation of magnetite (in pure form) was investigated by Grosvenor et al. (2004), Kamolpornwijit et al. (2004), and Karabelli et al. (2008). The latter has noted (on the basis of XRD analyses) a slow development of iron oxides, primarily in the forms of Fe_3O_4 and γ-Fe_2O_3. Kishimoto et al. (2011) reported that the iron hydroxide formed in earlier stages was finally oxidized and transformed into iron oxides.

In conclusion, based on the results of the XPS and XRD, the products formed on the surface of ZVI as a result of zinc removal from water were the following: (1) magnetite, in which zinc replaces iron creating $Zn_xFe_{3-x}O_4$—in acidic, neutral, and alkaline conditions and (2) lepidocrocite, mainly in acidic conditions. After the second batch tests, i.e., after inserting the samples into demineralised water, the same compounds remained on the ZVI.

Ferrous and ferric ions were formed as a result of the reactions 1, 2, and 3, while $Zn^{2+}{}_{(x)}Fe^{2+}{}_{(1-x)}Fe^{3+}{}_2O_4$ was formed in accordance with Eq. 4:

$$(1-x)\,Fe^{2+} + 2Fe^{3+} + xZn^{2+} + 2H_2O + O_2 \rightarrow Zn^{2+}{}_{(x)}\,Fe^{2+}{}_{(1-x)}\,Fe^{3+}{}_2O_4 + 4H^+ \quad (4)$$

where $x \leq 1$.

3.4 Specific Surface of Iron Samples

The values of specific surface areas obtained by means of multi-point BET method for the iron samples immersed in the zinc solutions with the pH values of 4.5, 7.0, and 9.5 (after the first batch tests) were 0.2704, 0.3324, and 0.2741 m^2/g, respectively, while these same parameters for the samples immersed in DW (after the second batch tests) equalled to 0.2482, 0.3521, and 0.2908 m^2/g. In case of the iron samples before the batch tests, the parameter amounted to 0.0302 m^2/g, which means that large amounts of chemical compounds have formed on the surface of the iron samples after the tests, especially at the initial pH = 7.0. It may be stated that after inserting the samples into DW, the surface area did not change. Based on these results, either the substances were not released into the water, or new compounds have been created on the surface of the samples.

3.5 pH_{pzc}

Figure 3 presents the results of the tests conducted for the determination of pH_{pzc}. The pH at the point of zero charge, determined by batch equilibrium method for the three iron samples (that have been used in the zinc solution at the initial pH = 4.5, 7.0, and 9.5) which were shaken in 0.01 M KNO_3 solutions, amounted to 7 (pH_f level at which a common plateau is reached, see the dashed arrows in Fig. 3). This leads to the conclusion that pH_{pzc} is independent from the initial pH of zinc solutions.

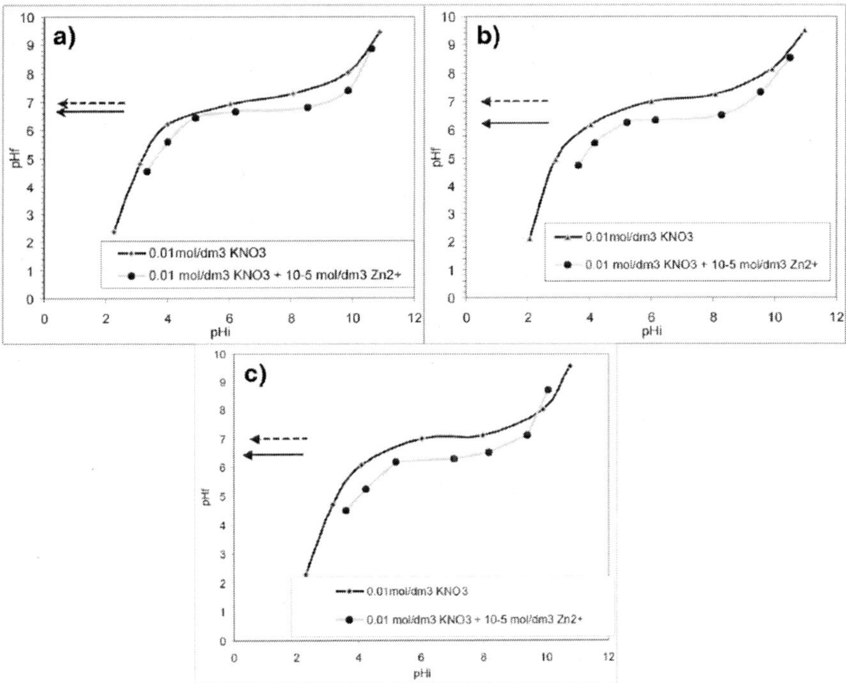

Figure 3 The final value of pH (pH_f) vs. the initial value of pH (pH_i) for the iron samples immersed (separately) in two solutions: 0.01 M KNO_3 and 0.01 M $KNO_3 + 10^{-5}$M Zn(II). Three samples have been used in each solution: **a** the one that was shaken in a solution with the initial pH = 4.5, **b** the one that was shaken in a solution with the initial pH = 7.0, and **c** the one that was shaken in a solution with the initial pH = 9.5

In the presence of zinc ions in solutions, pH_{pzc} decreases to the value of ca. 6.7, 6.3, and 6.4 (pH_f level at which a common plateau is reached, see the solid arrows in Fig. 3), respectively, for initial pH values of the solutions amounting to 4.5, 7.0, and 9.5. The decrease in pH_{pzc} is due to the specific adsorption of counter ions. The values of point of zero charge for iron oxides (hydroxide) are as follows: 7.8 for lepidocrocite γ-FeOOH, from 3.8 to 8.2 for magnetite, from 6.1 to 7.5 for maghemite γ-Fe_2O_3, from 5.5 to 9.3 for hematite α-Fe_2O_3, from 6.2 to 9.6 for goethite α-FeOOH, and from 6.0 to 6.5 for iron(II) hydroxide (Kosmulski 2011).

Based on the above, it may be stated that the sorption of Zn(II) on the surface of samples is not possible for the solution with the initial pH of 4.5, in case of which the final pH was 5.27 (see Table 1), while it may occur in case of the solutions with the initial pH of 7.0 and 9.5, for which the final pH was 6.31 and 6.85, respectively (see Table 1). The shells coating the iron samples following the tests exhibited nearly equal the amounts of negative and positive charges, as the pH of the solutions (6.31 and 6.85) was close to the pzc of the compounds which have formed on the surface of the iron samples. Since both the charges exist in the shells coating the samples at this pH, the sorption of Zn(II), as well as the negatively charged ions which were present in the solution, may occur at the surface of the samples.

In the research described in the papers by Kishimoto et al. (2011) and Rangsivek and Jekel (2005), the zinc removal process was also enhanced at higher pH values.

4 CONCLUSIONS

ZVI has the capacity to remove zinc from water in a wide range of pH values. The concentrations of zinc in water decreased much faster in higher pH levels of the solutions. At the same time, zinc was not released to demineralised water from the precipitates located on the ZVI (or was released to a small extent), as DW is not a reducing agent. The results of the XPS indicated, however, that some amount of zinc can be released into demineralised water. The specific surface of iron samples, in turn, did not change after inserting the samples into DW. In this case, no conclusion can be drawn. It can only be stated that zinc may be released into the DW to a very limited extent—if at all.

The products formed on the surface of the ZVI as a result of zinc removal from water were dependent on the initial pH values of the solutions. Under acidic conditions γ-FeO(OH) and $Zn_xFe_{3-x}O_4$ (where $x \leq 1$) were the main precipitates, while under alkaline conditions, the iron surface was covered with $Zn_xFe_{3-x}O_4$ and to a lesser extent with γ-FeO(OH). For lower initial values of pH, the presence of sulfates on the surface of the samples has also been identified. Similar observations to the ones described above were made for the second batch tests, in which the ZVI samples were submerged in demineralised water. To summarize, $Zn_xFe_{3-x}O_4$ was the main compound generated as a result of the removal of zinc in ionic form from water in a wide range of pH values.

The adsorption of Zn(II) on the surface of precipitates which have formed on the ZVI may occur as an additional process in neutral and high pH of solutions, since the pH was approximately 6.3–6.4 at point of zero charge for shells coating zero-valent iron. The final pH of these solutions reached the values of 6.31 and 6.85.

REFERENCES

1. Babić, B. M., Milonjić, S. K., Polovina, M. J., & Kaludierović, B. V. (1999). Point of zero charge and intrinsic equilibrium constants of activated carbon cloth. Carbon, 37, 477–481.
2. Bruzzoniti, M. C., & Fiore, S. (2014). Removal of inorganic contaminants from aqueous solutions: evaluation of the remediation efficiency and of the environmental impact of a zero-valent iron substrate. Water, Air, and Soil Pollution, 225(9), 1–14.
3. Cornell, R. M., & Schwertmann, U. (1996). The iron oxides-structure, properties, reactions, occurrence and uses (p. 573). New York: VCH Publishers.
4. Furukawa, Y., Kim, J. W., Watkins, J., & Wilkin, R. T. (2002). Formation of ferrihydrite and associated iron corrosion products in permeable reactive barriers of zero-valent iron. Environmental Science and Technology, 36(24), 5469–5475.
5. Grosvenor, A. P., Kobe, B. A., Biesinger, M. C., & McIntyre, N. S. (2004). Investigation of multiplet splitting of Fe 2p XPS spectra and bonding in iron compounds. Surface and Interface Analysis, 36, 1564–1574.
6. Kamolpornwijit, W., Liang, L., Moline, G. R., Hart, T., & West, O. R. (2004). Identification and quantification of mineral precipitation in Fe0 filings from a column study. Environmental Science and Technology, 38(21), 5757–5765.
7. Karabelli, D., Uzum, C., Shahwan, T., Eroglu, A. E., Scott, T. B., Hallam, K. R., & Lieberwirth, I. (2008). Batch removal of aqueous Cu2+ ions using nanoparticles of zero-valent iron: a study of the capacity and mechanisms of uptake. Industrial and Engineering Chemistry Research, 47, 4758–4764.
8. Kishimoto, N., Iwano, S., & Narazaki, Y. (2011). Mechanistic consideration of zinc removal by zero-valent iron. Water, Air, and Soil Pollution, 221(1-4), 183–189.
9. Kosmulski, M. (2011). The pH-dependent surface charging and points of zero charge: V. Update. Journal of Colloid and Interface Science, 35(1), 1–15.
10. Kowal, L. A., Świderska-Bróż, M., & Swiderska-Bróż, M. (1996). Oczyszczanie wody. Warszawa-Wrocław: PWN.
11. Li, X. Q., & Zhang, W. X. (2007). Sequestration of metal cations with zerovalent iron nanoparticles: a study with high resolution X-ray photoelectron spectroscopy (HRXPS). Journal of Physical Chemistry, 111(19), 6939–6946.
12. Morrison, S. J., Metzler, D. R., & Dwyer, B. P. (2002). Removal of As, Mn, Mo, Se, U, V and Zn from groundwater by zero-valent iron in a passive treatment cell: reaction progress modeling. Journal of Contaminant Hydrology, 56(1), 99–116.
13. Oh, B., Lee, J., & Yoon, J. (2007). Removal of contaminants in leachate from landfill by waste steel scrap and converter slag. Environmental Geochemistry and Health, 29(4), 331–336.
14. Rangsivek, R., & Jekel, M. R. (2005). Removal of dissolved metals by zero-valent iron (ZVI): kinetics, equilibria, processes and implications for stormwater runoff treatment. Water Research, 39, 4153–4163.
15. Roh, Y., Lee, S. Y., & Elless, M. P. (2000). Characterization of corrosion products in the permeable reactive barriers. Environmental Geology, 40, 184–194.
16. Ronald, E. (1993). Zinc hazard to fish, wildlife, and invertebrates: a synoptic review. Contaminant hazard reviews. Laurel, Maryland: U.S. Department of the Interior, Fish and Wildlife Service (10).
17. Suponik, T. (2015a). Zero-valent iron for inorganic contaminants removal from low pH water. Environment Protection Engineering, 41(1), 15–27.
18. Suponik, T. (2015b). Study of precipitates formed on the iron reactors following the removal of copper from water. Environment Protection Engineering, in press.
19. Suponik, T., & Blanko, M. (2014). Removal of heavy metals from groundwater affected by acid mine drainage. Physicochemical Probl of Miner Process, 50(1), 359–372.

20. Takeno, N. (2005). Atlas of Eh-pH diagrams; Intercomparison of thermodynamic databases. National Institute of Advanced Industrial Science and Technology, Geological Survey of Japan Open File Report No. 419, 102.

21. Wilkin, R. T., & Mcneil, M. S. (2003). Laboratory evaluation of zero-valent iron to treat water impacted by acid mine drainage. Chemosphere, 53, 715–725.

CHAPTER 18

Ground Water Quality Evaluation Near Mining Area and Development of Heavy Metal Pollution Index

Bably Prasad Puja Kumari Shamima Bano Shweta Kumari

Central Institute of Mining and Fuel Research, Barwa Road, Dhanbad 826015, Jharkhand, India

ABSTRACT

Opencast as well as underground coal mining are likely to disturb the underground water table in terms of quantity as well as quality. Added to this is the problem of leachates from the large number of industrial waste and overburden dumps that are in abundance in mining areas, reaching the ground water and adversely affecting its quality. Enhancement of heavy metals contamination of the ground water is one eventuality. In the present work, concentrations of 7 heavy metals have been evaluated at 20 important ground water sampling stations at Dhanbad township situated very near to Jharia coalfields. The concentration of heavy metals in general was found to be below the permissible levels although concentration of iron and manganese was found above the permissible limits at a few stations. These data have been used for the calculation of heavy metal pollution index (HPI). The HPI of ground water in total was found to be 6.8860 which is far below the critical index limit of 100 pointing to the fact that the ground water is not polluted with respect to heavy metals in spite of the prolific growth of mining and allied industrial activities near the town.

Keywords: Heavy metal pollution index Ground water Coal mining Heavy metals

INTRODUCTION

Opencast as well as underground coal mining disturb the underground water table in terms of its level, quantity as well as quality. Coal mining activity in India started decades back, since then the ground water is getting affected. Along with coal mining, leachates generated from large number of industrial waste and overburden dumps that are in abundance around the mining areas, may reach the ground water and may adversely affect its quality (Khan et al. 2005; Mohammad et al. 2010). Enhancement of heavy metals contamination of the ground water is one of the serious eventualities. Some of the heavy metals considered as micronutrients become detrimental to human health when their concentrations exceed the permissible level of drinking water (Prasanna et al. 2011). Thus, evaluation of heavy metals in ground water that is used for drinking purpose is of great significance from the human health point of view. In the present study, seven important heavy metals such as iron, manganese, lead, copper, cadmium, chromium and zinc have been evaluated in 20 ground water samples, obtained from different places of Dhanbad township which is located very near to Jharia coalfields, for 3 different seasons of the year 2011 and heavy metal pollution index has been evaluated to know the present status of overall pollution level of ground water with respect to heavy metals.

The pollution parameters monitored for the assessment of the quality of any system give an idea of the pollution with reference to that particular parameter only. Quality indices are useful in getting a composite influence of all parameters of overall pollution. Quality indices make use of a series of judgements into a reproducible form and compile all the pollution parameters into some easy approach. Several methods have been proposed to develop quality indices for estimation of characteristics of surface water with water quality parameters (Horton 1965; Lohani and Todno 1984; Tiwary and Mishra 1985; Joung et al. 1979; Landwehr 1979; Nishidia et al. 1982). In past years, the authors (Prasad and Jaiprakas 1999; Prasad and Bose 2001; Prasad and Kumari 2008; Prasad and Mondal 2008) have evaluated the heavy metals pollution in ground water and evaluated heavy metal pollution index (HPI).

Indexing approach

The HPI represents the total quality of water with respect to heavy metals. The proposed HPI is based on weighted arithmetic quality mean method and is developed on two basic steps: firstly, by establishing a rating scale for each selected parameter giving weightage to selected parameter (heavy metal) and, secondly, by selecting the pollution parameter on which the index is to be based. Rating system is an arbitrary value between 0 and 1, and its selection depends upon the importance of individual quality considerations in a comparative way, or it can be assessed by making values inversely proportional to the recommended standard for the corresponding parameter (Horton 1965; Mohan et al. 1996). In the present formula, unit weightage (W_i) is taken as value inversely proportional to the recommended standard (S_i) of the corresponding parameter. Iron, manganese, lead, copper, cadmium, chromium and zinc have been monitored for the model index application. The HPI model proposed is given by Mohan et al. (1996)

$$\text{HPI} = \frac{\sum_{i=1}^{n} W_i Q_i}{\sum_{i=1}^{n} W_i} \tag{1}$$

where Q_i is the sub index of the ith parameter. W_i is the unit weightage of ith parameter and n is the number of parameters considered.

The sub index (Q_i) of the parameter is calculated by

$$Q_i = \sum_{i=1}^{n} \frac{\{M_i(-)\,I_i\}}{(S_i - I_i)} \times 100 \tag{2}$$

where, M_i is the monitored value of heavy metal of ith parameter, I_i is the ideal value of ith parameter, S_i is the standard value of ith parameter. The sign (–) indicates the numerical difference of the two values, ignoring the algebraic sign.

Generally, pollution indices are estimated for any specific use of the water. The proposed index is intended for the purpose of drinking water. The critical pollution index value for drinking water is 100.

MATERIALS AND METHODS

Study area

Dhanbad, one of the biggest coal belts of India, is actively associated in mining activities for more than a century. The oldest geological formations of the Dhanbad district are composed of crystalline metamorphic rocks which belong to Dharwar system in Indian stratigraphy. In these formations are found rock types of both sedimentary and igneous origin. Course gritty soil is predominant in Dhanbad coal basin. This type of soil mixed with big fragment of rocks is formed from the weathering of pegmatites, quartz and conglomeratic sandstones. About 200 coal mines of Jharia coal fields, which produce prime coking coal, are located around Dhanbad township. The large scale industries like coal washing and coke making are located nearby. There is drinking water shortage in the township, especially in summer season. Population here is mainly depended on ground water sources. Taking all these into consideration, twenty ground water sampling points have been selected, which covered almost entire township of Dhanbad (Fig. 1).

Sampling and analysis

Out of 20, 4 samples have been taken from open well and 16 from hand pumps. The depth of open wells was 25–30 m and that of hand pumps 100–120 m. Sampling has been done for the month of May (summer) 2011, August (rainy) 2011 and December (winter) 2011 to get an idea of the seasonal variation of the heavy metal concentrations. All samples have been digested, concentrated and prepared for analysis by atomic absorption spectrophotometer (AAS) methods using model: M Series Thermo Fisher (Arnold et al. 1992). The digestion and concentration of water sample was carried out by reducing the volume of 1,000–50 ml by slow heating on a hot plate with addition of few millilitres of concentrated nitric acid. Slow heating was done to avoid metal loss due to evaporation. Analysis of these metals were performed by calibrating the instrument with different dilutions of standard solutions and analysis of Cu, Mn, Pb, Fe, Cd, Zn and Cr was performed at 324.7, 279.5, 217.0, 248.3, 228.8, 213.9 and 357.9 nm, respectively. Three replicates were run for each sample and the instrument was recalibrated after analysis of ten samples. The detection limit of AAS was well within the range for all elements analysed. Over all precision expressed as percent relative standard deviation was obtained for all the samples. Because the samples were concentrated during digestion, the result obtained was calculated to get the actual concentration.

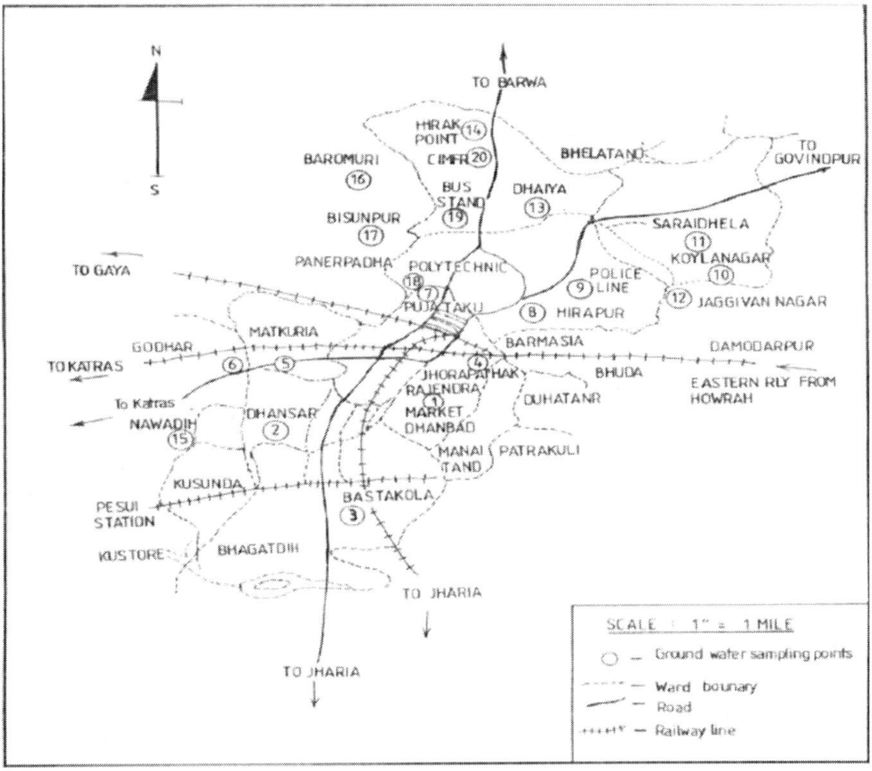

Figure. 1 Map showing water sampling points in Dhanbad township area

RESULT AND DISCUSSION

The evaluation of concentration of seven heavy metals such as Fe, Mn, Pb, Cu, Cd, Cr, and Zn in three seasons in ground water of Dhanbad town is listed in Tables 1 and 2. The mean concentrations of Fe, Mn, Pb, Cu, Cd, Cr, and Zn were 0.5416, 0.1662, 0.0038, 0.0084, 0.00041, 0.0072 and 0.2116 mg/l, respectively, which include total twenty ground water sampling points for three seasons of the year. From the results, it has been observed that concentrations of heavy metals such as Cu, Pb, and Zn were well below the permissible limits of Indian drinking water standard (IS: 10500). The concentration of Mn and Fe has been found more than the highest desirable limit of drinking water standard at many places, in all seasons, especially in hand pumps water samples. The concentration of Fe in summer has exceeded the highest permissible value of 1.5 mg/l at sampling points 8,9, and 17 and in winters at sampling points 8,16 and 17. This phenomenon may be due to corroded nature of iron pipes of hand pumps or due to presence of more iron in ground strata bearing the ground water. In general, highest concentration of Mn and Fe in hand pump water samples indicates the poor quality and maintenance of hand pumps. The open well water samples collected at sampling point 1, 2, 3, and 20 show that none of the water samples are having higher concentration of any heavy metals under study except Mn at sampling point 2 in summer season and Fe at sampling point 20 in rainy season. Since the weightage (W_i) given to Fe and Mn is very less, so in evaluation of HPI of ground water in total, these parameters do not contribute much on HPI value. Heavy metals like Pb, Cd, and Cr have been given no relaxation in drinking water standard and they have been given high weightage (W_i) value in HPI calculation. Hence even their smaller concentration present in water samples makes the water of poor quality and gives high values in HPI calculations. Concentration of Cu and Zn has not been found at any sampling point and at any season higher than the highest desirable limit of drinking water standard. Overall, the ground water of Dhanbad town with respect to heavy metals contamination is not a serious issue and ground water contamination of the township is not taking place due to severe coal mining activity near the township. The monitored data of ground water samples have been used to evaluate the HPI and assess the validity of the index model. The HPI has been determined by taking the mean concentration of heavy metals of all the three seasons using Eqs. 1 and 2. In Table 3 detailed calculation of pollution index with unit weightage (W_i), standard permissible value (S_i), monitored concentration value (M_i) and highest desirable value (I_i) has been done. The HPI calculated with mean concentration values of all metals including all sampling points of ground water for three seasons of the year 2011 comes out to be 6.8860 which is far below the critical index value of 100. The same study carried out earlier and HPI calculated for Dhanbad township ground water was 11.2531 (Prasad and Jaiprakas 1999). This indicated that even after continuous mining activity near the township for years, the ground water has not been contaminated by higher concentrations of heavy metals. The calculated HPI value indicates that, in general, the ground water is not contaminated with respect to heavy metals pollution. The heavy metal pollution index calculated for all 20 sampling locations and all seasons is given in Tables 4 and 5, respectively. This enables us to assess the quality of ground water at each sampling station which can be used to compare the quality with each other. Percentage deviation and mean deviation with mean HPI value are also calculated for each sampling location. At all sampling points, HPI calculated is below the critical index limit of 100 but at sampling point 6, 9, 11, 13 and 15 which are shallow hand pumps, the index value is high as compared to other sampling station. All other sampling locations indicate lower value of HPI as compared to mean HPI of 8.2091 (Table 4) and the percentage deviation is negative side. More negative deviation shows better quality of ground water and more positive deviation shows deteriorated quality of ground water. The lowest HPI value of 1.7094 has been obtained at sampling point 20 which is an open well and highest HPI value of 25.4608 at sampling point 11 which is a hand pump.

Table 1 Total heavy metals concentration (ppm) in ground water of Dhanbad town

Sl no.	Analysis of copper mean concentration 0.0084			Analysis of manganese mean concentration 0.1662			Analysis of lead mean concentration 0.0038			Analysis of iron mean concentration 0.5416		
	Summer season	Rainy season	Winter season	Summer season	Rainy season	Winter season	Summer season	Rainy season	Winter season	Summer season	Rainy season	Winter season
1. Open well	0.00288	0.00128	0.0020	0.00426	0.00334	0.0023	0.00546	BDL	0.0041	0.06662	0.06868	0.0269
2. Open well	0.00376	0.01016	0.0035	0.24986	0.07062	0.0833	0.00276	BDL	0.0002	0.02406	0.07568	0.0258
3. Open well	0.00386	0.00292	0.0032	0.03598	0.0943	0.0825	0.04874	0.00866	BDL	0.09322	0.04428	0.0211
4. Hand pump	0.03268	0.00854	0.0069	0.01146	0.03292	0.0475	BDL	0.02612	BDL	0.4181	0.1193	1.2607
5. Hand pump	0.00532	0.00706	0.0052	0.03468	0.10076	0.0515	0.00258	0.00194	0.0002	1.39866	0.11532	0.1219
6. Hand pump	0.00236	0.00256	0.0035	0.37404	0.28494	0.3351	BDL	0.00164	0.0005	0.43802	0.18056	0.0928
7. Hand pump	0.01072	0.00728	0.0060	0.1107	0.12938	0.1079	0.0009	0.011	0.0008	0.38542	0.342	0.1148
8. Hand pump	0.00904	0.00744	0.0048	0.0077	0.00844	0.0315	BDL	0.002	0.0005	1.667	0.80368	0.4548
9. Hand pump	0.01428	0.00642	0.0341	0.19608	0.20364	0.1758	BDL	0.00108	0.0130	1.98606	0.86664	5.5744
10. Hand pump	0.00128	0.0082	0.0052	0.00616	0.8954	0.0067	BDL	0.00404	0.0023	BDL	0.11726	0.0011
11. Hand pump	0.03148	0.00874	0.0258	1.24586	1.6652	1.6060	0.00154	0.0097	0.0002	0.0793	0.7385	0.3856
12. Hand pump	0.00164	0.00458	0.0123	0.00024	0.05184	0.0903	BDL	0.01198	0.0023	BDL	0.25904	0.4075
13. Hand pump	0.00594	0.0064	0.0068	0.02932	0.05458	0.0444	BDL	0.00194	BDL	0.164898	0.4932	0.1826
14. Hand pump	0.00292	0.00488	0.0047	0.09238	0.13662	0.0468	0.00038	0.01482	0.0050	0.0011	0.16778	0.2142
15. Hand pump	0.00818	0.0354	0.0043	0.00426	0.16624	0.1410	0.00142	0.00766	0.0015	0.55062	0.22474	0.1436
16. Hand pump	0.00582	0.0122	0.0153	0.00928	0.00934	0.0139	0.00044	0.0055	0.0046	0.68896	0.32732	2.6000
17. Hand pump	0.01396	0.00546	0.0087	0.01524	0.024	0.0216	0.00212	0.00048	0.0016	2.45064	0.4402	1.5932
18. Hand pump	0.0023	0.00232	0.0030	0.16788	0.18452	0.0421	BDL	0.00128	0.0017	0.37826	0.25584	0.3166
19. Hand pump	0.00782	0.00326	0.0067	0.1011	0.07128	0.0242	BDL	0.01136	0.0011	0.81092	0.66894	0.4460
20.Open well	0.00354	0.00622	0.0255	0.00848	0.06432	0.0637	0.00044	0.00084	0.0036	0.12258	0.35644	0.1258

Table 2 Total heavy metal concentration (ppm) in ground water of Dhanbad town

Sl no.	Analysis of cadmium mean concentration 0.00041			Analysis of zinc mean concentration 0.2116			Analysis of chromium mean concentration 0.0072		
	Summer season	Rainy season	Winter season	Summer season	Rainy season	Winter season	Summer season	Rainy season	Winter season
1. Open well	BDL	BDL	BDL	0.01518	0.02224	0.0152	BDL	BDL	BDL
2. Open well	0.0001	0.002	0.0004	0.09618	0.07168	0.1080	BDL	BDL	0.0006
3. Open well	BDL	BDL	0.0002	BDL	0.14132	0.1085	BDL	BDL	0.0023
4. Hand pump	BDL	0.00024	BDL	0.22426	0.14774	0.0988	0.0041	0.00708	0.0054
5. Hand pump	0.00084	BDL	BDL	0.17224	0.19492	0.0091	BDL	0.02488	0.0062
6. Hand pump	BDL	0.00562	BDL	0.01044	0.02196	0.0377	BDL	0.06444	0.0022
7. Hand pump	0.00194	BDL	BDL	0.02616	0.03656	0.0198	BDL	0.0198	0.0037
8. Hand pump	BDL	0.00066	BDL	0.28182	0.05214	0.0322	0.00154	0.01022	0.0044
9. Hand pump	BDL	0.00238	0.0003	0.01904	0.02052	1.5246	0.00028	0.01888	0.0094
10. Hand pump	0.00156	BDL	BDL	0.0376	2.7842	0.0545	BDL	0.01568	0.0021
11. Hand pump	BDL	0.00078	0.0002	0.0972	0.0386	0.0746	0.0066	0.04586	0.0131
12. Hand pump	BDL	0.00022	BDL	BDL	0.9836	0.4922	0.0012	0.04424	0.0057
13. Hand pump	0.00484	BDL	BDL	0.03118	0.03474	0.0207	0.00264	0.02152	0.0046
14. Hand pump	BDL	0.00004	BDL	0.03238	0.0362	0.0241	0.00362	0.01262	0.0029
15. Hand pump	0.0043	0.00068	BDL	0.07752	1.9594	0.1496	0.00158	BDL	0.0056
16. Hand pump	BDL	0.0004	BDL	0.17704	0.14574	0.4326	0.00118	0.00282	0.0057
17. Hand pump	BDL	0.00004	BDL	0.0227	0.02808	0.1496	0.00244	0.00992	0.0052
18. Hand pump	BDL	BDL	BDL	0.0687	0.02	0.0152	0.00118	0.00904	0.0032
19. Hand pump	BDL	BDL	BDL	0.46866	0.09144	0.1377	0.00228	0.0047	0.0062
20.Open well	BDL	BDL	BDL	0.07846	0.04796	0.3769	0.00286	0.00034	0.0024

Table 3 HPI calculation for ground water of Dhanbad town

Heavy metals	Mean value (ppb) M_i	Standard permissible value.(ppb) S_i	Highest desirable value (ppb) I_i	Unit weightage (W_i)	Sub index (Q_i)	$W_i \times Q_i$
Cu	8.4	1,500	50	0.0006	2.86	0.0017
Mn	166.2	300	100	0.0033	33.1	0.1092
Zn	211.6	15,000	5,000	0.00006	47.88	0.0028
Pb	3.8	50	–	0.02	7.6	0.152
Fe	541.6	1,000	300	0.001	34.51	0.0345
Cd	0.41	10	–	0.1	4.1	0.410
Cr	7.2	50	–	0.02	14.4	0.288

Mean HPI for all season

$\sum W_i = 0.14496$, $\sum W_iQ_i = 0.9982$, HPI = 6.8860

Table 4 Heavy metal pollution index of various sampling stations

Sl no.	Sampling point	Ground water source	HPI	% Deviation
1	Rajendra market	Open well	2.2544	−72.5377
2	Dhansar	Open well	6.7549	−17.7144
3	Bastakola	Open well	6.5569	−20.1264
4	Joraphatak	Hand pump	5.5905	−31.8987
5	Matkuria 1	Hand pump	5.9271	−27.7984
6	Matkuria 2	Hand pump	22.2770	+171.3690
7	Puja talkies	Hand pump	7.9932	−2.6300
8	Court more	Hand pump	4.8868	−40.4709
9	Police line	Hand pump	13.6258	+65.9840
10	Koyla nagar	Hand pump	8.4326	+2.7225
11	Steel gate	Hand pump	25.4608	+210.1533
12	Jagjivan nagar	Hand pump	7.0798	−13.7566
13	Dhaiya	Hand pump	14.5198	+76.8744
14	Hirak point	Hand pump	4.0044	−51.2199
15	Nawadih	Hand pump	13.1552	+60.2514
16	Baromuri	Hand pump	4.7178	−42.5296
17	Bisanpur	Hand pump	4.2025	−48.8068
18	Polytechnique	Hand pump	1.9161	−76.6588
19	Bus stand	Hand pump	3.1181	−62.0165
20	CIMFR	Open well	1.7094	−79.1767

Mean HPI = 8.2091

Table 5 HPI values at different seasons

Sampling month	Heavy metal pollution index	% Deviation with mean HPI value
Summer	6.2044	−10.8511
Rainy	11.4717	+64.8327
Winter	3.2028	−53.9801

Mean HPI = 6.9596

In Table 5, calculation of HPI has been done separately for three seasons May (summer), August (rainy) and December (winter) months. Rainy season HPI shows highest value of 11.4717 and percentage deviation is positive +64.8327 with mean HPI of 6.9596. Highest HPI in rainy season indicates that due to rain, many heavy metals from the surface of the ground may leach out to ground water. The HPI of summer and winter season is 6.2044 and 3.2028 which is lesser than the mean HPI and percentage deviation is −10.8511 and −53.9801, respectively. Lowest HPI in winter season with maximum negative percentage deviation indicates the best quality of ground water in winter season. This method has been found very useful to calculate the overall pollution level of ground water with respect to heavy metals.

CONCLUSION

The present mean HPI calculated for ground water of Dhanbad town has been found to be 6.8860 which is far below the critical index limit of 100. In our earlier study, which was carried out in 1997 for the same work, it was found that the mean HPI of Dhanbad township was 11.2531 (Prasad and Jaiprakas 1999). This indicates that after passing of several years still the ground water of the mining town is not polluted with respect to heavy metals. In addition, the several years of coal mining activity taking place near the township has not affected the ground water quality with respect to heavy metals. This method of evaluating the overall pollution level of ground water with respect to heavy metals is a very useful.

ACKNOWLEDGMENTS

Authors are grateful to Dr. Amalendu Sinha, Director of Central Institute of Mining and Fuel Research, Dhanbad, India for the support during the investigation period.

REFERENCES

1. Arnold EG, Lemore SC, Andrew DE, American Public Health Association (1992) Standard methods for the examination of water and wastewater APHA, 18th edn
2. Bhuiyan MAH, Islam MA, Dampare SB, Parvez L, Shigeyuki S (2010) Evaluation of hazardous metal pollution in irrigation and drinking water systems in the vicinity of a coal mine area of Northwest Bangladesh. J Hazard Mater 179:1065–1077
3. Horton RK (1965) An index number systems for rating water quality. J Water Pollut Control Fed 3:300
4. Joung HM, Miller WW, Mahammah CN, Gultjens JCA (1979) A generalised water quality index based on multivariate factor analysis. J Environ Qual 8:95
5. Khan R, Israili SH, Ahmad H, Mohan A (2005) Heavy metal pollution assessment in surface water bodies and its suitability for irrigation around the Neyveli lignite mines and associated industrial complex, Tamil Nadu, India. Mine Water Environ 24:155–161
6. Landwehr TM (1979) A statistical view of a class of water quality indices. Water Resour Res 15:460

7. Lohani BN, Todno M (1984) Water quality index for chao phraya river. J Environ Eng Div (ASCE) 110:1163–1176
8. Mohan SV, Nithila P, Reddy SJ (1996) Estimation of heavy metal in drinking water and development of heavy metal pollution index. J Environ Sci Health A31(2):283
9. Nishidia N, Miyai M, Tada F, Suzuki S (1982) Computation of index of pollution by heavy metals in river water. Environ Pollut 4:241
10. Prasad B, Bose JM (2001) Evaluation of heavy metal pollution index for surface and spring water near lime stone mining area of lower Himalayas. Environ Geol 41(1/2):183–188
11. Prasad B, Jaiprakas KC (1999) Evaluation of heavy metals in ground water near mining area and development of heavy metal pollution index. J Environ Sci Health (USA) A 34(1):91–102
12. Prasad B, Kumari S (2008) Heavy metal pollution index of ground water of an abandoned opencast mine filled with fly ash. Mine Water Environ 27(4):265–267
13. Prasad B, Mondal KK (2008) The impact of filling an abandoned open cast mine with fly ash on groundwater quality: a case study. Mine Water Environ 27:40–45
14. Prasanna MV, Chitambaram S, Hameed AS, Srinivasamoorthy K (2011) Hydrogeochemical analysis and evaluation of groundwater quality in the Gadilam river basin, Tamil Nadu, India. J Earth Syst Sci 120(1):85–98
15. Tiwary TN, Mishra M (1985) A preliminary assignment of water quality index to major Indian rivers. J Environ Prot 5:276

CHAPTER 19

A Semi-Infinite Interval-Stochastic Risk Management Model for River Water Pollution Control under Uncertainty

Jing Liu [1], Yongping Li [1,2,*], Guohe Huang [2] and Yurui Fan [2]

1 Department of Environmental Engineering, Xiamen University of Technology, Xiamen 361024, China; zyljing@126.com

2 Institute for Energy, Environment and Sustainable Communities, University of Regina, Regina, SK S4S 0A2, Canada

ABSTRACT

In this study, a semi-infinite interval-stochastic risk management (SIRM) model is developed for river water pollution control, where various policy scenarios are explored in response to economic penalties due to randomness and functional intervals. SIRM can also control the variability of the recourse cost as well as capture the notion of risk in stochastic programming. Then, the SIRM model is applied to water pollution control of the Xiangxihe watershed. Tradeoffs between risks and benefits are evaluated, indicating any change in the targeted benefit and risk level would yield varied expected benefits. Results disclose that the uncertainty of system components and risk preference of decision makers have significant effects on the watershed's production generation pattern and pollutant control schemes as well as system benefit. Decision makers with risk-aversive attitude would accept a lower system benefit (with lower production level and pollutant discharge); a policy based on risk-neutral attitude would lead to a higher system benefit (with higher production level and pollutant discharge). The findings can facilitate the decision makers in identifying desired product generation plans in association with financial risk minimization and pollution mitigation.

Keywords: decision making; financial risk management; functional interval; pollution control; stochastic with recourse; water quality

1. INTRODUCTION

Currently, the problem of water quality deterioration (e.g., water pollution) hinders socio-economic development and eco-environmental sustainability in many countries (e.g., China). According to the 2015 China Water Resources Bulletin, approximate 19.7% of the monitored river water (i.e., the total length of monitored river is 235×10^3 km) is in the worst two categories of water quality classification system (i.e., no longer fishable and of questionable agricultural value) [1]. Effective planning of water pollution control for watersheds plays an important role for national and/or regional sustainable development [2,3,4]. In fact, decisions in water pollution control are often made on the basis of uncertain information (i.e., various uncertainties) existing in system components and their interactions [5,6]. The major sources of uncertainty in water pollution control are the random characteristics of natural processes (e.g., precipitation and climate change) and stream conditions (e.g., stream flow, water supply, and point/nonpoint source pollution), the errors in estimated modeling parameters, and the imprecision of system objectives and constraints [7]. For example, pollutant discharge allowances are often affected by pollutant discharge rates (which influenced by random events such as temperature and precipitation) and decision makers' estimations; the allowable loads are not measured with certainty but in fact represent a probability distribution. Consequently, a number of mathematical techniques were launched to examine economic, environmental and ecological impacts of alternative pollution control actions under uncertainty, and thus aid the decision makers in generating effective water pollution control plans and policies [8,9,10,11,12,13,14,15,16,17].

Two-stage stochastic programming with recourse (TSP) is effective for problems where an analysis of policy scenarios is desired and uncertainties are expressed as random variables with known probability distributions [18,19]. The fundamental idea behind TSP is the concept of economic penalties with recourse that is the ability to take corrective actions after a random event occurs [20]. Previously, a number of researchers applied the TSP method to water pollution control [21]. Luo et al. [22] presented a simulation-based interval two-stage stochastic programming model for agricultural nonpoint source pollution control through land retirement under uncertain conditions, with the annual load of effluents before retirement being expressed as probability distributions. Harrison et al. [23] proposed a two-stage stochastic decision making programming method for problem of water quality pollution control. It was applicable to problems characterized by uncertainty in unobservable parameters; meanwhile, some problems were resolved upon observation of the outcome of the first-stage decision. Tavakoli [24] proposed a nonlinear two-stage stochastic fuzzy programming to tackle uncertainties described as fuzzy boundary intervals and probability distributions in decision making of optimal water allocation and pollutant load policies. Hu et al. [25] proposed a Bayesian-based two-stage inexact optimization method for supporting water pollution control under uncertainty; optimal production patterns of economic activities under different pollutant discharge allowance scenarios were generated. Generally, one major limitation of TSP is that it can only account for the expected second-stage cost without any consideration on the variability of the recourse values. For example, in water pollution control problems, one targeted benefit is firstly set, where one part is regular benefit from production generation and the other is in the case of excess pollutant discharge. If the event occurs, the penalty related to the pollutant mitigation might be tremendous such that the system benefit (i.e., the difference between the two parts) could be lower than the expected value. Although the event may happen associated with a low probability, it is unacceptable once it arises; any expected benefit below the targeted value is considered risky. The TSP model may become infeasible when the decision makers are risk averse under high variability conditions [26]. Ignoring risk might generate an unacceptable value for the objective function, which could lead to terrible outcomes for some extreme scenarios.

An effective tool that could help overcome the above shortcoming is financial risk management (i.e., risk below a certain targeted benefit), which was proposed to control the variability of the recourse cost, as well as capture the notion of risk in TSP [27,28]. The TSP model coupled with financial risk management enable decision makers to account for uncertainty in the evaluation and comparison of alternatives, while at the same time maximize the expected benefit and minimize the financial risk at every benefit level. Koppol and Bagajewicz [29] proposed a financial risk management based two-stage stochastic programming for the design of water utilization systems in process plants, where risk constraints were introduced to balance between cost and system reliability. Barbaro and Bagajewicz [30] presented a methodology which included financial risk management in the framework of two-stage stochastic programming for planning under uncertainty. Liu and Huang [31] proposed a dual interval two-stage restricted-recourse programming method to control the variability of the recourse cost of flood-diversion planning problem. Ahmed et al. [32] proposed financial risk management based two-stage stochastic programming for energy planning under uncertainties in demand and fuel price, in which the financial risk constraint were employed to choose the first stage decision in an optimal way without anticipation of future outcomes of uncertainties. Financial risk management is an advantageous measure to assess risk and obtain solutions appealing to decision makers. However, in practical water pollution control problems, various uncertain components may exist and may not be available as probability distribution and/or deterministic values, such as cost/benefit coefficients and pollutant discharge rates. Interval may be an appropriate format for dealing with uncertainties that cannot be quantified as distribution functions and/or deterministic values; moreover, the lower and upper bounds of interval numbers may be presented as functions instead of deterministic values when they vary with independent variables (e.g., time) [33]. Semi-infinite interval programming (SIP) can handle such dual uncertainties (i.e., functional intervals).

Therefore, this study aims at developing a semi-infinite interval-stochastic risk management (SIRM) model and applying it to water pollution control of the Xiangxihe watershed. The detailed tasks include: (i) integrating the concepts of financial risk management and functional interval parameters into a two-stage stochastic programming (TSP) framework, (ii) handling uncertainties presented as probability distributions and functional interval values, (iii) examining various scenarios that are related to economic penalties when the designed targets are violated, (iv) analyzing results of benefit, product generation and pollution control under a variety of risk levels, and (iv) assessing tradeoffs between risk and benefit on the basis of cumulative risk curves under different decision alternatives.

2. METHODOLOGY

Two-stage stochastic programming (TSP) reflects a tradeoff between pre-regulated strategies and the associated adaptive adjustments being approached to the expected system benefit [5]. In general, a TSP with recourse model can be formulated:

$$\max f = \sum_{j=1}^{n_1} c_j x_j - \sum_{j=1}^{n_2} \sum_{h=1}^{v} p_h e_j y_{jh} \tag{1}$$

subject to:

$$\sum_{j=1}^{n_1} a_{rj} x_j \leq b_r, r = 1, 2, ..., m_1 \tag{2}$$

$$\sum_{j=1}^{n_1} a_{tj} x_j + \sum_{j=1}^{n_2} a'_{tj} y_{jh} \geq \tilde{w}_h, t = 1, 2, ..., m_2; h = 1, 2, ..., v \tag{3}$$

$$x_j \geq 0, j = 1, 2, ..., n_1 \tag{4}$$

$$y_{jh} \geq 0, j = 1, 2, ..., n_2; h = 1, 2, ..., v \tag{5}$$

where xj represent the first-stage decision variables, which have to be decided before the actual realizations of the random variables; yjh denote the second-stage decision variables, which are related to the recourse actions against any infeasibilities due to particular realizations of uncertainties; cj is coefficient of the first-stage decision variable (xj) in the objective function; ej is coefficient of the recourse variable (yjh) in the objective function; arj is coefficient of the first-stage decision variable (xj) in constraint; br is right-hand side parameter in constraint; a'tj is coefficient of yjh in constraint *j*, which is associated with probability level ph. w˜h are random variables with probability levels ph, where h = 1, 2, ..., v and $\sum ph$=1. The objective function is composed of the benefit of first-stage decision minus the expected cost of the second stage under a series of constraints.

In water pollution control problems, when pollutant discharge is expressed as a random variable with known probability distribution, the actual benefit under pollutant discharge allowance scenarios may be unacceptably low, although the system expected benefit is maximized, due to the preceding TSP model's incapability in providing any control over the variability of the benefit under each scenario (h). Such a risk of loss (or potential regret from the decision makers) may be due to the above modeling situations where the variability of the second-stage cost is not considered except for its expected value [34]. Therefore, a concept of financial risk is introduced into the TSP problems, which is defined as the probability of not meeting the targeted benefit (η) in each individual scenario realization [29,30]:

$$\text{Risk}(x, \eta) = \sum_{h=1}^{v} P[B_h(x) < \eta] = \sum_{h=1}^{v} p_h z_h(x, \eta) \tag{6}$$

$$B_h(x) = \sum_{j=1}^{n_1} c_j x_j - \sum_{j=1}^{n_2} e_j y_{jh}, \; \forall j \tag{7}$$

$$z_h = \begin{cases} 1 \text{ if } \sum_{j=1}^{n_1} c_j x_j - \sum_{j=1}^{n_2} e_j y_{jh} \leq \eta \\ 0 \text{ otherwise} \end{cases} \tag{8}$$

where Bh(x) is the actual benefit (i.e., the benefit resulting after the uncertainty has been unveiled and scenario *h* has been realized). For any scenario, the benefit is either greater or equal than the targeted value; the corresponding probability *P[Bh(x)<η]=phzh(x, η)* is zero *(i.e., zh(x, η)*=0), or the benefit of the scenario is smaller than the target, rendering a probability of 1 *(i.e., zh(x, η)=1).* In order to control the financial risk of one TSP problem, Equations (2) to (4) can be converted into:

$$\sum_{j=1}^{n_1} c_j x_j - \sum_{j=1}^{n_2} e_j y_{jh} \geq \eta - UB_h z_h, \; \forall j; \; h = 1, 2, ..., v \tag{9}$$

$$\sum_{j=1}^{n_1} c_j x_j - \sum_{j=1}^{n_2} e_j y_{jh} \leq \eta + UB_h(1 - z_h), \; \forall j; \; h = 1, 2, ..., v$$

.(10)

$$\sum_{h=1}^{v} p_h z_h \leq \varepsilon, \; h = 1, 2, ..., v \tag{11}$$

$$z_h = \begin{cases} 1 \text{ if } \sum_{j=1}^{n_1} c_j x_j - \sum_{j=1}^{n_2} e_j y_{jh} \leq \eta \\ 0 \text{ otherwise} \end{cases} \tag{12}$$

where ε is the allowable risk exposure level. The integer variable *zh* should take a value of 1 if the benefit for scenario *h* is smaller than or equal to the targeted value *(η)* and a value of 0 otherwise. An upper-bound cost under each scenario (UB_h) is used for balancing benefit for scenario *h* and the target level (η) [35].

In many real-world problems, some parameters may be presented as interval numbers; meanwhile, the lower and upper bounds of them can rarely be acquired as deterministic values. Instead, they may be presented as functions of independent variables; this leads to functional intervals. For example, decision makers may estimate that the pollutant discharge allowance may vary with time (t). Different time corresponds to different lower and upper bounds of pollutant discharge allowance (e.g., [$a1+c1t$, $a2+c2t$]), resulting in infinite pollutant discharge allowance values. Semi-infinite interval programming (SIP) can handle such dual uncertainties. Therefore, through introducing concepts of SIP and financial risk management into TSP model, a semi-infinite interval-stochastic risk management (SIRM) model can be formulated as follows:

$$\text{Max } f^{\pm} = \sum_{j=1}^{n_1}\sum_{t=1}^{t_1} c_{jt}^{\pm} x_{jt}^{\pm} - \sum_{j=1}^{n_2}\sum_{h=1}^{v}\sum_{t=1}^{t_1} p_h e_{jt}^{\pm} y_{jht}^{\pm} \tag{13}$$

subject to:

$$\sum_{j=1}^{n_1} a_{rjt}^{\pm} x_{jt}^{\pm} \le b_{rt}^{\pm}, r = 1, 2, ..., m_1, t = 1, 2, ..., t_1 \tag{14}$$

$$\sum_{j=1}^{n_1} a_{kjt}^{\pm} x_j^{\pm} + \sum_{j=1}^{n_2} a'^{\pm}_{kjt} y_{jht}^{\pm} \ge \tilde{w}_{ht}^{\pm}(l), \; l = \{t | t \in [ll, lu]\}$$
$$k = 1, 2, ..., m_2; \; h = 1, 2, ..., v; \; t = 1, 2, ..., t_1 \tag{15}$$

$$\sum_{j=1}^{n_1}\sum_{t=1}^{t_1} c_{jt}^{\pm} x_{jt}^{\pm} - \sum_{j=1}^{n_2}\sum_{t=1}^{t_1} e_{jt}^{\pm} y_{jht}^{\pm} \ge \eta^{\pm} - UB_h z_h^{\pm}, \; \forall j; \; h = 1, 2, ..., v; \; t = 1, 2, ..., t_1 \tag{16}$$

$$\sum_{j=1}^{n_1}\sum_{t=1}^{t_1} c_{jt}^{\pm} x_{jt}^{\pm} - \sum_{j=1}^{n_2}\sum_{t=1}^{t_1} e_{jt}^{\pm} y_{jht}^{\pm} \le \eta^{\pm} + UB_h (1 - z_h^{\pm}), \; \forall j; \; h = 1, 2, ..., v; \; t = 1, 2, ..., t_1 \tag{17}$$

$$\sum_{h=1}^{v} p_h z_h^{\pm} \le \varepsilon^{\pm}, \; h = 1, 2, ..., v \tag{18}$$

$$z_h^{\pm} = \begin{cases} 1 \text{ if } \sum_{j=1}^{n_1} \sum_{t=1}^{t_1} c_j^{\pm} x_j^{\pm} - \sum_{j=1}^{n_2} \sum_{t=1}^{t_1} e_j^{\pm} y_{jh}^{\pm} \le \eta^{\pm} \\ 0 \text{ otherwise} \end{cases} \tag{19}$$

$$x_{jt}^{\pm} \ge 0, \; j = 1, 2, \ldots, n_1 \tag{20}$$

$$y_{jht}^{\pm} \ge 0, \; j = 1, 2, \ldots, n_2; \; h = 1, 2, \ldots, v \tag{21}$$

where superscripts '−' and '+' represent lower and upper bounds of the interval values; $\widetilde{w}_{ht}^{\pm}(l)$ denotes pollutant discharge allowance that vary with time; l denotes the length of time. In this study, a two-step solution process is proposed to transform the SIRM model into two deterministic submodels corresponding to the lower and upper bounds of the desired objective-function values. Figure 1 presents the overall flow chat of the methodology. The detailed solution method and procedure are presented in Appendix A to this paper.

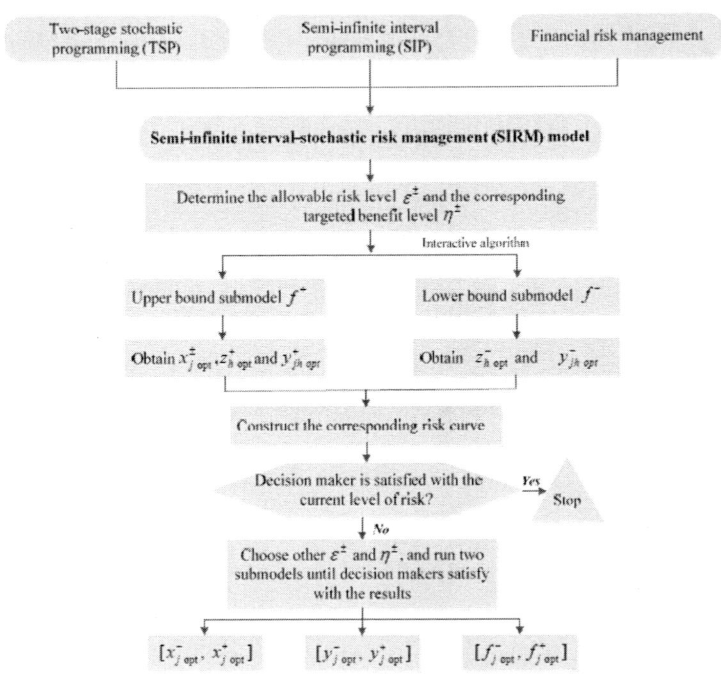

Figure 1. Overall flow chat of the methodology.

3. CASE STUDY

3.1. Study Area

Currently, water pollution problems due to point and nonpoint source pollution discharges have become more and more challenging in this watershed. Five chemical plants (i.e., GF, BSH, PYK, LCP, and XJLY), six phosphorus mining companies (i.e., XL, XH, XC, GP, JJW, and SJS), and four wastewater treatment plants (WTPs) (i.e., Gufu, Nanyang, Gaoyang, and Xiakou), four agricultural zones (AZ1 to AZ4) are the main pollutant discharge sources. These point and nonpoint sources scatter along a length of about 51 km river stretch which is segmented into five reaches (as shown in Figure 2). In the watershed, the main problems include: (1) the immoderate discharge of high-concentration phosphorus-containing wastewater and industrial soil wastes (i.e., chemical wastes, slags, and tailings) far exceed what can be decomposed by self-purification; (2) high potential for generating soil erosion and surface runoff due to the special geography and heavy rainfall; (3) large amounts of nutrient pollutants (in terms of phosphorus and nitrogen) in livestock wastewater and wastes are drained into the river by direct discharge or in rainfall. Effective water pollution control schemes are essential for the local sustainable development. In this study, a one-year planning horizon is selected and further sorted into two periods: dry season (i.e., November to May of the following year) and wet season (i.e., June to October) based on the specific growth periods of different crops. Wheat, potato, rapeseed, and alpine rice are identified as crops in dry season; second rice, maize, and vegetables are considered during wet season; citrus and tea grow up over the whole planning horizon.

3.2. Modelling Formulation

Then, the SIRM model is applied to optimize economic activities (i.e., industrial and agricultural activities) in the Xiangxihe watershed. The objective is to maximize the system benefit through identifying desired activities of chemical plants, phosphorus mining company, wastewater treatment plants, and crop farming. Biological oxygen demand (BOD), total nitrogen (TN) and total phosphorus (TP) are chosen as water quality indicators. The constraints involve relationships among economic activity, environmental restriction, and resources availability. The application of the proposed method is based on several assumptions: (1) the uncertainties, expressed as intervals and probability distributions, are independent; (2) the constraints are simplified with the respect to pollution loads rather than contaminant concentration, in order to prevent unnecessary complexity in

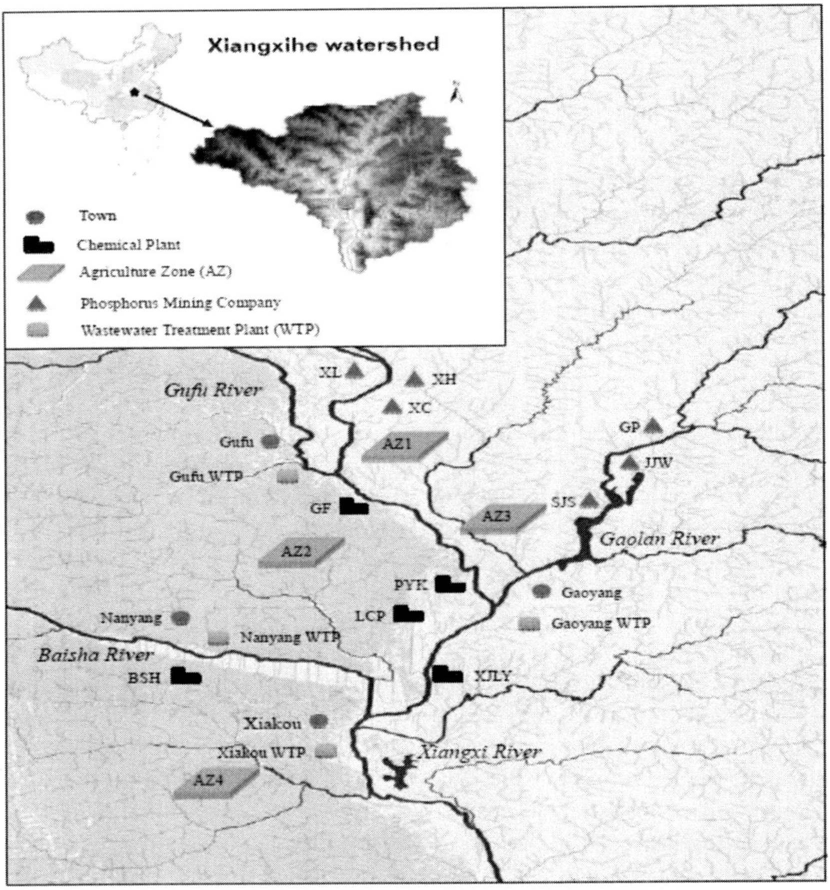

Figure 2. Study area. Note: GF, Gufu chemical plant; XL, Xinglong phosphorus mining company; XH, Xinghe phosphorus mining company; XC, Xingchang phosphorus mining company; BSH, Baishahe chemical plant; PYK, Pingyikou chemical plant; LCP, Liucaopo chemical plant; GP, Geping phosphorus mining company; JJW, Jiangjiawan phosphorus mining company; SJS, Shenjiashan phosphorus mining company; XJLY, Xiangjinlianying chemical plant.

the optimization procedure [36]. The SIRM model for supporting water pollution control in the Xiangxihe watershed can be represented as follows:

$$\begin{aligned}\text{Max } f^{\pm} &= L_t \cdot \sum_{i=1}^{5}\sum_{t=1}^{2} BC_{it}^{\pm} \cdot (PLC_{it}^{-} + x_{it} \cdot \Delta PLC_{it}^{\pm}) + L_t \cdot \sum_{n=1}^{6}\sum_{t=1}^{2} BP_{nt}^{\pm} \cdot (PLM_{nt}^{-} + y_{nt} \cdot \Delta PLM_{nt}^{\pm}) \\ &+ L_t \cdot \sum_{s=1}^{4}\sum_{t=1}^{2} BW_{st}^{\pm} \cdot (QW_{st}^{-} + u_{st} \cdot \Delta QW_{st}^{\pm}) + \sum_{j=1}^{4}\sum_{k=1}^{9}\sum_{t=1}^{2} BA_{jkt}^{\pm} \cdot (PA_{jkt}^{-} + w_{jkt} \cdot \Delta PA_{jkt}^{\pm}) \\ &- L_t \cdot \sum_{i=1}^{5}\sum_{t=1}^{2}\sum_{h=1}^{5} p_h \cdot PBC_{it}^{\pm} \cdot (PLC_{it}^{\pm} - APC_{ith}^{\pm}) \\ &- L_t \cdot \sum_{n=1}^{6}\sum_{t=1}^{2}\sum_{h=1}^{5} p_h \cdot PBP_{nt}^{\pm} \cdot (PLM_{nt}^{\pm} - APM_{nth}^{\pm}) \\ &- L_t \cdot \sum_{s=1}^{4}\sum_{t=1}^{2}\sum_{h=1}^{5} P_h \cdot PBW_{st}^{\pm} \cdot (QW_{st}^{\pm} - PW_{sth}^{\pm}) \\ &- \sum_{j=1}^{4}\sum_{k=1}^{9}\sum_{t=1}^{2}\sum_{h=1}^{5} p_h \cdot PBA_{jkt}^{\pm} \cdot (PA_{jkt}^{\pm} - PPA_{jkth}^{\pm})\end{aligned} \tag{22}$$

Constraints:

(1) Wastewater treatment capacity constraints:

$$PW_{sth}^{\pm} \cdot GT_{st}^{\pm} \le TPC_{st}^{\pm}, \ \forall s,\ t \tag{23}$$

$$APC_{ith}^{\pm} \cdot WC_{it}^{\pm} \le TPD_{it}^{\pm}, \ \forall i,\ t \tag{24}$$

(2) BOD discharge constraints:

$$APC_{ith}^{\pm} \cdot WC_{it}^{\pm} \cdot IC_{it}^{\pm} \cdot \left(1 - \eta_{BOD,it}^{\pm}\right) \le ABC_{ith}^{\pm}(l), \ \forall i,\ t \tag{25}$$

$$PW_{sth}^{\pm} \cdot GT_{st}^{\pm} \cdot BM_{st}^{\pm} \cdot \left(1 - \eta_{BOD,st}^{\pm}\right) \le ABW_{sth}^{\pm}(l), \ \forall s,\ t \tag{26}$$

(3) Nitrogen discharge constraints:

$$\sum_{k=1}^{9}\left(NS_{jk}^{\pm} \cdot SL_{jkt}^{\pm} + RF_{jkt}^{\pm} \cdot DN_{jkt}^{\pm} \cdot 10^{-5}\right) \cdot PPA_{jkth}^{\pm} \le MNL_{jth}^{\pm}(l), \ \forall j,\ t \tag{27}$$

(4) Phosphorus discharge constraints:

$$APC_{ith}^{\pm} \cdot \left[WC_{it}^{\pm} \cdot PCR_{it}^{\pm}\left(1 - \eta_{TP,it}^{\pm}\right) + ASC_{it}^{\pm} \cdot SLR_{it}^{\pm} \cdot PSC_{it}^{\pm}\right] \le APCM_{ith}^{\pm}(l), \ \forall i,\ t \tag{28}$$

$$\sum_{k=1}^{9}\left(PS_{jk}^{\pm}\cdot SL_{jkt}^{\pm}+DP_{jkt}^{\pm}\cdot RF_{jkt}^{\pm}\cdot 10^{-5}\right)\cdot PPA_{jkth}^{\pm}\leq MPL_{jth}^{\pm}(l),\ \forall j,\ t \tag{29}$$

(5) Soil loss constraints:

$$\sum_{k=1}^{9} SL_{jkt}^{\pm}\cdot PPA_{jkth}^{\pm}\leq MSL_{jth}^{\pm}(l),\ \forall j,\ t \tag{30}$$

(6) Financial risk management:

$$\begin{aligned}&\sum_{i=1}^{5}\sum_{t=1}^{2} BC_{it}^{\pm}\cdot PLC_{it}^{\pm}+\sum_{n=1}^{6}\sum_{t=1}^{2} BP_{pt}^{\pm}\cdot PLM_{pt}^{\pm}+\sum_{s=1}^{4}\sum_{t=1}^{2} BW_{st}^{\pm}\cdot QW_{st}^{\pm}\\&+\sum_{j=1}^{4}\sum_{k=1}^{9}\sum_{t=1}^{2} BA_{jkt}^{\pm}\cdot PA_{jkt}^{\pm}-\sum_{i=1}^{5}\sum_{t=1}^{2} PBC_{it}^{\pm}\cdot(PLC_{it}^{\pm}-APC_{ith}^{\pm})\\&-\sum_{n=1}^{6}\sum_{t=1}^{2} PBP_{nt}^{\pm}\cdot(PLM_{nt}^{\pm}-APM_{nth}^{\pm})-\sum_{s=1}^{4}\sum_{t=1}^{2} PBW_{st}^{\pm}\cdot(QW_{st}^{\pm}-PW_{sth}^{\pm})\\&-\sum_{j=1}^{4}\sum_{k=1}^{9}\sum_{t=1}^{2} PBA_{jkt}^{\pm}\cdot(PA_{jkt}^{\pm}-PPA_{jkth}^{\pm})\geq \eta^{\pm}-UB_{h}^{\pm}z_{h}^{\pm},\ \forall h\end{aligned} \tag{31}$$

$$\begin{aligned}&\sum_{i=1}^{5}\sum_{t=1}^{2} BC_{it}^{\pm}\cdot PLC_{it}^{\pm}+\sum_{n=1}^{6}\sum_{t=1}^{2} BP_{pt}^{\pm}\cdot PLM_{pt}^{\pm}+\sum_{s=1}^{4}\sum_{t=1}^{2} BW_{st}^{\pm}\cdot QW_{st}^{\pm}\\&+\sum_{j=1}^{4}\sum_{k=1}^{9}\sum_{t=1}^{2} BA_{jkt}^{\pm}\cdot PA_{jkt}^{\pm}-\sum_{i=1}^{5}\sum_{t=1}^{2} PBC_{it}^{\pm}\cdot(PLC_{it}^{\pm}-APC_{ith}^{\pm})\\&-\sum_{n=1}^{6}\sum_{t=1}^{2} PBP_{nt}^{\pm}\cdot(PLM_{nt}^{\pm}-APM_{nth}^{\pm})-\sum_{s=1}^{4}\sum_{t=1}^{2} PBW_{st}^{\pm}\cdot(QW_{st}^{\pm}-PW_{sth}^{\pm})\\&-\sum_{j=1}^{4}\sum_{k=1}^{9}\sum_{t=1}^{2} PBA_{jkt}^{\pm}\cdot(PA_{jkt}^{\pm}-PPA_{jkth}^{\pm})\leq \eta^{\pm}+UB_{h}^{\pm}(1-z_{h}^{\pm}),\ \forall h\end{aligned} \tag{32}$$

$$\sum_{h=1}^{5} p_h z_h^{\pm}\leq \varepsilon^{\pm},\ \forall h \tag{33}$$

$$z_h^{\pm}=0 \text{ or } 1,\ \forall h \tag{34}$$

(7) Cropland resources constraints:

$$\sum_{j=1}^{4}\sum_{k=1}^{6} PPA_{jkth}^{\pm} \geq MFP_{th}^{\pm}, \forall h, t = 1 \tag{35}$$

$$\sum_{j=1}^{4}\sum_{k=1}^{2} PPA_{jkth}^{\pm} + \sum_{j=1}^{4}\sum_{k=8}^{9} PPA_{jkth}^{\pm} \geq MFP_{th}^{\pm}, \forall h, t = 2 \tag{36}$$

$$PA_{jkt\,\min}^{\pm} \leq PPA_{jkth}^{\pm} \leq PA_{jkt}^{\pm} \leq PA_{jkt\,\max}^{\pm}, \forall j, t, k, h \tag{37}$$

(8) Industrial production scale constraints:

$$PLC_{it,\,\min} \leq APC_{ith}^{\pm} \leq PLC_{it}^{\pm} \leq PLC_{it,\,\max}, \forall i, t, h \tag{38}$$

$$QW_{st,\,\min} \leq PW_{sth}^{\pm} \leq QW_{st}^{\pm} \leq QW_{st,\,\max}, \forall s, t, h \tag{39}$$

$$PLM_{nt,\,\min} \leq APM_{nth}^{\pm} \leq PLM_{nt}^{\pm} \leq PLM_{nt,\,\max}, \forall n, t, h \tag{40}$$

(9) Non-negative constraints:]

$$APC_{ith}^{\pm}, PPA_{jkth}^{\pm}, PW_{sth}^{\pm}, APM_{nth}^{\pm} \geq 0, \forall i, j, s, n, k, t, h \tag{41}$$

The detailed nomenclatures for variables and parameters are provided in Appendix B. In the above modeling formulation, pollutant discharge allowances are selected as the key constraints and expressed as probability distributions (which are divided into functional intervals).

3.3. Data Analysis

The imprecise inputs are investigated through field surveys, statistical yearbooks, government reports, and literatures. They are presented as functional intervals, probability distributions, and interval numbers. Table 1 displays net benefits and penalties from industrial activities (chemical plant, water supply, and phosphorus mining company) that presented as intervals. Given a production target that is promised to each industrial activity, if the target is determined too high, the excessive pollutants will have to be mitigated in a more expensive way or discharged/drained into the stream (leading to penalties from the government). Table 2 lists the pollutant discharge allowances for different industrial activities which are presented as probability distributions. They are divided into a set of discrete functional intervals with probability distributions to approximate the stochastic values since the amounts of pollutant discharge are vary during the day in each season. The pollutant discharge allowances are estimated by the decision makers to fall within reasonable ranges. In this study, it is assumed that the lengths of dry and wet seasons are 213 and 152 days, respectively. Five scenarios for pollutant discharge allowance are generated, which are named as low, low-medium, medium, medium-high, and high, respectively.

Table 1. Net benefits and penalties from industrial activities.

Industrial Activities	Net Benefit		Penalty	
	Period 1	Period 2		
Chemical plant (RMB¥/t)				
Gufu (GF)	[791.3, 832.4]	[803.2, 864.7]	[823.6, 891.5]	[833.1, 910.4]
Baishahe (BSH)	[1217.4, 1493.8]	[1532.3, 1682.4]	[1357.4, 1634.1]	[1723.7, 1841.6]
Pingyikou (PYK)	[832.8, 912.4]	[943.5, 1032.4]	[861.5, 943.8]	[993.5, 1145.7]
Liucaopo (LCP)	[1532.6, 1843.5]	[1682.2, 1942.5]	[1735.7, 2011.1]	[1835.3, 2113.5]
Xiangjinlianying (XJLY)	[1843.3, 2174.3]	[1932.4, 2243.5]	[1992.3, 2294.5]	[2014.5, 2385.3]
Water supply (RMB¥/m^3)				
Gufu	[45.4, 52.7]	[58.8, 61.4]	[52.4, 58.9]	[63.7, 68.2]
Nanyang	[32.4, 38.5]	[41.4, 46.7]	[39.5, 43.1]	[47.2, 53.4]
Gaoyang	[48.3, 52.4]	[56.4, 59.4]	[55.5, 58.7]	[63.7, 70.1]
Xiakou	[42.1, 46.2]	[48.3, 53.4]	[49.2, 54.5]	[53.8, 59.1]
Phosphorus mining company (RMB¥/t)				
Xinglong (XL)	[167.3, 185.3]	[167.3, 194.2]	[196.7, 213.5]	[196.7, 221.3]
Xinghe (XH)	[154.5, 164.9]	[163.4, 173.4]	[178.4, 197.3]	[182.1, 207.3]
Xingchang (XC)	[161.5, 178.4]	[161.5, 178.4]	[188.4, 204.5]	[188.4, 204.5]
Geping (GP)	[166.4, 174.7]	[168.3, 179.4]	[189.2, 201.3]	[191.2, 207.3]
Jiangjiawan (JJW)	[161.4, 178.6]	[166.4, 181.4]	[188.3, 204.6]	[189.2, 211.3]
Shenjiashan (SJS)	[161.4, 183.4]	[171.9, 191.5]	[188.3, 211.3]	[198.2, 201.4]

4. RESULTS AND DISCUSSION

4.1. Benefit and Cost

In this study, a number of financial risk levels from 0 to 1 (ε) were taken into account. Figure 3 displays the results of various benefits (i.e., targeted benefit, system benefit, and regular benefit) and cost (i.e., penalty) under different risk levels. An increase in the targeted benefit implies the raised product generation meeting economic requirement in advance as the $\boldsymbol{\varepsilon}$ level is raised; an increase in the penalty for excessive pollutant discharge would then exist in case of low pollutant discharge allowance event occurring. In addition, any change in targeted benefit and $\boldsymbol{\varepsilon}$ level would lead to different system benefits; the trend of the **f±** variation along with the $\boldsymbol{\varepsilon}$ level would illustrate a balance between economic objective and violation risk. For instance, when $\boldsymbol{\varepsilon} = 0.12$ (with $\boldsymbol{\eta}^{\pm}$ = RMB¥ [100,600] × 10^6), the system benefit would be RMB¥ [121.9, 646.3] × 10^6; when $\boldsymbol{\varepsilon} = 0.64$ (with $\boldsymbol{\eta}^{\pm}$ = RMB¥ [160,800] × 10^6), the system benefit would be RMB¥ [160.8, 800.6] × 10^6. An increased $\boldsymbol{\varepsilon}$ level (a raised risk) corresponds to a decreased strictness for the constraints and thus an increased system benefit; such an increased value, however, would be associated with a raised constraint-violation risk.

Table 2. Allowable pollutant discharges for industrial activities.

Industrial Activities	Period 1	Period 2
Allowable phosphorus discharge for each chemical plant (kg/day)		
Gufu (GF)	[0.60 + 0.002*l*, 1.05 + 0.002*l*]	[0.70 + 0.001*l*, 1.15 + 0.001*l*]
Baishahe (BSH)	[195.63 + 0.089*l*, 432.56 + 0.089*l*]	[162.64 + 0.086*l*, 407.87+ 0.086*l*]
Pingyikou (PYK)	[30.49 + 0.09*l*, 70.71+ 0.09*l*]	[50.03 + 0.013*l*, 90.30 + 0.013*l*]
Liucaopo (LCP)	[124.7 + 0.046*l*, 294.7 + 0.046*l*]	[213.40 + 0.043*l*, 409.5 + 0.043*l*]
Xiangjinlianying (XJLY)	[126.22 + 0.042*l*, 321.76 + 0.042*l*]	[174.4 + 0.039*l*, 364.8 + 0.039*l*]
Allowable phosphorus discharge for each mine company (kg/m³)		
Xinglong (XL)	[47.41 + 0.043*l*, 123.49 + 0.043*l*]	[78.31 + 0.41*l*, 191.59 + 0.041*l*]
Xinghe (XH)	[23.68 + 0.044*l*, 86.87 + 0.044*l*]	[39.19 + 0.42*l*, 86.87 + 0.042*l*]
Xingchang (XC)	[30.54 + 0.038*l*], 62.75, + 0.038*l*]	[52.74 + 0.034*l*, 115.76 + 0.034*l*]
Geping (GP)	[48.39 + 0.012*l*, 109.36 + 0.012*l*]	[70.03 + 0.011*l*, 173.95 + 0.011*l*]
Jiangjiawan (JJW)	[25.91 + 0.014*l*, 40.13 + 0.014*l*]	[38.49 + 0.013*l*, [87.51 + 0.013*l*]
Shenjiashan (SJS)	[20.23 + 0.011*l*, 42.3*l* + 0.011*l*]	[28.19 + 0.009*l*, [82.6*l* + 0.009*l*]
Allowable BOD discharge for each WTP (kg/day)		
Gufu	[70 + 0.003*l*, 110 + 0.003*l*]	[75 + 0.002*l*, 115 + 0.002*l*]
Nanyang	[3 + 0.0001*l*, 8 + 0.0001*l*]	[4 + 0.0009*l*, 9 + 0.0009*l*]
Gaoyang	[10 + 0.0003*l*, 25 + 0.0003*l*]	[12 + 0.0002*l*, 27 + 0.0002*l*]
Xiakou	[15 + 0.0002*l*, 28 + 0.0002*l*]	[17 + 0.0001*l*, 30 + 0.0001*l*]

4.2. Risk Analysis

The concept of financial risk was employed to manage the economic loss risk related to the production policy. Figure 4 presents the variation of benefit corresponding to each risk level and pollutant discharge allowance scenario. It is indicated that, under a given scenario, benefit would increase with the $\boldsymbol{\varepsilon}$ level. For instance, when pollutant discharge allowance is low, the system benefit would be RMB¥ [124.8, 685.6] × 10^6 under $\varepsilon = 0.36$ and RMB¥ [173.7, 820.1] × 10^6 under $\boldsymbol{\varepsilon} = 0.88$, respectively. The higher risk level corresponds to a higher targeted benefit and thus a higher production level, leading to a higher benefit. A straightforward way of evaluating the tradeoff between risk and benefit is to use the cumulative risk curve. Figure 5 illustrates the lower and upper bound cumulative risk curves under different risk levels, respectively. Under a fixed $\boldsymbol{\varepsilon}$ level, the cumulative risk curve indicates the level of incurred risk for the generation target at each benefit level (which is the difference between regular benefit and penalty under a designed pollutant discharge allowance scenario). The cumulative risk curves would monotonically increase because they are cumulative probability functions; therefore, the risk is given by the cumulative probability. The risk area would increase as the $\boldsymbol{\varepsilon}$ level is raised. Generally, decisions at a lower $\boldsymbol{\varepsilon}$ level (i.e., risk averse) would lead to an increased reliability in fulfilling the system requirements but with a lower system benefit; conversely, a strong desire to obtain high system benefit (i.e., risk neutral or risk prone) would yield a raised risk of violating the constraints. It is revealed that a tradeoff between system benefit and system failure risk exists in the decision processes.

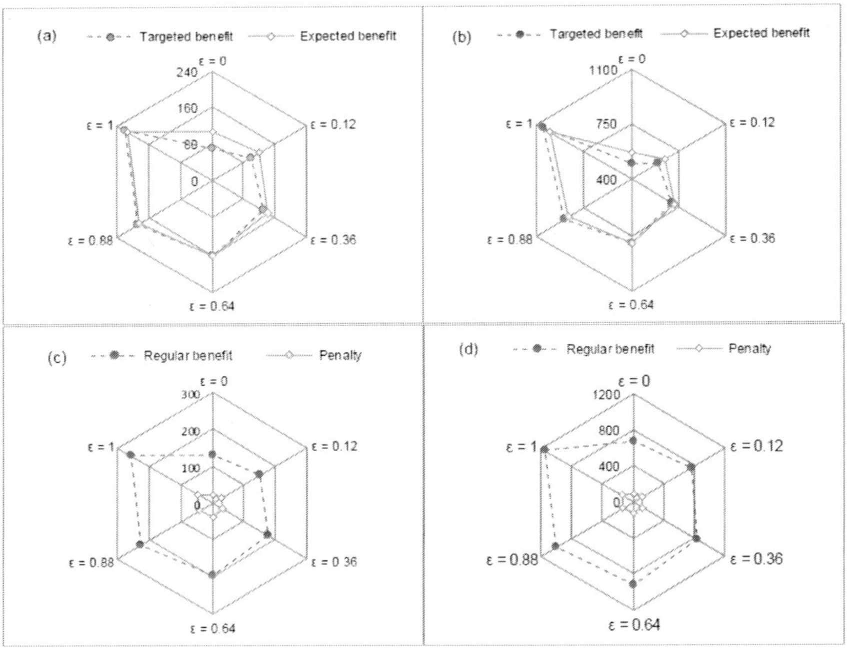

Figure 3. Benefits and costs under different risk levels. Note: (a) lower-bound targeted benefit and system benefit, (b) upper-bound targeted benefit and system benefit, (c) lower-bound regular benefit and penalty, (d) upper-bound regular benefit and penalty; unit: 10^6 RMB¥.

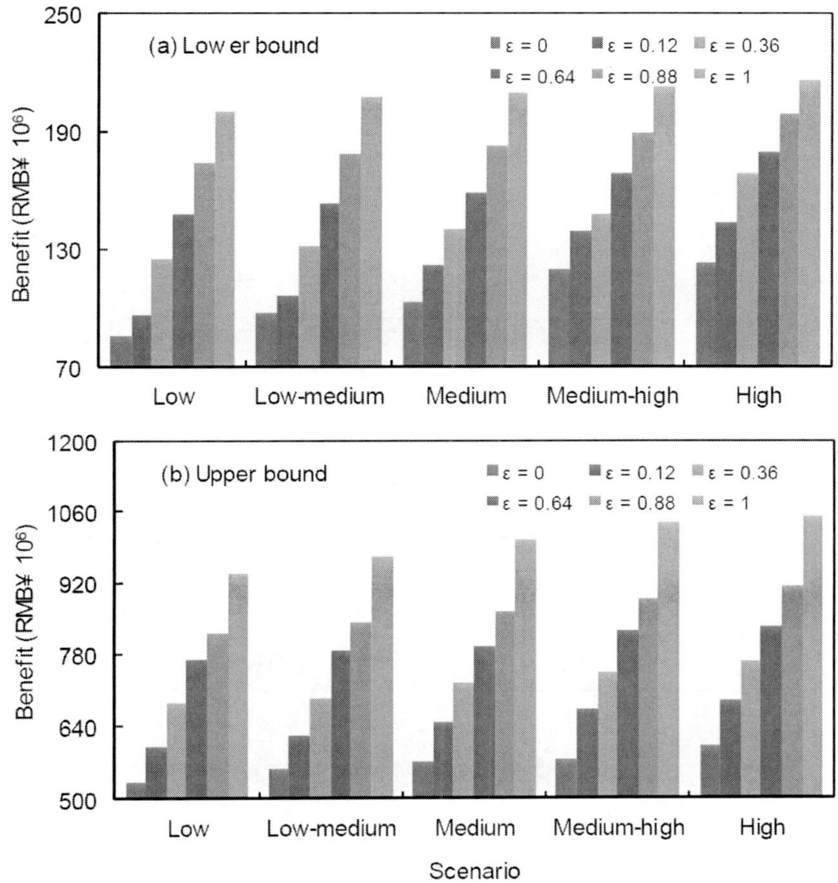

Figure 4. Benefit corresponding to each scenario and risk level. (a) Lower bound and (b) Upper bound.

4.3. Proportion of Benefit under Different Scenarios

Figure 6 presents proportion of benefit from all activities (including chemical plant, phosphorus mining company, water supply, and crop farming) under different scenarios. As shown in Figure 6, over the planning horizon, benefits from chemical plant and phosphorus mining company would be ranked as the first and the second proportion. For instance, under high scenario, the proportions would be [19.5, 33.3]% and [41.6, 44.9]%, respectively. This is due to the fact that the there are plenty of mineral resources (particularly phosphate ore) in the watershed; meanwhile, unit benefit of mineral related production is high (leading to phosphorus related industry-oriented pattern). In addition, crop farming would produce less than 3% of the benefit due to its low crop yield and limited arable land. Under low scenario, it would only produce [0.4, 1.4]% of the total benefit. However, crop farming is an important income source for rural households, which accounts for about 73% of the total population. Thus, how to enhance crop yield would be a huge concern to the decision makers.

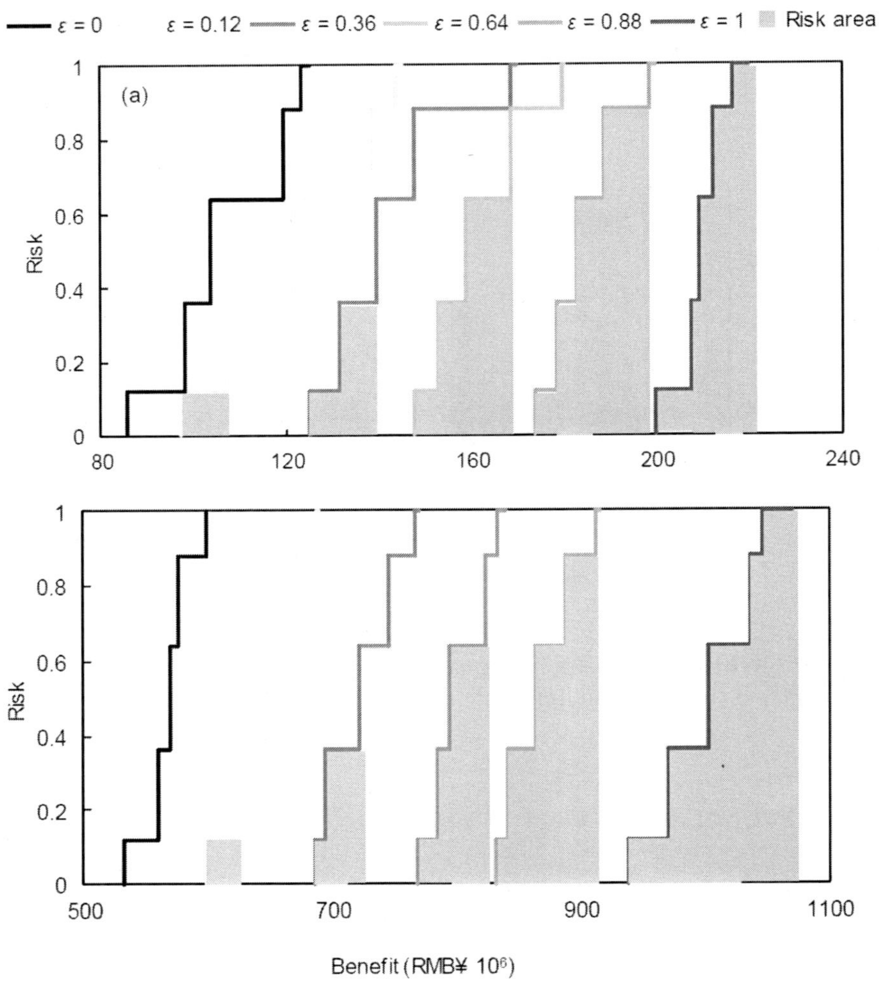

Figure 5. Cumulative risk curve ((a) Lower bound and (b) Upper bound).

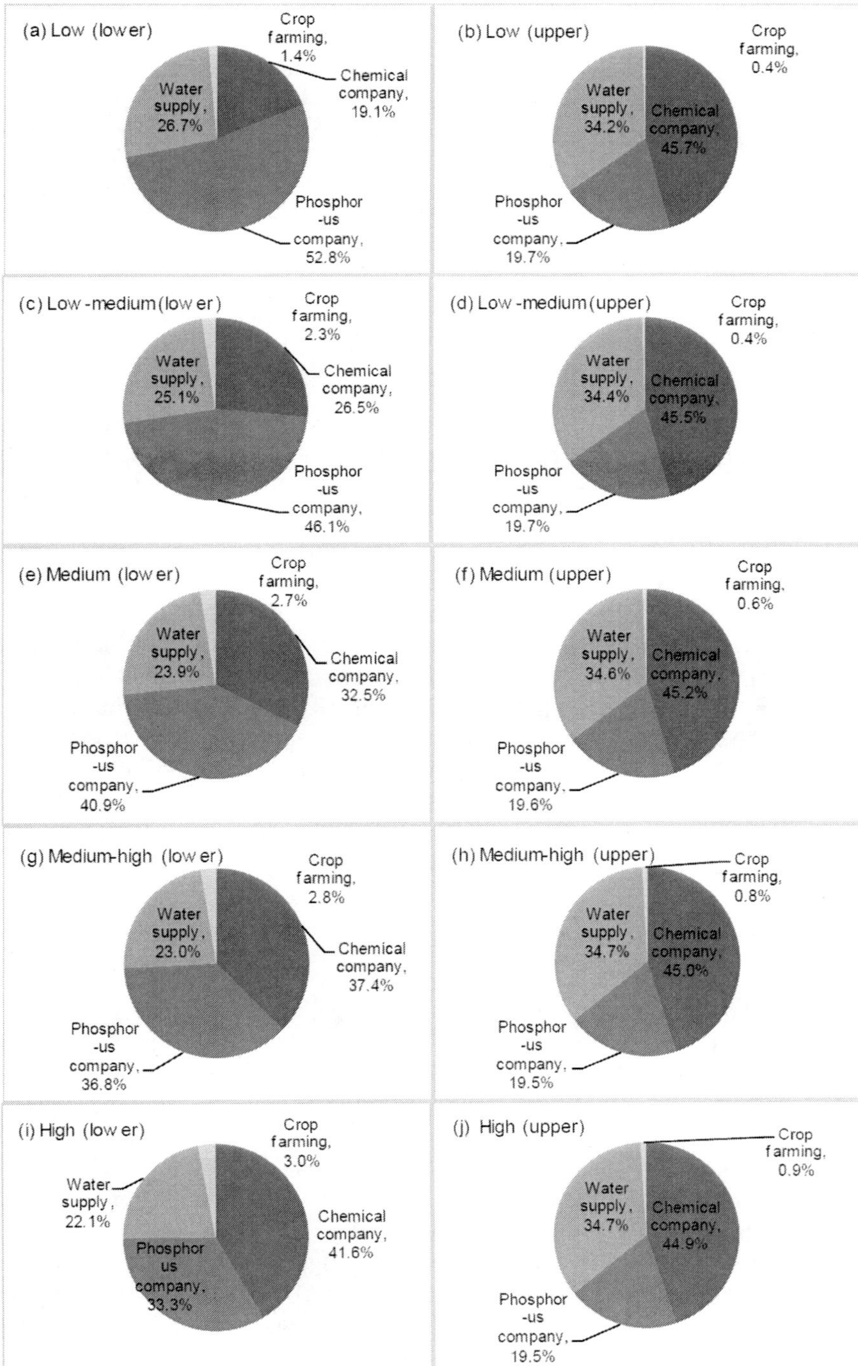
(a) Low (lower)
Crop farming, 1.4%
Chemical company, 19.1%
Water supply, 26.7%
Phosphor-us company, 52.8%
(b) Low (upper)
Crop farming, 0.4%
Water supply, 34.2%
Chemical company, 45.7%
Phosphor-us company, 19.7%
(c) Low-medium (lower)
Crop farming, 2.3%
Chemical company, 26.5%
Water supply, 25.1%
Phosphor-us company, 46.1%
(d) Low-medium (upper)
Crop farming, 0.4%
Water supply, 34.4%
Chemical company, 45.5%
Phosphor-us company, 19.7%
(e) Medium (lower)
Crop farming, 2.7%
Water supply, 23.9%
Chemical company, 32.5%
Phosphor-us company, 40.9%
(f) Medium (upper)
Crop farming, 0.6%
Water supply, 34.6%
Chemical company, 45.2%
Phosphor-us company, 19.6%
(g) Medium-high (lower)
Crop farming, 2.8%
Water supply, 23.0%
Chemical company, 37.4%
Phosphor-us company, 36.8%
(h) Medium-high (upper)
Crop farming, 0.8%
Water supply, 34.7%
Chemical company, 45.0%
Phosphor-us company, 19.5%
(i) High (lower)
Crop farming, 3.0%
Water supply, 22.1%
Chemical company, 41.6%
Phosphor us company, 33.3%
(j) High (upper)
Crop farming, 0.9%
Water supply, 34.7%
Chemical company, 44.9%
Phosphor-us company, 19.5%

Figure 6. Proportion of benefit under different scenarios. (a) Lower bound under low scenario; (b) Upper bound under low scenario; (c) Lower bound under low-medium scenario; (d) Upper bound under low-medium scenario; (e) Lower bound under medium scenario; (f) Upper bound under medium scenario; (g) Lower bound under medium-high scenario; (h) Upper bound under medium-high scenario; (i) Lower bound under high scenario and (j) Upper bound under high scenario.

4.4. Optimal Production Target and Actual Production Level

Table 3 provides the optimal production targets under different $\boldsymbol{\varepsilon}$ levels. Results obtained reveal that different risk levels lead to changed optimal production targets. For example, in period 1, the total targets of all chemical plants would change from 1279.4 t/day ($\boldsymbol{\varepsilon} = 0$) to 1618.5 t/day ($\boldsymbol{\varepsilon} = 1$). The increment of target is mainly derived from raised targeted benefits along with risk attitudes of decision makers. Figure 7 shows the corresponding results of actual production levels within the optimal production targets. It is indicated that the actual production levels would increase with the raised risk levels and pollutant discharge allowance scenarios. For instance, under low scenario ($\boldsymbol{t} = 1$), production level of chemical plant would change from [412.4, 733.9] t/day ($\boldsymbol{\varepsilon} = 0$) to [794.5, 883.7] t/day ($\boldsymbol{\varepsilon} = 1$). Under a fixed $\boldsymbol{\varepsilon}$ level of 1, production level of chemical plant would be [681.4, 832.7] t/day (with low-medium scenario) and [1278.2, 1743.2] t/day (with high scenario). Results reveal that such changes would be mainly associated with risk attitudes of decision makers and allowable pollutant discharges.

4.5. Excess Pollutant Discharge

Figure 8 describes the excess BOD, TN, and TP discharges under different ε levels. Different ε levels would result in different production levels, leading to different pollutant loadings. Results indicate that the excess BOD, TN, and TP discharges would increase as ε level is raised. For instance, under medium scenario, the total amounts of excess BOD, TN, and TP discharges would respectively approach [12.9, 18.1] t, [49.6, 62.5] t, and [65.2, 85.7] t (i.e., tonne) under $\varepsilon = 0.12$, while they would be [24.1, 28.9] t, [71.2, 90.6] t, and [86.6, 112.4] t under $\varepsilon = 0.88$, respectively. This is due to the fact that higher ε levels correspond to optimal production targets of economic activities, leading to higher pollutant discharges. Decision makers could justify pollutants control schemes according to the risk levels. Besides, chemical plant would be the main BOD and TP dischargers (associated with its high pollutant discharge rate and production level), while crop farming would be the main TN contributor (due to heavy runoff and soil loss). Thus, advanced wastewater treatment technologies (e.g., tertiary wastewater treatment and depth processing technologies) should be recommended to further improve pollutant removal efficiency. Moreover, some ecologically engineered management measures (e.g., constructed wetland, buffer stripes, and river channel enhancement zone) can be taken to mitigate soil loss and nutrient losses.

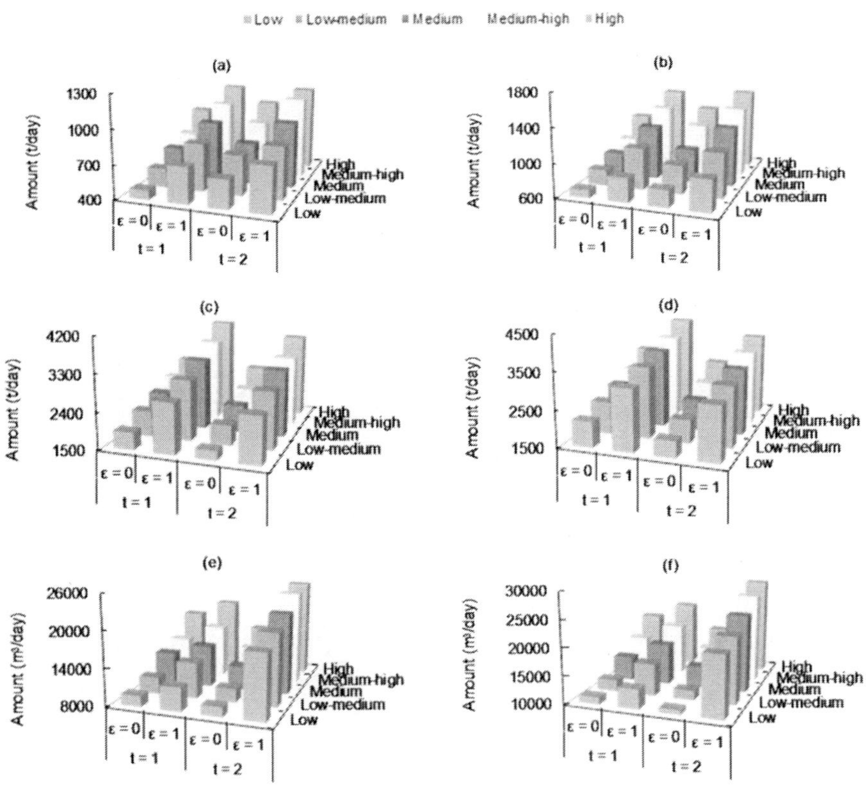

Figure 7. Actual production levels of industrial activities. Note: (a) lower-bound production level of chemical plant; (b) upper-bound production level of chemical plant; (c) lower-bound production level of phosphorus mining company; (d) upper-bound production level of phosphorus mining company; (e) lower-bound water supply and (f) upper-bound water supply.

Table 3. Optimal production targets under different ε levels.

Activity	Designed Target		$\varepsilon = 0$		$\varepsilon = 1$	
	$t = 1$	$t = 2$	$t = 1$	$t = 2$	$t = 1$	$t = 2$
Production level of chemical plants (t/day)						
$i = 1$	[21.8, 28.6]	[17.6, 28.3]	28.6	28.3	28.6	28.3
$i = 2$	[457.5, 573.7]	[461.6, 625.6]	573.7	461.6	573.7	625.6
$i = 3$	[69.5, 112.6]	[62.8, 110.5]	69.5	110.5	112.6	110.5
$i = 4$	[347.2, 526.8]	[327.6, 552.4]	347.2	552.4	526.8	552.4
$i = 5$	[260.4, 376.8]	[256.7, 373.7]	260.4	256.7	376.8	373.7
Production level of phosphorus mining companies (t/day)						
$n = 1$	[679.4, 1435.6]	[736.5, 1324.5]	679.4	736.5	1435.6	1324.5
$n = 2$	[246.7, 424.5]	[194.6, 412.5]	424.5	412.5	424.5	412.5
$n = 3$	[257.4, 436.4]	[227.5, 415.3]	436.4	415.3	436.4	415.3
$n = 4$	[385.3, 783.5]	[325.6, 704.5]	783.5	704.5	783.5	704.5
$n = 5$	[348.6, 568.4]	[287.4, 503.5]	568.4	503.5	568.4	503.5
$n = 6$	[495.6, 689.4]	[437.8,612.7]	689.4	437.8	689.4	612.7
Quantity of water supply to towns (m^3/day)						
$s = 1$	[9921.5, 17533.5]	[10385.7, 21756.4]	9921.5	10385.7	17533.5	21756.4
$s = 2$	[729.5, 1232.7]	[798.5, 1463.6]	1232.7	1463.6	1232.7	1463.6
$s = 3$	[2804.5, 3424.5]	[2964.5, 3853.7]	3424.5	3853.7	3424.5	3853.7
$s = 4$	[1428.7, 2104.7]	[1564.6, 2746.4]	2104.7	2746.4	2104.7	2746.4

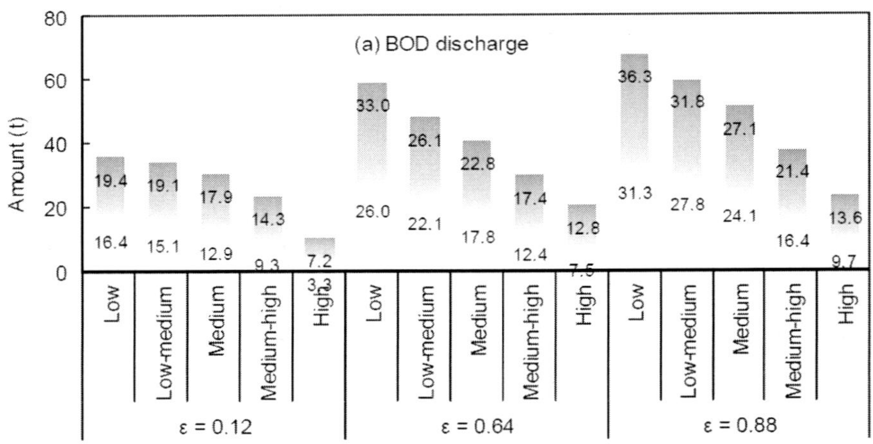

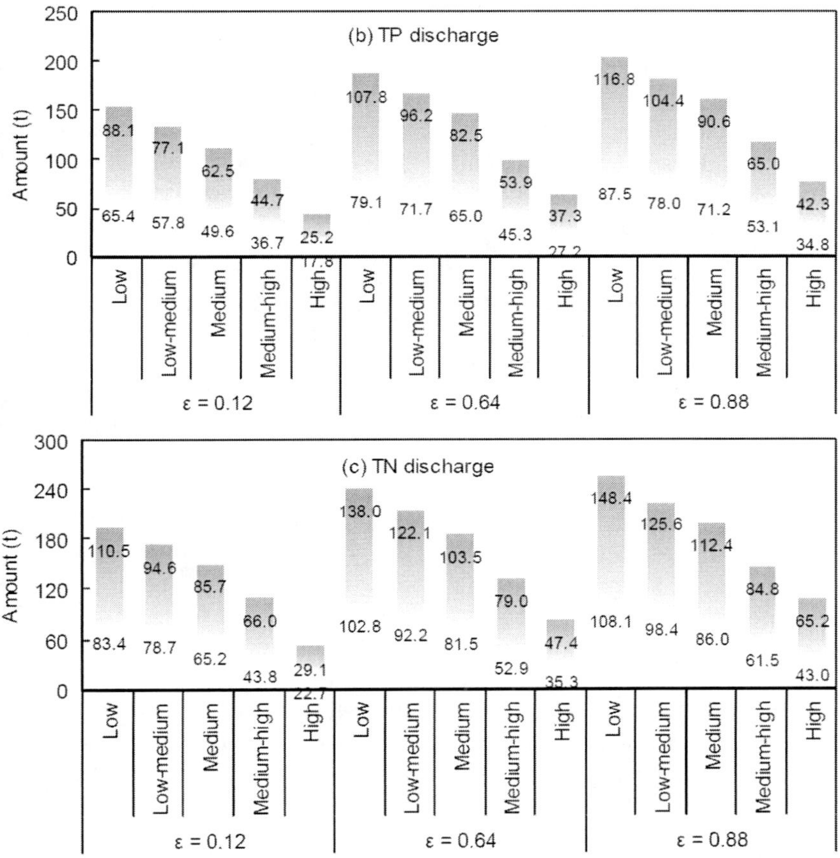

Figure 8. Excess Pollutant Discharge. (**a**) BOD discharge; (**b**) TP discharge and (**c**) TN discharge.

5. CONCLUSIONS

In this study, a semi-infinite interval-stochastic risk management (SIRM) model has been proposed through integrating the concepts of financial risk management and functional interval parameters into a two-stage stochastic programming (TSP) framework. The SIRM model can handle uncertainties described as probability distributions and functional interval values. Economic penalty is imposed when random variables are violated after the realization of designed targets. Decision makers' attitudes towards system risk are reflected using a risk management measure by controlling the variability of the second-stage cost (i.e., penalty). The SIRM model enables decision makers to account for uncertainty in alternative evaluation and comparison, and help maximize the expected benefit as well as minimize the financial risk at every benefit level. The SIRM model has been applied to water quality management in the Xiangxihe watershed. A number of pollutants (i.e., BOD, TN, TP) generated by industrial and agricultural activities have been considered. Interval solutions with different risk levels and allowable pollutant discharge scenarios concerning optimal production target, actual production level, and excess pollutant discharge have been generated through solving the model.

Results disclose that uncertainties and risks have significant effects on the watershed's production generation pattern and pollutant control schemes as well as the system benefit and economic penalty. It is revealed that the optimal production target and actual production level would increase with a raised risk level (i.e., a risk-neutral attitude to the system failure); under each ε level, more pollutant would be generated due to the increased allowable pollutant discharges. Under a fixed ε level, higher allowable pollutant discharge corresponds to a higher production level, resulting in a higher benefit. Moreover, under a given scenario, benefit would increase with the ε level; higher risk level corresponds to a higher targeted benefit and thus a higher production level, leading to a higher benefit. In summary, decisions at a lower ε level (i.e., risk averse) would lead to an increased reliability in fulfilling the system requirements but with a lower system benefit; conversely, a strong desire to obtain high system benefit (i.e., risk neutral or risk prone) would yield a raised risk of violating the constraints.

On the other hand, results disclose that chemical plant and phosphorus mining company would be the main income source of the study area (occupying more than 60% of the total benefit), while chemical plant would be the main BOD and TP dischargers (account for more than 72% and 75% of the total BOD and TP dischargers, respectively). Although crop farming would produce less than 3% of the total benefit, it would be the main TN contributor due to heavy runoff and soil loss (crop farming discharge 82% of the total TN discharge). Abatement of pollutant contaminations from industrial and agricultural activities is important for the river pollution control. However, the implementation of management practices for water pollution control can have potentials to affect the local economic income. There is a tradeoff between the economic income and pollution control. Several suggestions that could help local authority generate desired decision schemes in association with both maximizing economic income and reducing water pollution are: (i) wastewater treatment technologies (e.g., tertiary treatment and depth processing technologies) would be enhanced to improve the pollutant removal efficiency; (ii) definite land management alternatives suitable to legally manage the farmland need to be enacted for controlling soil erosion; (iii) some ecologically engineered management measures (e.g., constructed wetland, buffer stripes, and river channel enhancement zone) can also be taken to mitigate nitrogen and phosphorus losses.

The first attempt to employ the SIRM model to support water quality management of a watershed system demonstrates its applicability. However, the SIRM model still has space for further improvement. Firstly, it has difficulties in tackling uncertainties expressed as possibility distributions (i.e., fuzzy uncertainties). For example, decision makers' estimations of pollutant discharge allowances are objectively determined relying on some available historical data; the observed values of them may be ambiguous that can be estimated as fuzzy sets, leading to more complex uncertainties (e.g., intervals with fuzzy-random boundary). The shortcoming can be overcome by incorporating fuzzy mathematical programming into the SIRM framework. Secondly, the effects of the uncertainties as well as their interactions on the water quality management system performance should be further disclosed. Multi-level factorial design can effectively identify the important parameters (or factors) and detect their interactions on modelling outputs. Thirdly, decision support regarding pollution concentration management could be further provided by incorporating certain water quality simulation models into the SIRM framework, which can effectively reflect dynamic interactions between pollutant loading and water quality.

Acknowledgements: This research was supported by the National Key Research Development Program of China (2016YFA0601502 and 2016YFC0502800), and the 111 Project (B14008). The authors are grateful to the editors and the anonymous reviewers for their insightful comments and suggestions.

APPENDIX A. SOLUTION METHOD

In SIRM model, the first-stage decision variables $x_j^{\pm}$ in the model are expressed as intervals, which should be identified before the random variables are disclosed. Consequently, decision variables ***uj*** are introduced to identify an optimized set of the first stage values for supporting the related policy analyses.

Let $x_j^{\pm} = x_j^- + \Delta x_j u_j$, where $\Delta x_j = x_j^+ - x_j^-$ and $u_j \in [0, 1]$. Then, a two-step process is proposed to transform the model into two deterministic submodels that are associated with the lower and upper bounds of the desired objective. Since the objective is to be maximized, the submodel corresponding to ***f+*** (i.e., most desirable objective) can be firstly formulated:

$$\max f^+ = \sum_{j=1}^{n_1} c_j^+ (x_j^- + \Delta x_j u_j) - \sum_{j=1}^{k_2} \sum_{h=1}^{v} p_h e_j^- y_{jh}^- - \sum_{j=k_2+1}^{n_1} \sum_{h=1}^{v} p_h e_j^- y_{jh}^+ \tag{A1}$$

subject to:

$$\sum_{j=1}^{n_1} \left| a_{rj} \right|_+ \mathrm{Sign}(a_{rj}^+)(x_j^- + \Delta x_j u_j) \le b_r^+, \forall r \tag{A2}$$

$$\sum_{j=1}^{n_1} \left| a_{rj} \right|_+ \mathrm{Sign}(a_{rj}^+)(x_j^- + \Delta x_j u_j) + \sum_{j=1}^{k_2} \left| a'_{tj} \right|_+ \mathrm{Sign}(a'^{+}_{tj}) y_{jh}^- + \sum_{j=k_2+1}^{n_2} \left| a'_{tj} \right|_- \mathrm{Sign}(a'^{-}_{tj}) y_{jh}^+ \ge \tilde{\omega}_h^-, \forall t, h \tag{A3}$$

$$\sum_{j=1}^{n_1} c_j^+ (x_j^- + \Delta x_j u_j) - \sum_{j=1}^{k_2} p_h e_j^- y_{jh}^- - \sum_{j=k_2+1}^{n_1} p_h e_j^- y_{jh}^+ \ge \eta^+ - UB_h z_h^-, \forall j, h \tag{A5}$$

$$\sum_{h=1}^{v} p_h z_h^- \le \varepsilon^+, \forall h \tag{A6}$$

$$z_h^- = \begin{cases} 1 \text{ if } \sum_{j=1}^{n_1} c_j^+ (x_j^- + \Delta x_j u_j) - \sum_{j=1}^{k_2} p_h e_j^- y_{jh}^- - \sum_{j=k_2+1}^{n_1} p_h e_j^- y_{jh}^+ \ge \eta^+ \\ 0 \text{ otherwise} \end{cases} \tag{A7}$$

$$x_j^- + \Delta x_j u_j \ge 0, \ j = 1, 2, \ldots, n_1 \tag{A8}$$

$$y_{jh}^{-} \geq 0,\ j = 1,\ 2,\ ...,\ k_2;\ \forall h \tag{A9}$$

$$y_{jh}^{+} \geq 0,\ j = k_2 + 1,\ k_2 + 2,\ ...,\ n_2;\ \forall h \tag{A10}$$

where u_j, z_h^{-}, y_{jh}^{-}, and y_{jh}^{+} are decision variables; $y_{jh}^{-}(\ j = 1,\ 2,\ ...,\ k_2)$ are decision variables with positive coefficients in the objective function; $y_{jh}^{+}(j = k_2 + 1,\ k_2 + 2,\ ...,\ n_2)$ are decision variables with negative coefficients. Solutions of $y_{jh}^{-}(\ j = 1,\ 2,\ ...,\ k_2),\ y_{jh}^{+}(j = k_2 + 1,\ k_2 + 2,\ ...,\ n_2),\ u_j$, and z_h^{-} can be obtained through solving submodel (A1).The optimized first-stage variables are $x_{j\ \mathrm{opt}}^{\pm} = x_j^{-} + \Delta x_j u_{j\ opt}$ **(*j*=1, 2, ..., *n1*)**. Then, the submodel corresponding to the lower-bound objective-function value (f^{-}) is:

$$\max f^{-} = \sum_{j=1}^{n_1} c_j^{-}(x_j^{-} + \Delta x_j u_{j\ \mathrm{opt}}) - \sum_{j=1}^{k_2}\sum_{h=1}^{v} p_h e_j^{+} y_{jh}^{+} - \sum_{j=k_2+1}^{n_1}\sum_{h=1}^{v} p_h e_j^{+} y_{jh}^{-} \tag{A11}$$

subject to:

$$\sum_{j=1}^{n_1} \left|a_{rj}\right|_{-}\mathrm{Sign}(a_{rj}^{-})(x_j^{-} + \Delta x_j u_{j\ \mathrm{opt}}) \leq b_r^{-},\ \forall r \tag{A12}$$

$$\sum_{j=1}^{n_1} \left|a_{rj}\right|_{-}\mathrm{Sign}(a_{rj}^{-})(x_j^{-} + \Delta x_j u_{j\ \mathrm{opt}}) + \sum_{j=1}^{k_2} \left|a'_{tj}\right|_{-}\mathrm{Sign}(a'^{-}_{tj}) y_{jh}^{+} + \sum_{j=k_2+1}^{n_2} \left|a'_{tj}\right|_{+}\mathrm{Sign}(a'^{+}_{tj}) y_{jh}^{-} \geq \widetilde{w}_h^{+},\ \forall t,\ h \qquad \text{A(13)}$$

$$\sum_{j=1}^{n_1} c_j^{-}(x_j^{-} + \Delta x_j u_{j\ \mathrm{opt}}) - \sum_{j=1}^{k_2} p_h e_j^{+} y_{jh}^{+} - \sum_{j=k_2+1}^{n_1} p_h e_j^{+} y_{jh}^{-} \geq \eta^{-} - UB_h z_h^{+},\ \forall j,\ h \qquad \text{A(14)}$$

$$\sum_{j=1}^{n_1} c_j^-(x_j^- + \Delta x_j u_{j\,\text{opt}}) - \sum_{j=1}^{k_2} p_h e_j^+ y_{jh}^+ - \sum_{j=k_2+1}^{n_1} p_h e_j^+ y_{jh}^- \le \eta^- + UB_h(1 - z_h^+),\ \forall j,\ h$$

A(15)

$$\sum_{h=1}^{v} p_h z_h^+ \le \varepsilon^-, \forall h$$

A(16)

$$z_h^- = \begin{cases} 1 \text{ if } \sum_{j=1}^{n_1} c_j^-(x_j^- + \Delta x_j u_{j\,\text{opt}}) - \sum_{j=1}^{k_2} p_h e_j^+ y_{jh}^+ - \sum_{j=k_2+1}^{n_1} p_h e_j^+ y_{jh}^- \ge \eta^- \\ 0 \text{ otherwise} \end{cases}$$

A(17)

$$y_{jh}^+ \ge y_{jh\,\text{opt}}^-, j = 1, 2, ..., k_2; \forall h$$

A(18)

$$0 \le y_{jh}^- \le y_{jh\,\text{opt}}^+, j = k_2 + 1, k_2 + 2, ..., n_2; \forall h$$

A(19)

$$z_h^- \le z_{h\,\text{opt}}^+, \ \forall h$$

A(20)

where z_h^-, $y_{jh}^+(j = 1, 2, ..., k_2)$, and $y_{jh}^-(j = k_2 + 1, k_2 + 2, ..., n_2)$ are decision variables. Solutions of z_h^-, $y_{jh}^+(j = 1, 2, ..., k_2)$, and $y_{jh}^-(j = k_2 + 1, k_2 + 2, ..., n_2)$ can be obtained through solving submodel (A2). Through integrating the solutions of the two sets of submodels, the solution for the objective function value can be obtained. The detailed computational processes for solving the SIRM model can be summarized as follows:

Step 1. Formulate the SIRM model.

Step 2. Transform the SIRM model into two submodels, based on an interactive algorithm.

Step 3. Formulate and solve the upper bound submodel corresponding to $|f^+$, obtain $x_{j\,\text{opt}}^\pm$ and $y_{jh\,opt}^+$, and calculate f_{opt}^+.

Step 4. Formulate and solve the f^- submodel base on the solutions obtained through step 3, obtain $y_{jh\,opt}^-$, and calculate f_{opt}^-.

Step 5. Construct the corresponding risk curve. If the decision maker is satisfied with the current level of risk then stop, otherwise go to step 6.

Step 6. Let the allowable risk level from the decision makers be $\varepsilon^{\pm}$ and the corresponding aspiration benefit level be $\eta^{\pm}$.

Step 7. Formulate the SIRM problem with the objective being maximized under chosen $\varepsilon^{\pm}$ and $\eta^{\pm}$.

Step 8. If the decision maker is satisfied with the solutions, then stop. Otherwise, go to Step 6 for the next level.

Appendix B. Nomenclatures of SIRM Model

i	chemical plant, 1 = Gufu (GF), 2 = Baishahe (BSH), 3 = Pingyikou (PYK), 4 = Liucaopo (LCP), 5 = Xiangjinlianying (XJLY)
j	agricultural zone, and j = 1, 2, 3, 4
k	main crop, 1 = citrus, 2 = tea, 3 = wheat, 4 = potato, 5 = rapeseed, 6 = alpine rice, 7 = second rice, 8 = maize, 9 = vegetables
n	phosphorus mining company; 1 = Xinglong (XL), 2 = Xinghe (XH), 3 = Xingchang (XC), 4 = Geping (GP), 5 = Jiangjiawan (JJW), 6 = Shenjiashan (SJS)
s	town, 1 = Gufu, 2 = Nanyang, 3 = Gaoyang, 4 = Xiakou
t	planning time period, 1 = dry season, 2 = wet season
L_t	length of period (day)
l	independent variable representing time with the range of [0, 180]
h	allowable pollutant discharge scenario; h = 1, 2, 3
p_h	probability of occurrence allowable pollutant discharge level h (%);
$BC_{it}^{\pm}$	net benefit from chemical plant i in period t (RMB¥/t);
$PLC_{it}^{\pm}$	production level of chemical plant i in period t (t/day);
$BP_{nt}^{\pm}$	average benefit for per unit phosphate ore (RMB¥/t);
$PLM_{nt}^{\pm}$	production level of phosphorus mining company n during period t (t/day);
$BW_{st}^{\pm}$	net benefit from water supply to municipal uses (RMB¥/m^3);
$QW_{st}^{\pm}$	quantity of water supply to town s in period t (m^3/day);
$BA_{jkt}^{\pm}$	average benefit for agricultural product (RMB¥/t);
$PA_{jkt}^{\pm}$	planning area of crop k in agricultural zone j during period t (ha);
$PBC_{it}^{\pm}$	economic cost when the production targets from plant i in period t are not met (RMB¥/t);
$APC_{ith}^{\pm}$	actual production level of plant i under scenario h in period t (t/day);
$PBP_{nt}^{\pm}$	economic cost when the production targets from company n in period t are not met (RMB¥/t);
$APM_{nth}^{\pm}$	actual production level of company n under scenario h in period t (t/day);
$PBW_{st}^{\pm}$	economic cost when the water supply targets for town s in period t are not met (RMB¥/m^3);
$PW_{sth}^{\pm}$	actual water supply level of town s under scenario h in period t (m^3/day);
$PBA_{jkt}^{\pm}$	economic cost when the plant targets of crop farm j in period t are not met (RMB¥/ha);
$PBA_{jkth}^{\pm}$	actual plant area of crop farm j under scenario h in period t (ha);
$WC_{it}^{\pm}$	wastewater generation rate of chemical plant i during period t (m^3/t)
$GT_{st}^{\pm}$	wastewater generation rate at town s during period t (m^3/m^3)
$TPC_{st}^{\pm}$	capacity of wastewater treatment capacity (WTPs) (m^3/day)
$TPD_{it}^{\pm}$	capacity of wastewater treatment capacity (chemical plants) (m^3/day)
$IC_{it}^{\pm}$	BOD concentration of raw wastewater from chemical plant i in period t (kg/m^3)
$\eta_{BOD,it}^{\pm}$	BOD treatment efficiency in chemical plant i during period t (%)
$ABC_{ith}^{\pm}(l)$	allowable BOD discharge for chemical plant i under scenario h in period t (kg/day)
$\mathrm{BM}_{\mathrm{st}}^{\pm}$	BOD concentration of municipal wastewater at town s during period t (kg/m^3)
$\eta_{BOD,st}^{\pm}$	BOD treatment efficiency of WTPs at town s during period t (%)
$ABW_{sth}^{\pm}(l)$	allowable BOD discharge for WTPs at town s during period t (kg/day)
$NS_{jk}^{\pm}$	nitrogen content of soil in agricultural zone j planted with crop k (%)

$SL^{\pm}_{jkt}$	average soil loss from agricultural zone j planted with crop k in period t (t/ha)
$RF^{\pm}_{jkt}$	runoff from agricultural zone j with crop k in period t (mm)
$DN^{\pm}_{jkt}$	dissolved nitrogen concentration in runoff from agricultural zone j planted with crop k in period t (mg/L)
$MNL^{\pm}_{jth}(I)$	maximum allowable nitrogen loss in agricultural zone j under scenario h during period t (t)
$PCR^{\pm}_{it}$	phosphorus concentration of raw wastewater from chemical plant i in period t (kg/m^3)
$\eta^{\pm}_{TP,it}$	phosphorus treatment efficiency in chemical plant i in period t (%)
$ASC^{\pm}_{it}$	amount of slag discharged by chemical plant i in period t (kg/t)
$SLR^{\pm}_{it}$	slag loss rate due to rain wash in chemical plant i during period t (%)
$PSC^{\pm}_{it}$	phosphorus content in slag generated by chemical plant i in period t (%)
$APC^{\pm}_{ith}(I)$	allowable phosphorus discharge for chemical plant i under scenario h in period t (kg/day)
$PCM^{\pm}_{st}$	phosphorus concentration of municipal wastewater at town s in period t (kg/m^3)
$\eta^{\pm}_{TP,st}$	phosphorus treatment efficiency of WTP at town s in period t (%)
$APW^{\pm}_{sth}(I)$	allowable phosphorus discharge for WTP at town s under scenario h in period t (kg/day)
$WPM^{\pm}_{nt}$	wastewater generation from phosphorus mining company n in period t (m^3/t)
$MWC^{\pm}_{nt}$	phosphorus concentration of wastewater from mining company n in period t (kg/ m^3)
$\eta^{\pm}_{TP,st}$	phosphorus treatment efficiency of WTP at town s in period t (%)
$APW^{\pm}_{sth}(I)$	allowable phosphorus discharge for WTP at town s under scenario h in period t (kg/day)
$WPM^{\pm}_{nt}$	wastewater generation from phosphorus mining company n in period t (m^3/t)
$MWC^{\pm}_{nt}$	phosphorus concentration of wastewater from mining company n in period t (kg/ m^3)
$\eta^{\pm}_{TP,nt}$	phosphorus treatment efficiency in mining company n (%)
$ASM^{\pm}_{nt}$	amount of slag discharged by mining company n during period t (kg/t)
$PCS^{\pm}_{nt}$	phosphorus content in generated slag (%)
$SLW^{\pm}_{nt}$	slag loss rate due to rain wash (%)
$APCM^{\pm}_{nth}(I)$	allowable phosphorus discharge for mining company n under scenario h during period t (kg/day)
$PS^{\pm}_{jk}$	phosphorus content of soil in agricultural zone j planted with crop k (%)
$SL^{\pm}_{jkt}$	average soil loss from agricultural zone j planted with crop k in period t (t/ha)
$DP^{\pm}_{jkt}$	dissolved phosphorus concentration in runoff from agricultural zone j with crop k (mg/L)
$MPL^{\pm}_{jth}(I)$	maximum allowable phosphorus loss in agricultural zone j under scenario h in period t (t/ha)
$MSL^{\pm}_{jth}(I)$	maximum allowable soil loss agricultural zone j under scenario h in period t (t)
$\eta^{\pm}$	targeted benefit level of economic activities (RMB¥)
$z^{\pm}_{h}$	integer variable, which would take a value of zero if the benefit for each economic activity under scenario *h* is greater than or equal to the target level ($\eta^{\pm}$) and a value of one otherwise
UB_h	upper bound benefit under each scenario *h* (RMB¥)
$\varepsilon^{\pm}$	desired risk exposure level of economic activity
$MFP^{\pm}_{th}$	the government requirement for minimum area of farmland during period t (ha);
$PLC^{\pm}_{it,\min}$	minimum production level of chemical plant i in period t (t/day)
$PLC^{\pm}_{it,\max}$	maximum production level of chemical plant i in period t (t/day)
$QW^{\pm}_{st,\min}$	minimum quantity of water supply to town s in period t (m^3/day)
$QW^{\pm}_{st,\max}$	maximum quantity of water supply to town s in period t (m^3/day)
$PLM^{\pm}_{nt,\min}$	minimum production level of phosphorus mining company n during period t (t/day)
$PLM^{\pm}_{nt,\max}$	maximum production level of phosphorus mining company n during period t (t/day)

REFERENCES

1. Ministry of Water Resources, People's Republic of China. *China Water Resources Bulletin*; China Water Power Press: Beijing, China, 2017.
2. Cederkvist, K.; Jensen, M.B.; Ingvertsen, S.T.; Holm, P.E. Controlling stormwater quality with filter soil—Event and dry weather testing. *Water* **2016**, *8*, 349.

3. He, J. Probabilistic evaluation of causal relationship between variables for water pollution control. *J. Environ. Inform.***2016**, *28*, 110–119.
4. Toyosada, K.; Otani, T.; Shimizu, Y.; Managi, S. Water quality study on the hot and cold water supply systems at Vietnamese Hotels. *Water* **2017**, *9*, 251.
5. Li, Y.P.; Zhang, N.; Huang, G.H.; Liu, J. Coupling fuzzy-chance constrained program with minimax regret analysis for water pollution control. *Stoch. Environ. Res. Risk Assess.* **2014**, *28*, 1769–1784.
6. Li, Y.P.; Huang, G.H. Fuzzy-stochastic-based violation analysis method for planning water resources management systems with uncertain information. *Inf. Sci.* **2009**, *179*, 4261–4276.
7. Li, Y.P.; Huang, G.H. Two-stage planning for sustainable water-quality management under uncertainty. *J. Environ. Manag.* **2009**, *90*, 2402–2413.
8. Liu, J.; Li, Y.P.; Huang, G.H.; Nie, S. Development of a fuzzy-boundary interval programming method for water pollution control under uncertainty. *Water Resour. Manag.* **2014**, *29*, 1169–1191.
9. Kerachian, R.; Karamouz, M. A stochastic conflict resolution model for water pollution control in reservoir-river systems. *Adv. Water Resour.* **2007**, *30*, 866–882.
10. Li, Y.P.; Huang, G.H.; Yang, Z.F.; Nie, S.L. IFMP: Interval-fuzzy multistage programming for water resources management under uncertainty. *Resour. Conserv. Recycl.* **2008**, *52*, 800–812.
11. Ocampo-Duque, W.; Osorio, C.; Piamba, C.; Schuhmacher, M.; Domingo, J.L. Water quality analysis in rivers with non-parametric probability distributions and fuzzy inference systems: Application to the Cauca River, Colombia. *Environ. Int.* **2013**, *52*, 17–18.
12. Mohammad, R.N.; Najmeh, M. Water quality zoning using probabilistic support vector machines and self-organizing maps. *Water Resour. Manag.* **2013**, *27*, 2577–2594.
13. Chen., C.F.; Tsai, L.Y.; Fan, C.H.; Lin, J.Y. Using exceedance probability to determine total maximum daily loads for reservoir water pollution control. *Water* **2016**, *8*, 541.
14. Hwang, S.A.; Hwang, S.J.; Park, S.R.; Lee, S.W. Examining the relationships between water shed urban land use and stream water quality using linear and generalized additive models. *Water* **2016**, *8*, 155.
15. Ji., Y.; Huang, G.H.; Sun, W.; Li, Y.F. Water pollution control in a wetland system using an inexact left-hand-side chance-constrained fuzzy multi-objective approach. *Stoch. Environ. Res. Risk Assess* **2016**, *30*, 621–633.
16. Zolfagharipoor, M.A.; Ahmadi, A. A decision-making framework for river water pollution control under uncertainty: Application of social choice rules. *J. Environ. Manag.* **2016**, *183*, 152–163.
17. Chen, Y.; Zou, R.; Han, S.; Bai, S.; Faizullabhoy, M.; Wu, Y.Y.; Guo, H.C. Development of an integrated water quality and macroalgae simulation model for Tidal Marsh eutrophication control decision support. *Water* **2017**, *9*, 277.
18. Li, Y.P.; Huang, G.H.; Huang, Y.F.; Zhou, H.D. A multistage fuzzy-stochastic programming model for supporting sustainable water resources allocation and management. *Environ. Model. Softw.* **2009**, *24*, 786–797.
19. Nematian, J. An extended two-stage stochastic programming approach for water resources management under uncertainty. *J. Environ. Inform.* **2016**, *27*, 72–84.
20. Huang, G.H.; Loucks, D.P. An inexact two-stage stochastic programming model for water resources management under uncertainty. *Civ. Eng. Environ. Syst.* **2000**, *17*, 95–118.
21. Zhang, N.; Li, Y.P.; Huang, W.W.; Liu, J. An inexact two-stage water pollution control model for supporting sustainable development in a rural system. *J. Environ. Inform.* **2014**, *24*, 52–64.

22. Luo, B.; Li, J.B.; Huang, G.H.; Li, H.L. A simulation-based interval two-stage stochastic model for agricultural non-point source pollution control through land retirement. *Sci. Total Environ.* **2006**, *361*, 38–56.
23. Harrison, K.W. Two-stage decision-making under uncertainty and stochasticity: Bayesian Programming. *Adv. Water Resour.* **2007**, *30*, 641–664.
24. Tavakoli, A.; Nikoo, M.R.; Kerachian, R.; Soltani, M. River water pollution control considering agricultural return flows: Application of a nonlinear two-stage stochastic fuzzy programming. *Environ. Monit. Assess.* **2015**, *187*, 158.
25. Hu, X.H.; Li, Y.P.; Huang, G.H.; Zhuang, X.W.; Ding, X.W. A Bayesian-based two-stage inexact optimization method for supporting stream water pollution control in the three gorges reservoir region. *Environ. Sci. Pollut. Res.* **2016**, *23*, 9164–9182.
26. Ahmed, S.; Sahinidis, N.V. Robust process planning under uncertainty. *Ind. Eng. Chem. Res.* **1998**, *37*, 1883–1892.
27. Li, Y.P.; Huang, G.H. Interval-parameter robust optimization for environmental management under uncertainty. *Can. J. Civ. Eng.* **2009**, *36*, 592–606.
28. Khan, U.T.; Valeo, C. Short-term peak flow rate prediction and flood risk assessment using fuzzy linear regression. *J. Environ. Inform.* **2016**, *28*, 71–89.
29. Koppol, A.P.R.; Bagajewicz, M.J. Financial risk management in the design of water utilization systems in process plants. *Ind. Eng. Chem. Res.* **2004**, *42*, 5249–5255.
30. Barbaro, A.; Bagajewicz, M.J. Managing financial risk in planning under uncertainty. *Process Syst. Eng.* **2004**, *50*, 963–989.
31. Liu, Z.F.; Huang, G.H. Dual-interval two-stage optimization for flood management and risk analyses. *Water Resour. Manag.* **2009**, *23*, 2141–2162.
32. Ahmed, S.; Elsholkami, M.; Elkamel, A.; Du, J.; Ydstie, E.B.; Douglas, P.L. Financial risk management for new technology integration in energy planning under uncertainty. *Appl. Energy* **2014**, *128*, 75–81.
33. Zhu, Y.; Li, Y.P.; Huang, G.H. Planning carbon emission trading for Beijing's electric power systems under dual uncertainties. *Renew. Sustain. Energy Rev.* **2013**, *23*, 113–128.
34. Felfel, H.; Ayadi, O.; Masmoudi, F. Multi-objective stochastic multi-site supply chain planning under demand uncertainty considering downside risk. *Comput. Ind. Eng.* **2016**, *102*, 268–279.
35. McCray, A.W. *Petroleum Evaluations and Economic Decisions*; Prentice Hall: Englewood Cliffs, NJ, USA, 1975; ISBN 0136622136.
36. Chen, H.W.; Chang, N.B. Decision support for allocation of watershed pollution load using grey fuzzy multobjective programming. *J. Am. Water Resour. Assoc.* **2006**, *42*, 725–745.

Index